Environmental Tax Policy Review of Andalusia

OECD
BETTER POLICIES FOR BETTER LIVES

Please cite this publication as:
OECD (2023), *Environmental Tax Policy Review of Andalusia*, OECD Publishing, Paris, https://doi.org/10.1787/fe6d8b45-en.

ISBN 978-92-64-46283-0 (print)
ISBN 978-92-64-53269-4 (pdf)
ISBN 978-92-64-41121-0 (HTML)
ISBN 978-92-64-82411-9 (epub)

Photo credits: Cover © davidionut/Shutterstock.com.

Corrigenda to publications may be found on line at: www.oecd.org/about/publishing/corrigenda.htm.

Foreword

Environmental issues related to climate change, air pollution, waste and water management are key concerns in the Autonomous Region of Andalusia, Spain. Greenhouse gas emissions directly contribute to the global threat of climate change and air pollution has significant localised impacts on human health and the environment. Water pollution increases water treatment costs and degrades water quality, which is particularly concerning as water scarcity becomes more severe with climate change. A notable share of waste is disposed of in landfills and the extraction of raw materials damages ecosystems and generates greenhouse gas emissions and air pollution.

The report assesses how the Andalusia tax legal framework aligns with principles of good environmental tax policy, including a key focus on external (environmental) cost management. Andalusia has extensive responsibilities in taxation and environment policy, and therefore has scope to mitigate negative environmental outcomes while maintaining (or raising) tax revenues. As taxes can reflect the external costs of production and consumption activities on the environment, aligning taxes more closely with marginal external costs can improve market efficiency and move environmental outcomes towards socially optimal levels. The analysis also considers policy objectives beyond the efficient management of external costs, such as revenue raising and managing the distributional consequences from taxation, and considers other market-based or regulatory instruments, which may be better suited to address environmental concerns where taxation is less effective or where setting the “right” level of taxation would be too complex.

The analysis is guided by Andalusia’s environmental policy goals, existing legislation and the region’s legal competencies within the multi-level governance framework of Spain. The report provides the Andalusian government with strategic recommendations to use its tax competencies to pursue environmental goals in key environmental areas, namely greenhouse gas emissions and air pollution, water usage and pollution as well as waste and circular economy.

The report contains five parts. Part I describes the multi-level governance framework and tax competences in Spain and Andalusia. Part II analyses greenhouse gas emissions and air pollution focusing on stationary sources (industry and electricity, Section 3) and non-stationary sources (road transport, Section 4). Part III assesses water usage and pollution, Part IV assesses waste and circular economy, and Part V discusses the taxation of tourist stays. Parts II to IV begin by identifying the legal scope for action at the regional level. Parts II to V then assess how existing environmentally related taxes and fees applicable in Andalusia align with principles of good environmental tax policy and additional policy goals, and then close by providing strategic reform options in Andalusia. Detailed national and regional case studies are provided in the Annex to Parts II to V and underpin the strategic recommendations.

The project was funded by the European Union via the Technical Support Instrument, and implemented by the OECD, in cooperation with the Directorate-General for Structural Reform Support of the European Commission.

Acknowledgements

The report *Environmental Tax Policy Review of Andalusia* is the final report of the project on "Technical support for an integral reform of the environmental tax legal framework of the Autonomous Region of Andalusia" funded by the European Union via the Technical Support Instrument, and implemented by the OECD, in cooperation with the Directorate-General for Structural Reform Support (DG REFORM) of the European Commission.[1] The project seeks to provide strategic recommendations to Andalusia to plan potential adjustments to its environmentally relevant tax legal framework, with a view to improving regional environmental outcomes and strengthening contributions to national and global performance.

The project was led by the OECD's Centre for Tax Policy and Administration (CTPA) and carried out in partnership with the OECD Environment Directorate (ENV) and the OECD's Centre for Entrepreneurship, SMEs, Regions and Cities (CFE).

The project and final report were led by Luisa Dressler under the responsibility of Kurt Van Dender at CTPA. Anasuya Raj (CTPA) conducted the assessment in the areas of greenhouse gas emissions and air pollution for stationary sources (electricity and industry) and for water pollution and usage. Luisa Dressler (CTPA) carried out the analysis on greenhouse gas emissions and air pollution in road transport. The case study work and the analysis of the legal framework for stationary sources, road transport and water were conducted by Isabelle Chatry, Kate Power and Margaux Vincent (CFE), with the support of Francisco Martes Porto Macedo. Anna Martínez Guallar, Verónica Martínez Sánchez and Ignasi Puig Ventosa from *ENT Environment & Management* provided the legal and economic analysis for circular economy and waste management under the guidance of Peter Börkey and Frithjof Laubinger (ENV). Frithjof Laubinger (ENV) led the assessment in the area of the taxation of tourist stays, in collaboration with Luisa Dressler and Anasuya Raj (CTPA).

A Scientific Committee accompanied the project and provided valuable input and feedback at various stages: Francisco Adame Martínez (University of Seville), Maria Garcia Valiñas (University of Oviedo), Xavier Labandeira (University of Vigo), Marina Rosales Martinez (University of Seville) and Jaime Vallés-Giménez (University of Zaragoza).

The publication and dissemination of the report was coordinated by Bethany Millar-Powell under the supervision of Assia Elgouacem, with the valuable support of Karena Garnier, Nathalie Lagorce, Michael Sharatt, Violet Sochay and Carine Tyler.

The authors wish to thank Javier Jiménez López (General Secretariat of Finance, Region of Andalusia) and Bogdan Tasnadi (DG REFORM, European Commission) for their continuous support throughout the project. The results of the project were discussed with the Region of Andalusia at a workshop in December 2022. The Government of Andalusia assisted with the provision of data and shared suggestions, inputs and comments at various stages. The authors are also grateful to the following colleagues from the Region of Andalusia. From the Regional Ministry of Economy, Finance and European Funds: Javier Jiménez López, Javier Chavarría Fuentes, Alberto González González (General Secretariat of Finance), María del Robledo Romero Rus, María José López Romeu, Álvaro Yanes Elejalde, Fernando Alpresa Gutiérrez, Inmaculada Bravo Falcón, María Inmaculada Fernández Bernárdez, Desiré Benítez Zamora (General

Secretariat of Finance; Directorate General for Taxation, Financing, Financial Relations with Local Corporations and Gambling), Enrique Arcos Vargas (General Secretariat of Finance; Andalusian Tax Agency), Macarena Hernández Salmerón (General Secretariat of Economy), Sonia Fernández Boniquito, and Martín Manzanera (General Secretariat of Economy; Institute of Statistics and Cartography of Andalusia). From the Regional Ministry of Sustainability, Environment and Blue Economy: Elena Ortega Díaz (General Secretariat for Sustainability, Environment and Blue Economy) and David Fernández Guerra (General Secretariat for Sustainability, Environment and Blue Economy; Directorate General for Environmental Sustainability and Climate Change). From the Regional Ministry of Agriculture, Fisheries, Water and Rural Development: Pedro Gómez Galán (General Secretariat for Water). From the Regional Ministry of Industrial Policy and Energy: Joaquín Villar Rodríguez, Marcos Borrero Gaviño (Andalusian Energy Agency). From the Regional Ministry of Health and Consumption: Luis Ángel Moya Ruano (Directorate General for Public Health and Pharmaceutical Regulation).

The authors thank OECD colleagues for their valuable inputs in the analysis: Konstantinos Theodoropoulos and Kurt Van Dender (CTPA), Xavier Leflaive and Rodrigo Pizarro (ENV), and Ander Eizaguirre, Andres Fuentes Hutfilter and Oriana Romano (CFE).

Table of contents

FIGURES

TABLES

Follow OECD Publications on:

https://twitter.com/OECD

https://www.facebook.com/theOECD

https://www.linkedin.com/company/organisation-eco-cooperation-development-organisation-cooperation-developpement-eco/

https://www.youtube.com/user/OECDiLibrary

https://www.oecd.org/newsletters/

This book has... StatLinks

A service that delivers Excel® files from the printed page!

Look for the StatLink at the bottom of the tables or graphs in this book. To download the matching Excel® spreadsheet, just type the link into your Internet browser or click on the link from the digital version.

Executive Summary

Environmental concerns related to climate change, air pollution, waste and water management compel the Autonomous Region of Andalusia to make use of its extensive legal competences to implement policy measures to mitigate such negative environmental outcomes. The *Environmental Tax Policy Review of Andalusia* supports the government of Andalusia in developing plans for potential reforms to its environmentally relevant tax legal framework, with a view to improving regional environmental outcomes and strengthening the region's contributions to national and global environmental performance. The report is the outcome of the project on "Technical support for an integral reform of the environmental tax legal framework of the Autonomous Region of Andalusia" funded by the European Union (EU) via the Technical Support Instrument, and implemented by the OECD, in cooperation with the European Commission.

The report provides strategic recommendations for environmentally related tax reform in the areas of greenhouse gas (GHG) emissions and air pollution, with a focus on electricity, industry (stationary sources) and road transport; in the area of water usage and pollution; and on waste and circular economy.

The recommendations derive from a thorough review of the legal framework at the regional, national and EU levels, as well as Andalusia's tax competences. The report takes into account the multilevel governance framework of Spain, which is a quasi-federal country within the EU and has three tiers of subnational government benefitting from different levels of constitutionally recognised autonomy. The first tier of the subnational governance structure is composed of 17 autonomous communities, the second tier is made up of 50 provinces, and the third tier comprises 8 131 municipalities and two autonomous cities. Andalusia is an autonomous community that is governed under the common regime and benefits from some tax autonomy (though less than the two regions governed by the foral regime). Andalusia can establish own-source taxes, apply a surcharge on centrally levied taxes (with some limitations), and has some discretion over assigned taxes, which are centrally levied taxes where autonomous communities receive a share of the revenues and have control over some elements of the tax design (e.g. exemptions). As own-source taxes must be based on a taxable event that is not already subject to tax by the central government or the municipalities, to avoid double-taxation, the report also includes a review of existing environmentally related taxes at the different levels of governance in Spain.

The recommendations take into consideration how existing taxes and fees applicable at the Andalusian level align with general principles of sound environmental tax policy. Concrete practical examples support the assessment of specific instruments and their designs. The Andalusian Climate Plan (PAAC) is one example of the pioneering role that Andalusia plays amongst the Spanish autonomous communities in this respect by being the first autonomous community in Spain to develop a regional Climate Change Strategy in 2002. More recently, the region has been seeking to focus on aligning the regional tax framework with its environmental and climate strategies.

Environmental taxes can reflect the external costs of production and consumption activities on the environment. Integrating environmental costs into market prices creates incentives that influence economic agents' decision making and reduce pollution in a cost-effective manner. Aligning taxes more closely with marginal external costs will improve market efficiency and move environmental outcomes towards socially optimal levels. Analysing the alignment of the Andalusian tax legal framework with this principle of external cost management is an objective of the project.

This report focuses on assessing how current taxes applicable in Andalusia align with external costs and the principles of good environmental tax policy more generally, based on options that have been identified as feasible for the region from a legal perspective. On several occasions, the analysis goes one step further and considers policy objectives beyond the efficient management of external costs, such as revenue raising and the distributional consequences of taxation. In particular, the report considers how environmentally related tax policy in Andalusia may contribute to specific policy goals, such as those set out in the Andalusian Climate Action Plan – with its overall GHG emissions reduction target of 39% by 2030 (relative to 2005) - and the Andalusian Strategy for Air Quality – to support the elaboration of air quality improvement plans by local governments.

Where relevant, the analysis discusses taxation in relation to other market-based or regulatory instruments. Other instruments may better suit specific contexts, e.g. when there is a risk that reactions to prices and taxes will be limited due to limited behavioural responses or because no alternatives are available, or when setting the 'right' tax rate would be too complex.

The analysis considers recent policy developments and draws on new analysis of environmental costs. The *Committee of experts to prepare the White Paper on the tax reform*, established by the Spanish Treasury, recently published the "White Book for the reform of the tax system and its adaptation to the reality of the 21st century". The White Book includes a diagnosis of the Spanish tax system, including in respect of environmental taxation, and provides detailed proposals for tax reform. In addition, the present report draws on a recent European Commission (EC) report on "Green taxation and other economic instruments: Internalising environmental costs to make the polluter pay", which estimates the cost of various forms of environmental damage, including those covered in the report. The EC report finds that across EU Member States, the external costs generated by the various forms of environmental damage significantly outweigh the revenues raised through tax and other instruments and tax rates are not aligned with the marginal external cost; with 16 out of 27 of Member States having internalisation rates below 50%.

Andalusia is a pioneer in establishing strategies to reduce GHG emissions and air pollution, though improvements could be made to its main emissions pricing tool. Andalusia was one of the first autonomous regions to introduce a tax on greenhouse gas and air pollutant emissions when it introduced the Tax on the Emission of Gases into the Atmosphere (IEGA) in 2003. The IEGA covers CO_2 emissions and two important air pollutants, nitrogen oxide (NOx) and sulphur oxide (SOx). It currently exempts other pollutants as well as emissions from landfill, the combustion of biomass and biofuel, and facilities for the intensive rearing of animals.

Andalusia could reform the IEGA to strengthen the price signal and cover emissions that are currently excluded from the taxable base. The IEGA currently bundles all pollutants into a single tax base, which hinders its ability to send clear price signals. In addition, the tax does not cover some important pollutants (e.g. PM emissions, NH_3) and sectors (e.g. agriculture). To strengthen the price signal, taxes could apply separately to each type of emission and could act as a price floor to the EU ETS for CO_2 emissions, given the significant overlap between facilities covered by the IEGA and the EU ETS. Andalusia could also consider broadening the scope of the tax to address the harmful effects of emissions not currently covered, for example by extending the taxable base to other pollutants, such as fine particular matter (PM) emissions, and other sectors, such as waste management and the agriculture sector. The analysis also finds that distributional considerations could be addressed through complementary policy instruments, including revenue recycling in the form of direct support to firms for the adoption of abatement technologies.

In the area of road transport, no specific tax instrument applicable in Andalusia incorporates air pollution and congestion costs, despite their significance at the subnational level. The local and regional governance levels are well-equipped to implement congestion charges in urban areas where benefits are likely most important. Implementing congestion pricing at the regional or local level will help manage local congestion problems and improve local air quality. In addition, Andalusia could consider introducing a regional tax on vehicle emissions or a feebate (i.e. that penalises higher emissions vehicles and subsidises zero-emissions vehicles) that accounts for both GHG emissions (of which 99% were CO_2 emissions in 2019) and other air pollutants arising from road transport. To avoid supporting efficient internal combustion engines that still emit CO_2 and air pollution, favourable tax treatment should be targeted to zero-carbon emissions vehicles only. As aligning vehicle tax rates with external cost estimates of emissions is administratively complex, an alternative could be to vary tax rates according to environmental indicators such as the Euro emissions standards for vehicles and to increase their stringency over time. Caution should be exercised when implementing a tax on vehicle emissions at the regional level as it would increase the compliance burden for vehicle owners, who are already subject to a vehicle registration tax at the national level and a vehicle circulation tax at the municipal level.

To contribute to the targets of climate neutrality and reducing mobility-related GHG emissions, reform efforts to decarbonise road transport could be prioritised by focussing on substantially increasing the share of zero-carbon vehicles. In the context of national and EU targets for climate neutrality by 2050 and the Andalusian goal to reduce mobility-related emissions by 30-43% in 2030 (compared to 2005), tax and non-tax policy tools can play an important role. Pioneering countries, like Norway, have been successful in advancing the decarbonisation of their fleet by setting clear targets of zero-emissions vehicles and using additional steering instruments, including taxation. Andalusia could set a clear target indicating the share of electric vehicles in total future car sales to provide a tangible milestone and certainty to economic actors. Taxation could also play an important role in the decarbonisation process, for example through consistent fuel excise and carbon pricing, favourable taxation of zero-emissions vehicles and carefully designed tax incentives to purchase zero-emissions vehicles. This would need to be accompanied by complementary tools such as emissions regulations, investment in charging infrastructure and clean electricity generation and good communication. Co-ordination between the regional, national and EU-level is important to ensure the effectiveness of such policies.

Risks linked to climate change call for a focus on demand-side instruments to address water scarcity. Water scarcity has traditionally been addressed in Spain through supply-side infrastructure (e.g. dams, wells, inter-basin water transfers), but climate change is placing renewed focus on demand-side instruments (e.g. levies, regulation). A water abstraction levy, which would reflect the environmental costs arising from the process of extracting water from a natural source, is one option to address water scarcity and the environmental externalities arising from water use. Non-price tools such as water allocation regimes could also be considered if water users are generally unresponsive to pricing.

Currently, water pricing relates to supply cost recovery and does not capture the environmental externality costs that arise in the context of water use, including harm to ecosystems and reduced potential for carbon sequestration. A water abstraction charge has the potential to encourage sustainable use of water and to price environmental externalities. However, there are political economy and practical barriers to the introduction of such a levy, particularly given the special status of water as a human right and lack of information about water demand responsiveness. Alternatives to a water abstraction levy include water use regulation and water allocation regimes; the latter can be more effective in the case of water use, given the generally low responsiveness of water users to prices. Other non-pricing tools to reduce environmental harm arising from water abstraction include mechanisms at the water user association level to monitor the informal extraction of water (i.e. through wells), which covers a non-negligible share of agricultural water use. Finally, ensuring policy coherence as well as setting clear policy goals and priorities is key for achieving water use sustainability and fairness without prejudice to other policy areas. In particular, it is important to pay careful attention to coherence between agricultural policies and water use concerns.

There are a number of tools available to help address externalities arising from water pollution, which causes damage to health and ecosystems and increases water treatment costs. Andalusia currently levies a pollution control fee for urban and industrial use. However, this fee does not cover pollution arising from the agricultural sector, which is the main sector responsible for aquifer pollution. Unlike water pollution arising from point sources, where the pollution levels can be determined by a regular measurement of discharges, most water pollution in the agricultural sector arises from diffuse sources and is difficult to measure. As an alternative to the pollution levy, a tax on polluting inputs could be a means to capture the environmental and economic impact of this source of pollution. Levying a tax on pesticides and fertilisers, which are responsible for a significant share of water pollution in the agricultural sector, would allow the region to address the related environmental and economic externalities.

Tax instruments may be more efficient than the current limits that apply to fertiliser and pesticide use, but complementary policies may be needed given evidence of low responsiveness of farmers to input taxes and political barriers. As the tax base for input taxes is likely highly mobile, whereby farmers could source their inputs from a neighbouring region with lower tax rates, coordination with other Autonomous Communities would be key and could increase responsiveness levels. Political economy barriers may be addressed through the use of revenues to support the best performers and through careful tax design, as evidence shows that aligning the tax liability with the environmental damage increases the chances of the tax being perceived as fair.

Pricing instruments can also help reduce waste generation, avoid incentivising waste imports and increase circularity of materials. The disposal tax on hazardous waste introduced at the national level in 2022 (which is currently being implemented and would replace regional-level taxes) applies lower tax rates to landfill of hazardous waste than those that currently apply in Andalusia. Given Andalusia is already managing more hazardous waste than it generates, allowing the effective taxation of landfill to decrease could lead to a surge of hazardous waste imports into the region. Andalusia could apply a surcharge to the national tax rate for hazardous waste so the combined regional and national rates match the existing level of taxation in Andalusia. Additionally, the proposed regional increase of the national tax rate on Construction and Demolition Waste (CDW) disposal, together with an aggregates extraction tax (see following paragraph), could increase the circularity of the building sector, which is one of the sectors with greatest material use and low material recovery rates.

As Andalusia is a significant producer of raw mining materials, there is scope to encourage materials recycling and reduce the extraction of raw materials. Forty percent of the Spanish mining production value comes from Andalusia. A tax on virgin aggregates is one option to reduce the consumption of raw materials in favour of more material recycling and thereby reduce environmental impacts related to extraction. To avoid imports of raw materials from bordering regions that apply no taxes on raw materials extraction, this regional tax should not exceed EUR 3 per tonne. Given the lack of information on differentiated environmental impacts by type of material extracted, which prevents the implementation of a detailed Pigouvian tax, two alternative approaches could be considered: an *ad valorem* tax on the value of the raw materials extracted and an *ad quantum* tax on the quantity of raw materials extracted. An *ad quantum* tax would be more straightforward, as it better reflects the environmental impact, is simpler to administer, and has more predictable tax revenues. As an *ad quantum* flat tax rate applied to all raw materials would have a greater relative impact on cheaper materials, materials could be grouped by market price and a tax schedule with three tax brackets applied to the different groups (i.e. lower tax rate for cheaper materials). More analysis is needed to understand the effects of such a tax on competition.

Technical summary

The *Environmental Tax Policy Review of Andalusia* report supports the government of Andalusia in developing plans for potential reforms to its environmentally related tax legal framework, with a view to improving regional environmental outcomes and enhancing national and global environmental performance. The report is the outcome of a project on "Technical support for an integral reform of the environmental tax legal framework of the Autonomous Region of Andalusia" funded by the European Union (EU) via the Technical Support Instrument (TSI), and implemented by the OECD, in cooperation with the European Commission.

The report provides strategic recommendations for environmentally related tax reform in three key areas; greenhouse gas (GHG) emissions and air pollution (with a focus on stationary sources of emissions and on road transport), water usage and pollution, and waste and circular economy. The recommendations derive from a thorough review of the legal framework at the regional, national and EU level and from an assessment of how existing taxes and fees applicable at the Andalusian level align with principles of good environmental tax policy. Concrete practical examples underpin the assessment of specific instruments and their design.

This technical summary presents key findings and strategic recommendations from the report.

The multi-level governance framework in Spain and Andalusia

Spain is a quasi-federal country with a three-tier system of subnational government whose autonomy is constitutionally recognised. The first tier of the subnational governance structure is composed of 17 autonomous communities, the second tier is made up of 50 provinces, and the third tier comprises 8 131 municipalities and two autonomous cities (Ceuta and Melilla). Spain also has an asymmetric system of subnational governance. Andalusia is one of the 15 autonomous communities that fall under a "common regime", while the Basque Country and Navarra fall under a "foral regime" that provides them with special responsibilities and more fiscal autonomy.

In addition to the Constitution of Spain, each autonomous community is governed by a Statute of Autonomy through which the central government may transfer some of its responsibilities. The Andalusia Statute of Autonomy (*Estatuto de Autonomia de Andalucia*) was adopted in 1981 and revised in 2007 and provides for the transfer of powers, including revenue raising capacities, from the central to the regional government. This Statute also provides full guarantee and protection of local autonomy and is complemented by the organic 1985 Law regulating the basis of local administration (*Ley reguladora de las bases del régimen local (LBRL)*, which sets the framework of the local government system. Local and regional responsibilities are defined in accordance with the principle of subsidiarity, where decisions should be taken at the lowest decision-making level possible. The legislation of higher levels of government must ensure that lower levels of government have the right to intervene in matters that affect their interests and have adequate powers, in accordance with the principles of decentralisation, proximity, effectiveness and efficiency.

As Spain is one of the most decentralised countries in the OECD, subnational governments assume significant responsibilities for public spending and service provision. Subnational governments were responsible for almost half (47.3%) of total public spending in 2020, amounting to 24.8% of GDP, above the OECD average (respectively 36.6% and 17.1%). The autonomous communities may assume both exclusive and shared responsibilities with the central government. Exclusive responsibilities must not fall under the central government's remit and are listed in the Constitution of Spain. Environmental protection is a regional responsibility (though legislation must be in line with national and EU legislation), which provides Andalusia with extensive obligations in the areas of environment and climate. The region also has responsibilities in areas related to the green transition, such as transport, economic development, agriculture and forestry, water management, regional planning and housing. Responsibilities may differ across autonomous communities as the exact allocation of responsibilities is determined by each community's Statute of Autonomy. Municipalities are in charge of waste collection and treatment, drinking water supply system and urban environmental protection, as outlined in the 1985 LBRL. This differs across municipalities; for example, waste treatment and urban environmental protection may revert to regions for very small municipalities.

Tax revenue accounted for 37.5% of total subnational government revenues in 2020 in Spain. Autonomous communities can establish own-source taxes, apply a surcharge on centrally levied taxes (with some limitations), and have some discretion over assigned taxes (e.g. exemptions). The central government is responsible for establishing and administering assigned taxes, while the revenue is wholly or partially shared with the autonomous communities. By contrast, own-source taxes are created by the autonomous communities and must be based on a taxable event that is not already subject to tax by the central government or by municipalities (e.g. the Tax on Gas Emissions into the Atmosphere, the Tax on Discharges into Coastal Waters or the Single-Use Plastic Bag Tax in Andalusia). If the taxable event is already subject to tax at a lower level of government, the government that established the new tax must compensate the lower level of government for the revenue loss. Similar to autonomous communities, municipalities can finance their responsibilities through their own taxes and assigned taxes from the autonomous communities and the central government.

Parts of the Spanish tax system are currently under review. The report includes proposals from the "White Book for the reform of the tax system and its adaptation to the reality of the 21st century" published in March 2022 by the *Committee of experts to prepare the White Book on the tax reform*, established by the Spanish Treasury (Comité de personas expertas (2022[1]), Labandeira (2022[2])). This White Book analyses the tax system as a whole and considers topics such as environmental taxation, corporate taxation and property taxation.

Stationary sources of GHG emissions and air pollutants

Over the past two decades Andalusia has been reducing GHG emissions, which are directly responsible for climate change. Climate change is an existential threat that is increasingly impacting ecosystems and people's lives. Since the global warming impact of GHGs[2] is independent of where the emissions occur, a concerted effort across countries and all levels of government is needed to address the threat of climate change. In Andalusia, carbon dioxide (CO_2) represented 80% of GHG emissions in 2019 and have steadily declined from about 75 $MtCO_2e$ in 2007 to about 54 $MtCO_2e$ in 2019. The main sectors responsible for CO_2 emissions are the electricity (29%), industry (24%) and road transport (31%) sectors, though the main sources for other GHGs differ. Agriculture is the main sector responsible for methane (CH_4) emissions (representing 95% of total CH_4 emissions in Andalusia) and nitrous oxide (N_2O) emissions (86%).

Air pollutant emissions have direct and localised effects on human health and on the environment but have generally followed a downward trend over time in Andalusia. Air pollutants might indirectly impact climate change, but the direct impact of air pollutants is often local, and their harmfulness generally depends on local conditions, such as population density and weather conditions. [3] The OECD's *Air pollution effects* indicator estimates that in 2019, exposure to fine particulate matter ($PM_{2.5}$) caused 190 deaths per 1 000 000 inhabitants in Spain (OECD, 2022[3]). Evidence shows that beyond the health and environmental impacts, air pollution may also have detrimental effects on the economy by reducing worker productivity, increasing public health expenditure or causing loss of crop yield. In Andalusia, air pollutant emissions followed a downward trend between 2003 and 2019, with reductions of up to 80% for SO_2. Ammonia (NH_3) emissions experienced a much smaller decrease of 6% over the same period, declining significantly until 2011 and then increasing to 2019. In 2019, the main anthropogenic sources of air pollutants in Andalusia differed across emissions; SO_2 arose principally from the industry sector (about 51%), maritime traffic (22%) and electricity production (20%). Nitrogen oxide (NOx) emissions arise principally from road traffic (28%), agriculture (26%), maritime traffic (13%) and electricity production (11%) and $PM_{2.5}$ from buildings (39%) and agriculture (32%).

Andalusia has established multiple instruments to reduce GHG emissions and air pollution, which align with national and EU level action. Andalusia has pioneered regional-level action on climate change; the 2002 Andalusian Strategy on Climate Change was the first initiative of its kind in Spain. The autonomous community has since adopted several measures to strengthen climate change action, including its latest Andalusian Climate Action Plan (PAAC) in 2021, based on the 2018 Andalusian Law on Measures Against Climate Change and the Transition Towards a New Energy Model (Law 8/2018). The main objective of the PAAC is to reduce GHG by 39% by 2030 compared to 2005 and is composed of three programmes (mitigation, adaptation and communication). The PAAC aligns with national and EU-level action on climate change, specifically the Spanish Climate Change and Energy Transition Law (Law 7/2021) and the European Climate Law (EU Regulation 2021/1119). To complement the PAAC and reduce the emissions of other atmospheric pollutants, Andalusia adopted the Strategy for Air Quality in 2020, based on the Andalusian Law 7/2007 and in line with the Spanish National Program on Atmospheric Pollution Control (Royal Decree 818/2018) and EU Directive 2016/2284. The strategy aims to support the preparation of air quality improvement plans by local governments and provides a comprehensive assessment of air quality at the local level, including the main pollutants emitted by sectors.

The Andalusian Tax on the Emission of Gases into the Atmosphere (IEGA) is an important element of regional action on stationary sources of GHG and air pollutant emissions. Introduced in 2003, the IEGA covers emissions from CO_2 and from two important air pollutants, nitrogen oxyde (NOx) and sulfur oxyde (SOx). The IEGA covered about 70 installations in 2019, of which about 40% were electricity providers or autoproducers of electricity (i.e. some industrial firms) and 60% were in the manufacturing industry. Slightly more than 80% that were covered by the IEGA were also subject to the EU ETS. To calculate the IEGA liability, CO_2, NOx and SOx emissions are each adjusted by a reference value, which is similar to the threshold levels set in the European Pollutant Emission Register (EPER) Decision. The resulting adjusted CO_2, NOx and SOx emissions are added up to form one taxable base, referred to as polluting units, which are subject to a progressive tax rate schedule, with marginal tax rates ranging from 0 to EUR 14,000 per polluting unit. Exemptions apply to emissions from landfills and facilities for the intensive rearing of animals as well as those from the combustion of biomass and biofuel.

The IEGA follows good administrative practice and is more comprehensive than pricing tools that only cover CO_2 emissions. The IEGA determines which entities are covered by the tax through physical characteristics (e.g. production capacity and storage, level of thermal power) instead of emission thresholds. This allows entities to be clearly identified and avoids relying on emissions reporting, which would create a high administrative burden on installations and the verifying entities. The IEGA also played a pioneering role in air pollutant emissions pricing, with Andalusia being one of the first regions to establish such a tax, and represents a more comprehensive approach to emissions taxation in its effort to cover CO_2 emissions as well as NOx and SOx emissions.

The design of the IEGA is complex, which risks muting its price signal and providing unintended incentives. The IEGA currently bundles all three gases into a single tax base and applies to polluting units rather than quantities of pollutants emitted (e.g. in tonnes). This reduces the salience of the tax as it makes it difficult for firms to know how much they are taxed on each type and unit of emissions. Rather than incentivising reductions across all types of emissions, the combined tax base allows firms to offset increases in one type of emissions by reducing another type. Moreover, the reference values used to adjust the quantities of emissions are not based on emission limit values or the relative harmfulness of the different types of emissions; rather they are based on reporting thresholds intended to cover most emissions at a limited administrative burden (European Commission, 2017[4]).

The IEGA sets a very low CO_2 price signal compared to recommended price levels for the transition to net-zero. The average marginal rate is EUR 0.036/tCO_2 for installations covered by EU ETS and IEGA and almost all installations covered only by the IEGA fall below the IEGA exemption threshold and have a tax liability of EUR 0. For installations covered by both the IEGA and the EU ETS, the price signal from the IEGA adds very little to EU ETS permit prices, which in 2019 averaged EUR 25/tCO_2. Moreover, the low price level of the IEGA does not help reach the standard low-end benchmark price level needed to trigger meaningful abatement efforts, which is EUR 30/tCO_2 in 2021 and EUR 60/tCO_2 in 2025. It also does not help attain the EUR 100/tCO_2 Social Cost of Carbon retained by a recent European Commission study (European Commission, 2021[5]) either.

The tax rates on NOx and SOx that apply through the IEGA are at the lower range of observed rates worldwide, but they are similar to those observed in other parts of Spain. Benchmark price levels are rarely discussed in the case of air pollutants, as these are location specific. Indeed, where the objective relates to target reduction levels, the cost and local availability of abatement technologies should be considered as this will affect how reactive firms are to the tax. If the objective relates to costing the externalities that arise from air pollution then the local population, climate, weather and environmental considerations should be accounted for.

As GHG and air pollutant emissions from stationary sources are already taxed in Andalusia, the main legal possibility for tax reform is to broaden the scope of the IEGA. The tax base of the IEGA can be expanded to cover other pollutants (e.g. particulate matter (PM) emissions, NH_3) and other industrial or productive activities (e.g. waste management, poultry) that are currently exempt. Other possible improvements include simplifying the calculation of the tax value and updating the tax with the current regulatory framework at the national level. As suggested by the White Book for Tax Reform in Spain, the design of this tax would also benefit from harmonisation between regions that levy an equivalent tax.

Based on the assessment, the following strategic recommendations are proposed (Box 1).

Box 1. Strategic recommendations – Stationary sources

General recommendations regarding the Andalusian Tax on the Emission of Gases into the Atmosphere (IEGA)

Separating the **IEGA into three tax schedules, each applied per tonne of emission of CO_2, nitrogen oxide (NOx) or sulphur oxides (SOx)** would make the price signal more salient, and would enable a better alignment of price levels with environmental costs and mitigation targets.

Applying flat IEGA tax rates would ensure coherent abatement incentives, ensuring that the price signal aligns with the environmental costs of each unit of emissions rather than increasing with total emissions as currently.

Compared to the progressive tax rates that currently apply, **equity concerns for firms of different sizes are better dealt with by complementary instruments, which can be direct or indirect**. Indirect support could include a **time-progressive phase-in of the tax base and rates** to give firms the opportunity to make the necessary investments. Direct support could include **subsidies for green technology adoption**.

Subsidies for firms should be carefully designed to ensure they are properly targeted (e.g. tailored to firm size) and are effective in addressing affordability and competitiveness concerns. The revenue implications of such subsidies need to be considered; for example, they could be implemented using the general budget or the revenue from green taxes (revenue recycling).

Reform options for Greenhouse Gas (GHG) emissions

Given that GHG emissions are a global issue, **the regional level may not be the most suitable** governance level for regulation in this area. In terms of effectiveness, climate change and GHG mitigation are best dealt with at a national or even supranational level. Indeed, this enables emissions cuts where they are the cheapest at a much larger scale and can help avoid carbon leakage.

Reform options for air pollutant emissions

Air pollutant emissions (e.g. NOx, SOx, ammonia (NH_3), particulate matter (PM) emissions) are principally a local issue, which makes them **a suitable target for regional level action**.

If the **objective is to reach specific air pollution reduction targets**, tax rates would need to be **set with reference to local air pollution reduction targets** and with reference to **available mitigation technologies and costs** (i.e. where mitigation is relatively inexpensive, lower tax rates may be sufficient to incentivise behavioural shifts).

If the objective is to **reflect external costs from pollution in tax rates**, Andalusia could consider **including population density in the calculation of tax rates**. This would better align price levels with health costs (which are higher in more populated areas) and possibly discourage firms from settling in densely populated areas going forward. A similar measure applies in Chile (Box 3.7)

An **extension of the tax to cover PM emissions** from industrial and electricity sector stationary sources could be considered. This would be relatively straightforward to implement given that NOx and SOx emissions are already measured and taxed. Moreover, this would deal with one of the most harmful air pollutants for human health.

The current IEGA exemption for emissions of NOx, SOx and PM arising from biofuels could be removed. Indeed, while biomass might be carbon-neutral over the life cycle, it is a highly emitting fuel in terms of air pollutants.

Reform options for emissions from the farming sector

An extension of the tax to the farming sector would entail **extending the coverage to other air pollutants, such as NH_3, and to other GHGs, such as nitrous oxide (N_2O) and methane (CH_4)**. The agricultural sector in Andalusia is responsible for a low share of CO_2 emissions but is the main source of CH_4 and N_2O emissions (95% and 86% respectively). The sector is also responsible for the air pollutants NOx, PM and NH_3 emissions.

The **emissions measurement methods should be adapted** to capture emissions from the farming sector. Farm-level emissions, of which a low share arises from fuel use, cannot be measured in the same way as firm-level emissions. Such a reform therefore requires emissions the measurement to capture farm-level emissions.

Extending taxation of GHGs and air pollutants to the farming sector would require **dialogue and engagement with stakeholders**, **proposals for and existence of alternatives**, and **support for farmers in the transition**. Andalusia could adopt a similar approach to the New Zealand 2022 proposal for taxing farm-level emissions (see Part I) and should ensure dialogue with farmers highlights the benefits that better air quality and mitigated climate change would have on their sector and employees.

Greenhouse gas emissions and air pollutants arising from road transport

Road transport is a major source of GHGs and local air pollutants due to the combustion of fossil fuels in vehicles and causes a range of additional external costs unrelated to fossil fuel use. Road transport is responsible for external costs due to the combustion of fossil fuels in vehicles, as well as potentially substantial costs related to accidents, congestion, noise and road damage. In Andalusia, CO_2 emissions from road transport increased by 12% between 2011 and 2019 but remained stable as a share of the region's total CO_2 emissions over the same period (around 28%). In contrast, air pollutant emissions followed a downward trend over the same period; in 2019 road transport accounted for 12.6% of nitrogen oxide (NOx) emissions (compared to 34.7% in 2011), 2.7% of fine particulate matter ($PM_{2.5}$) emissions (22.7% in 2011), and 3.2% of particulate matter (PM_{10}) emissions (21.5% in 2011). Although air pollution from vehicles has been decreasing, it remains an important source of pollution, especially in urban environments.

The Andalusian Climate Plan (PAAC) sets out the region's objective to reduce GHG emissions from road transport and complements national and EU-level action. The PAAC establishes a reduction target between 30% and 43% by 2030 compared to 2018 levels for transport and mobility. In addition, the Spanish Climate Change Law (Law 7/2021) sets out the national objective to reach a zero GHG emissions fleet of passenger cars and light commercial vehicles by 2050, in line with EU Regulation 2019/631 setting CO_2 emission performance standards for such vehicles. To help autonomous communities promote electrified transportation, the central government launched the Moves III Plan in 2021 (Royal Decree 266/2021). The Spanish Climate Change Law also promotes the adoption of sustainable urban mobility plans by 2023 for large municipalities with the aim to reduce emissions from mobility. This includes, among others, the development of low-emission zones, the improvement of the public transport network and its electrification.

Private drivers and vehicle owners in Andalusia are currently liable for national and local taxes and charges, but these do not fully reflect environmental externalities from private road transport. At the national level, drivers are subject to an excise duty on fuels and are liable for the national vehicle registration tax (a one-off tax paid on the first registration of a vehicle). In addition, municipality-specific annual circulation taxes apply. However, there is currently no direct taxation on emissions from non-stationary sources in Andalusia and drivers currently do not face road tolls or congestion charges.[4]

Well-designed tax policy can contribute to pricing the costs from environmental damage caused by drivers and vehicle owners. Pricing can improve transport-related decision-making and environmental outcomes by reflecting the costs from environmental damage. Different tax measures account better for different types of external costs. For example, vehicle taxes can reflect external costs related to average vehicle characteristics and fuel excise taxes can capture external costs related to fuel types. When these costs vary with the location and time of driving or depend on population and ecosystem exposure, distance-based or location-specific charges (i.e. road tolls or congestion charges) are better suited. Beyond taxation, non-price policies like fuel or emission standards are another option to achieve environmental policy goals.

Fuel and carbon taxes are well suited to account for the external costs related to CO_2 emissions, as CO_2 emissions are proportional to fuel consumption; but the Region has little room for manoeuvre as they are set at the national and EU level. The current Spanish fuel excise tax rates exceed a low-end estimate of climate costs today. This does not mean the rates are necessarily too high, as climate cost estimates are highly uncertain and external costs from fuel use are broader than climate costs. Fuel excise tax rates vary across fuel types and users and are not based on carbon content, which leads to an unequal treatment of taxpayers and potential distortions. However, Andalusia has no direct control over these rates that are regulated at the national level in alignment with the EU Energy Tax Directive.

Vehicle taxes can reflect the average emissions profile of a vehicle but are currently not used for this purpose in Andalusia. Vehicle taxes can account to some extent for the range of health and environmental impacts arising from vehicles, as they can reflect those impacts linked to average vehicle characteristics. Such taxes can have unintended effects however; for example, the Spanish registration tax varies by vehicle type and CO_2 profile of a car, but as it does not account for air pollution profiles it has the potential to stimulate the sale of diesel cars, despite the relatively large negative impact of diesel on health and the environment through air pollution. Moreover, current municipal-level annual ownership taxes account for neither CO_2 emissions nor air pollutants. While well-designed vehicle taxes have the potential to reflect a range of vehicle emissions types, they are less efficient in targeting driving-related and location-specific external costs, such as congestion and population exposure to air pollution.

Distance-based fees or congestion charges are currently not used in Andalusia, although they have the potential to deliver more efficient road transport if carefully designed. Andalusia does not currently levy distance-based or congestion charges, though these can usefully reflect driving-related external costs like congestion and accidents, which are not covered in fuel or vehicle taxes. Benefits from distance-based charging are also evident in terms of their revenue stability, as driving likely adjusts less quickly to pricing and taxation than energy use. While distance-based charging has several downsides, such as their administrative and implementation costs and privacy concerns related to data collection, technological progress is reducing costs and may remedy privacy concerns. Potential distributional concerns may be alleviated by using part of the revenue for public transport improvements or direct transfers to low-income drivers that have no alternative to driving in the short run.

An alternative means to charge distances driven is to take odometer readings, though additional information would be needed to fully reflect the external costs. Distance-based charges relying on odometer readings assess distances travelled by a vehicle without collecting detailed information on when and where the driving took place. The downside of these types of charges is they cannot vary with location and congestion levels and would only cover cars registered in Andalusia. Nevertheless, implementing odometer readings aligns better with external cost management than having no distance-related charging at all.

Synergies and coordination with other levels of government and between tax and non-tax policy instruments is crucial for successful environmental policy. In the climate context, for example, Andalusia can act in coordination local, national and supranational governments and draw on other instrument in the climate policy toolkit in addition to taxation. Other relevant climate policy instruments at the Spanish national level or at the EU level include the Spanish National Fund for the Sustainability of the Electricity System (FNSSE), national and European regulations on emissions of air pollutants, the European regulation on GHG emissions from vehicles, Euro standards, the EU Energy Tax Directive, and the EU Emissions Trading System, including the potential extension to road transport. These policies are currently being reviewed with the intention of increasing their environmental policy stringency, particularly in relation to carbon neutrality. If EU or national level policies become more ambitious, the scope for regional level activity declines. Nonetheless, it will be important for Andalusia to adapt the use of taxation as the policy mix and regulatory approaches evolve in the future; for example implementing low emission zones and integrating the upcoming Euro 7 standard.

From a legal perspective, Andalusia has the possibility to establish several new taxes since there is currently no direct taxation on emissions from non-stationary sources in Andalusia. Andalusia has several options to target emissions and air pollution, given the region's competencies (e.g. areas related to environmental protection and transport) and given existing tools at the national and EU levels (e.g. fuel excise taxes) and at the municipal level (e.g. circulation taxes). Reforms could include the introduction of a tax on the emissions from mechanical traction vehicles (i.e. motor vehicles). Additional legal possibilities include the creation of a congestion charge for polluting vehicles that circulate in central urban areas, aligned with the potential creation of the Low Emission Zones, and encouraging the central government to update its vehicle registration tax (e.g. fewer exemptions, updating the tax rate bands more regularly to account for technological advances).

Based on the assessment, the following strategic recommendations on tax reform are proposed (Box 2).

Box 2. Strategic recommendations – Road transport

General assessment of the road transport tax framework

On an external cost basis, **the tax framework applicable to drivers and passenger car owners in Andalusia could be improved**. Fuel taxes, for example, apply heterogeneously and are not based on carbon content. No specific tax instrument applies to incorporate costs from air pollution and congestion, despite their significance for and variation at the local level.

The sub-national level is well-placed to manage pricing of air pollution and congestion. Taxes (or feebates) targeted to the emissions of vehicles or congestion pricing in urban centres can help manage local congestion problems and improve local air quality.

Focusing tax reform on reflecting external costs can fail to address additional policy considerations, such as the transition to a zero-emission transport sector or other key transport and environmental fiscal policy goals, such as managing distributional consequences or revenue raising.

Engaging in tax reform can take up significant administrative resources and political capital, **the government is therefore encouraged to decide on a ranking of policy objectives before starting a comprehensive reform process**. If the main goal of the Andalusian government is decarbonisation in the road transport sector, administrative resources may better be used to design a reform that strongly encourages zero-carbon vehicles to enter the fleet, instead of engaging in marginal but burdensome reform that aligns the tax framework with external cost estimates.

Reform options for a cleaner passenger vehicle fleet

A combination of tax elements may help Spain and Andalusia to push for a clean passenger vehicle fleet in the context of existing and future policy mix.

- **Consistent fuel excise and carbon pricing will align climate incentives across the economy and provide strong signals that fossil fuels are not the future.** These considerations are relevant for the national or the EU level.
- **Vehicle taxes could be reformed to cover average CO_2 emissions jointly with air pollution profiles of vehicles**. Israel is a practice case of a vehicle tax with broad coverage of CO_2 and air pollutants that tracks their external costs in detail (see Box 4.2). It would be important to favour zero-emission vehicles only. A downside of fine-tuning tax rates to different emissions and external costs is its administrative complexity.

  An alternative could be to vary tax rates according to environmental indicators such as the Euro emissions standards for vehicles[5] that increase in stringency over time. The vehicle tax could also be transformed into a feebate that charges a fee on dirty and large vehicles and subsidises zero-emissions vehicles of regular size (France uses feebates for example (see Box 4.6)). Whether such reform would best happen at the national or regional level depends on advances of vehicle tax reform at the Spanish level.

The creation of an additional tax on emissions from vehicles at the regional level, as implemented in Catalonia, requires careful thinking. From a taxpayer perspective, the compliance burden may increase significantly when vehicle owners are subject to three different, but similar taxes, i.e. the existing national tax on vehicle registration, the potential new regional tax on emissions from vehicles and the existing municipal tax on circulation of vehicles.

- **Subsidies and tax incentives provided through the corporate or the personal income tax system can further support the adoption of clean vehicles but could have important budgetary implications**. They can help overcome consumer myopia, financial constraints and other barriers that prevent households from making the relevant investments. But these tools lead to forgone revenue or expenditures that need to be assessed. They also risk predominantly benefitting richer households. A means-tested approach directed towards low- and middle-income buyers may overcome such shortcomings.
- **Congestion pricing could be implemented at the regional or local level to help manage local congestion problems, while improving local air quality.** The local governance level may be well-equipped to implement congestion charges in urban areas where effects are likely most important. This can be done through combining information of density in different cities with traffic and air pollution data.
- **Preparing to use distance-based charges can be an alternative option.** If not pursued at the national level, local level action in this area can bring local benefits, such as better traffic management, reduced congestion, fewer accidents, lower air pollution and additional revenue.[6]

Independently of their principal objective (i.e. external cost management or decarbonisation of transport sector), the above points highlight that such tax policy choices have budgetary and distributional impacts that should be considered when designing environmental policy. It is also important to consider that electrifying the car fleet can only be successful if accompanied by significant investment in charging infrastructure for electric vehicles and will contribute to the net-zero transition only if electricity production is decarbonised.

Water usage and pollution

Water scarcity is a growing concern worldwide, in Spain and in Andalusia. Spain has historically high temporal and spatial variability in water resources, and Andalusia is one of the driest regions in the country. Traditionally, the Spanish government has mainly dealt with scarcity in regions such as Andalusia through supply-side instruments (e.g., dams, reservoirs, inter-basin water transfers). However, in the coming decades, freshwater availability is projected to decrease globally and drought cycles to increase. Climate change models project warming temperatures, increased variability in precipitation patterns, and more frequent and extreme weather events (OECD, 2020[6]). This calls for additional efforts through demand-side instruments, which include pricing and taxation.

Water pollution is caused by many factors and may originate from urban, industrial and agricultural users. Water pollution generates a range of external costs; it poses risks to human health and ecosystems, increases the costs of water use by raising treatment costs and increases water scarcity by reducing the quantity of water that is safe to use. In the urban and industrial sectors, water pollution is mainly due to wastewater and direct industrial discharges and arises from point sources, defined as direct discharges into water bodies at a discrete location, such as pipes and ditches from sewage treatment plants and industrial sites. Water pollution from the agricultural sector mainly arises from sedimentation and pesticides use as well as certain practices of nutrient use, animal feeding, livestock grazing and irrigation. These diffuse sources of water pollution are defined as indirect discharges to receiving water bodies, via overland flow and subsurface flow to surface waters, and leaching through the soil structure to groundwater. In recent years there has also been a growing focus on contaminants of emerging concern (CECs) from households, businesses and farmers (e.g., pharmaceuticals, industrial and household chemicals, personal care products, nanopesticides and nanomedicines) and the use of intrants in agriculture, in particular pesticides and fertilisers.

Legislation at the regional, national and EU levels sets out water-related environmental objectives and regulates responsibilities between different levels of government. The main legislation on water in Andalusia is the Andalusian Water Law (Law 9/2010), which establishes a set of environmental objectives and regulates the responsibilities between the autonomous community and local governments with the aim to achieve water protection and sustainable water usage. The law is in line with the Spanish Water Law (Royal Legislative Decree 1/2001), which determines river basin districts as the basic managerial units of Spanish water resources, and the European Water Framework Directive (2000/60/EC), which provides an integrated framework for the protection and sustainable use of water within the EU. The Andalusian Water Law regulates the organisation of the river basin district authorities and their management plans, the supply and sanitation system of urban water use, and the revenue earmarked for infrastructure and public service provisions, among others. In 2020, the Andalusian government also launched the Andalusian Water Pact to promote a participatory process on water-related issues. Additionally, the Royal Decree 47/2022 on water diffuse pollution produced by nitrates from agricultural sources provides River Basin Authorities with the possibility to establish limits on new water concessions and other activities that may result in nitrate contamination.

In Andalusia a number of pricing instruments apply to water users. The main water users in Andalusia are agriculture (about 80%, above the world average of 70%) and urban users. Industry that is not connected to the grid represents a much smaller share of water use in most Andalusian river basins. Several instruments are in place to directly price water use of urban users, while only one instrument prices water use from agriculture directly. These instruments are implemented at the national, regional or local level. The national-level instruments principally address service-cost recovery and environmental costs related to the installation of water extraction activities. The Andalusian and local charges address service-cost recovery as well as affordability and sustainable use criteria to a certain extent.

Pricing of water usage should satisfy several environmental and economic criteria. Addressing one of the criteria, however, does not guarantee that the other criteria will be addressed and there might be trade-offs between the different objectives. From an economic and environmental perspective, five key criteria should be considered (no order of importance) when designing a price for water usage:

- **Cost of service recovery** (i.e., water prices cover the full current and future supply, administrative and governance costs of water use and guaranty financial sustainability);
- **Universal access and affordability** (i.e. access to a minimum level of water for everyone);
- **Promotion of sustainable water use for human populations** (i.e. adapted to the issue of scarcity and avoid potential overuse and water losses in networks);
- **Internalisation of externalities caused to ecosystems** (e.g., depriving fish of their habitat, decreasing availability of water as a support to wetlands or for healthy vegetation);
- **Equity** (i.e. pricing ensures that the burden falls in an equitable way on users, for example avoiding misalignment between the costs that users generate and the price that they bear).

Water pricing in Andalusia could be better aligned with the good practices outlined above. Water use from agriculture is subject to irrigation charges, which principally seek to cover service-related costs and address equity between agricultural users. Urban users are subject to pricing instruments that deal with service-related costs as well as seek to ensure affordability and sustainable use. Water pricing currently does not account for the environmental externalities linked to water use.

Equity across water users could be improved in Andalusia, including across agriculture and urban users as well as among urban users. Agricultural water users only pay for the supply cost of the water used for irrigation, which is distributed amongst users according to the *share* of volume used, but does not directly depend on the volume *level* itself. Urban users, on the other hand, pay a fee that directly depends on their water use. Moreover, agricultural users sometimes also use water directly from wells, and then face no other cost than their private costs of extraction. The differential rates and coverage observed, however, may be justified by other reasons, such as accounting for the pass-through of price increases for farmers to food prices for households, or the different demand responsiveness of users. The design of the main water use fee in Andalusia, i.e. the improvement fee, can also create equity issues among urban users belonging to households of different sizes, due to the fixed exemption per household (rather than an exemption per person), the progressive fee structure, and the higher thresholds for the progressive fees that apply to large households.

Taxation could better address the external costs arising from water pollution in Andalusia. Water pollution damages human health and ecosystems (environmental externalities) and gives rise to economic externalities. For example, groundwater provides a non-negligible share of drinking water to both humans and animals, so the higher its pollution level, the higher water treatment costs are. For urban and industrial use, the main external costs from water pollution are currently addressed in Andalusia by the national level pollution control fee on discharges of water from those users. However, no pollution tax or fee applies in the agricultural sector, even though it is the main sector responsible for aquifer pollution today.

The main legal possibility identified at the regional level for water usage is the creation of a levy for water abstraction for agricultural and industrial purposes. A levy for water abstraction would reflect the environmental costs arising from the process of taking or extracting water from a natural source. Additional legal possibilities exist at the national level, such as developing incentive mechanisms on sustainable groundwater abstraction for Irrigation Communities or creating a tax that covers water-related environmental costs arising from tourism (this could also cover other environmental costs such as electricity and waste).

Regarding water pollution, the creation of a tax to disincentivise the use of pesticides and fertilisers at the national level constitutes the main legal possibility. A tax on the use of pesticides and fertilisers would reflect the environmental impact of water pollution from the agricultural sector, which can be difficult to target where water pollution arises from diffuse sources. The White Book for Tax Reform in Spain also proposed the creation of a national tax on the nitrogen content of fertilisers used in agriculture, combined with a VAT increase for these products.

Based on the assessment the following strategic recommendations are proposed (Box 3).

Box 3. Strategic recommendations – Water usage and pollution

Reform options for water usage

Given the **special status of water as a human right**, the price of water for different users is determined by government. As a result, the usual market dynamics that would increase prices and decrease demand when water supply is low are not in play. While this ensures water remains accessible and affordable, it fails to send a price signal to encourage users to reduce water consumption when supply is low. **Further government intervention may therefore be needed to promote sustainable water use.**

To better balance key criteria for water use pricing such as cost-recovery and financial sustainability, equity, affordability and sustainable use, the government of Andalusia is encouraged to **set clear objectives** and **acquire additional information on water supply costs and water demand responsiveness to prices** (price elasticity of demand). This could also allow for reflecting environmental costs of water abstraction (i.e. the costs on the ecosystem) in prices.

With respect to the **formal extraction of water** (e.g. through concessions) for urban, industrial and agricultural uses, **an abstraction charge at the Andalusian-level** could be put in place, to align with sustainable use goals of Andalusia. This is also a recommendation for the national level of the 2022 White Book for Tax Reform in Spain. France and Estonia have such taxes (Box 6.4).

Water allocation regimes are also an option and if designed properly can be more effective than pricing in the case of water use, given the generally low responsiveness of water users to prices. They can also allow a clearly identifiable share of water to be dedicated to the environment. Water markets based on such regimes exists in Australia (see Part II). However, these come with high administrative costs and may trigger unintended effects, such as lower return flows to water bodies.

With respect to the **informal extraction of water** (through wells, that may be legal or not), which covers a non-negligible share of agricultural water use, **monitoring mechanisms could be put in place at the user association level** (also referred to as Irrigation Communities). Monetary fines, for example, could be put in place if the groundwater body to which these wells are attached reaches poor quantitative status. This latter mechanism, however, would fall within the jurisdiction of Spain.

Evidence finds that **pricing policy effectiveness is enhanced when combined with non-pricing policies**. Accompanying measures, such as public awareness campaigns about water scarcity, information on water fees themselves aimed at increasing their salience or smart metering devices can contribute to increasing responsiveness.

Reform options for water pollution

The region could consider introducing a **price on polluting inputs, such as pesticides and fertilisers,** which are both responsible for an important share of water pollution. Currently, no price applies on water pollution from agriculture, and instead of applying a price to diffuse pollution directly, taxes on pesticides and fertilisers could target the quantities purchased of a specific product and tax rates could depend on their respective environmental impact. Pesticides on the European market are already risk assessed by the European Chemicals Agency, so defining products to be targeted by the tax and grouping them into different rate bands would be relatively straightforward. Norway has such a tax (Box 6.7).

However, evidence points to low responsiveness of farmers to input taxes. This stresses the **importance of complementary policies**, which can help farmers reduce pesticide use without risking an important decrease in yield or income, and ensuring the broader policy environment is aligned with water protection objectives (e.g., policies that promote quality of agricultural production over quantity).

If action at the regional level is envisaged, coordination with other Autonomous Communities is key to avoid farmers buying their input provisions (which constitutes a mobile tax base) from neighbouring regions with no input tax.

Advances in nutrient pollution modelling can provide an opportunity **to tax diffuse pollution outputs directly**, rather than taxing inputs that can only serve as proxies. This is particularly the case of fertiliser use, which is harmful to the environment if over-applied and not necessarily from first application. Such an approach could increase the efficiency and fairness of water pollution taxes, by **promoting a tax which would more closely align with direct environmental damage**.

General considerations

Introducing or reforming fees for water usage or pollution can involve **political economy sensitivities**. Political barriers may be addressed through **better communication on evidence-based results of pollution pricing mechanisms and earmarking of revenues**. For example, the revenue from such taxes could be used to accompany farmers in their transition to more sustainable agricultural practices.

This can also help increase responsiveness as well as address affordability issues and sustain the economic well-being of farming in Andalusia, which is a key sector of the region.

Price-based mechanisms are generally more cost-efficient, as they encourage abatement where costs are the lowest and provide continued incentives, for example. However, given the little knowledge there is about demand elasticities for water use and polluting inputs and given the high temporal variability in water supply, **regulation as a policy tool might in some cases be better suited to the context of water use and water pollution**. Regulations on input use already exist and they can be made more stringent. Regulation on water use, especially during summer months can also help address such issues.

Finally, ensuring **policy coherence** and setting **clear policy goals and priorities** is key in achieving water use and pollution sustainability and fairness without prejudice to other policy areas, in particular economic viability of rural communities. In this respect, **long-term and short-term goals and sustainability should be carefully assessed**. This trade-off can be illustrated by the recent take-off of avocado cultivation in Andalusia, which is substantially more profitable in the short-term than historic olive cultivation but less environmentally (and economically) sustainable in the long-term as much more water demanding.

Waste management and circular economy

The share of waste recycled in Spain is below the EU average and does not meet EU recycling targets. In 2020, Spain recycled only 36% of its Municipal Solid Waste (MSW) while the EU-27 average was 48%. Spain did not meet the EU recycling targets of 2020 and without further action it will be a challenge to meet future EU targets related to recycling and waste management. It is therefore imperative to investigate reform options in Andalusia to reduce waste generation and encourage a shift towards a more circular economy.

A substantial share of non-hazardous waste in Andalusia is disposed of in landfill and there is significant uncontrolled disposal of waste from the construction sector. In 2018, 18.34 million tonnes of non-hazardous waste were generated in Andalusia. This arose primarily from waste management facilities and water treatment (31%), municipal waste (27%), and construction and demolition waste (22%). More than a quarter of non-hazardous waste was landfilled, while 31% was recycled in Andalusia. Waste from Construction and Demolition Waste (CDW) poses a unique challenge; while authorised facilities achieve high recycling rates, up to 30% of CDW in Spain is uncontrolled and some is deposited in unauthorised places or their fate remains unknown. Accounting for this uncontrolled CDW, Andalusia is unlikely to reach the objective of 70% of non-hazardous CDW destined for reuse or recovery as established in the National Waste Framework Plan (PEMAR) for the year 2020.

A substantial share of hazardous waste disposed of in Andalusia is imported. In 2018, Andalusia generated 327,646 tonnes of hazardous waste, nearly half of which came from the waste recovery sector (26%) and the extractive and metallurgic industry (22%). The amount of hazardous waste treated and/or disposed of by Andalusia was around two and a half times the hazardous waste generated in the region, as some was imported from other countries or autonomous communities, and some was treated more than once (i.e. there are primary and secondary destinations for hazardous waste). The import of waste is partly due to Andalusia's (and Spain's) available waste treatment facilities and landfills, competitive pricing for waste disposal, and low population density.

Andalusia is a significant producer of raw mining materials, which causes environmental damage. Forty percent of the Spanish mining production value comes from Andalusia and includes fuels, metallic minerals, industrial minerals, ornamental rocks and quarry products. For instance, the province of Almeria within the region of Andalusia concentrates around 60% of the gypsum extracted in Spain. The extraction of raw materials (also called virgin aggregates) has environmental impacts, such as soil degradation, damage to ecosystem functions and air pollution from fine particles, as well as greenhouse gas emissions from energy use. Reducing the consumption of virgin aggregates and increasing the use of recycled aggregates would reduce environmental impacts related to extraction.

Traditionally, environmental taxes related to resources and waste are predominantly levied at the regional level, but in 2022 the Spanish waste law implemented nationwide waste taxes. At the regional level, Andalusia currently applies a waste landfill tax, as well as a tax on single use plastic bags and at municipal level, waste charges apply to households and businesses. In April 2022 a new national Spanish waste law (Spanish Law 7/2022), which is currently being implemented, introduced two nationwide waste taxes: one on non-reusable plastic packaging and one on landfilling, incineration and co-incineration of wastes. While the new Spanish Waste Law applies lower taxes on landfill compared to Andalusia's current landfill tax on hazardous waste, the Law allows autonomous communities to implement a surtax on the national tax rates. The law also foresees the implementation of a Deposit Refund System for beverage containers if Spain does not meet the collection target of 70% for bottles by 2023 as established in the EU Single Use Plastics Directive.

Recent taxes established at the national level and EU directives have implications for Andalusia's incumbent fiscal legislation and require action to ensure coherence. The existing waste disposal tax of Andalusia applies different categorisation of waste types and higher tax rates than the national tax. As a result, waste disposal would be taxed at a lower rate under the new national tax than under the existing regional tax. In addition, the transposition of the EU Single-Use Plastic Directive into Spanish law (Spanish Royal Decree 239/2018) bans the types of plastic bags that are taxed under the Andalusian plastic bag tax and leaves this regional tax without a tax base. A reform of Andalusia's tax regime is therefore necessary to ensure coherence between regional and national tax policies, as well as to better internalise external environmental costs related to waste management and resource use, including emissions to air or leachate to soil and water.

The main legal possibilities identified at the regional level for waste management and resource use is a surtax on the national waste tax for specific waste streams and taxation of aggregate material extraction. A surtax on the new national waste tax would preserve the existing level of taxation of hazardous waste disposal and maintain the current incentive scheme. Construction and demolition waste are not taxed at the regional level but will fall within the scope of the new national-level law on waste disposal, allowing Andalusia to implement a surtax. The extraction of raw materials is currently not subject to taxation, but environmental taxes on material extraction falls within the regional governments competencies and would help incentivise material recovery, recycling and the use of secondary materials.

Based on the assessment, the following two strategic recommendations for tax reforms are proposed to address the circularity of different economic sectors whilst increasing waste prevention and recycling (Box 4).

Box 4. Strategic recommendations – waste management and circular economy

Reform options for waste disposal taxes

Even with the current regional landfill tax rate for hazardous waste, Andalusia is receiving substantial amounts of hazardous waste and imports would likely increase if tax rates were to be lowered to the levels set by the Spanish Waste Law. The import of waste is partly due to Andalusia's (and Spain's) available waste treatment facilities and landfills, competitive pricing for waste disposal, and low population density.

To avoid an additional influx in hazardous waste imports, **Andalusia should apply a surtax to the national waste disposal tax rate to match the level of the current regional waste disposal tax.** The recommended surtax on the national tax to maintain current tax levels in Andalusia would avoid a surge in waste imports and maintain incentives for material recovery.

To increase material recovery for Construction and Demolition Waste (CDW), it is recommended to **increase landfill tax rates for this waste stream to EUR 5 per tonne (with pre-treatment) and EUR 3 per tonne (without pre-treatment).** As the national tax rate for CDW is low compared to regional taxes applied by other autonomous communities, a surtax on the national tax for CDW is recommended to increase incentives for material recovery in the sector.

Reform options for aggregates extraction taxes

Andalusia could introduce an environmental tax on aggregates to reduce the consumption on virgin aggregates and favour material recycling.

The tax design of such an aggregate tax requires **careful consideration**:

- In order to avoid imports of aggregate material from bordering regions, which apply lower or no aggregate taxes, the tax rate should not exceed 3 EUR per tonne.
- Since no studies are available for Andalusia that assess differentiated environmental impacts of extraction activities by type of material (making it difficult to implement a detailed Pigouvian tax), a straightforward option would be to apply an *ad quantum* flat rate to all aggregates. The drawback of this option is, however, that it would represent a greater relative impact on cheaper materials. For instance, an ad quantum tax of 1.35 EUR/t represents up to 91% of the price of the cheapest aggregates (e.g. clay and loam) and only 11% for the most expensive aggregate (e.g. siliceous sand). As a consequence, impacts on demand will also vary greatly in such a tax setting.
- Alternatively an *ad valorem tax* of 10% of the aggregate value could be charged, which would result in equal impacts on demand for different aggregates. This tax option would however result in less tax revenues and arguably in low-value materials being taxed insufficiently to effectively incentivise a shift to the use of secondary materials.

Based on these considerations, **an intermediate option is recommended, which would take the form of a differentiated ad quantum tax with three tax brackets based on an aggregate's market price.**

In addition, **the tax rate could be differentiated according to the location of the extractive activity** to account for higher environmental externalities in areas with high natural capital. Whilst this further differentiation may lead to greater sector acceptability, it would also complicate implementation.

The taxation of tourist stays

The tourism industry is one of the major economic sectors in the Andalusian economy but generates external environmental costs and places pressure on local infrastructure. While tourism makes important contributions to the regional economy, making up 13% of the region's economy and 14% of its employment, the sector also generates external costs such as pollution, noise and congestion, and overuse of ecosystems such as national parks and beaches. In addition, tourism affects infrastructure needs, as the seasonal inflow of tourists requires investments to increase the capacity of infrastructure such as roads and waste beyond what is required for local residents.

Existing pricing tools may not reflect the external costs or increased infrastructure needs linked to tourism. Some tourists will face a lower water abstraction charge than local residents, as hotels are considered non-residential urban users and are not subject to the fixed charge or progressive tax rates that apply to residents. As the tax rates applied to non-residential water users do not rise with use, water pricing for the tourism industry may not encourage sustainable use of water. Similarly, rental cars are currently exempt from the national registration tax, which means tourist drivers may also contribute less than local residents to the construction and maintenance of road transport infrastructure. This preferential tax treatment raises concerns regarding visitors' contribution to the external costs they generate and equity in the tax treatment of local residents and visitors.

Some European countries and cities levy tourist taxes to fund municipal expenses, but these taxes typically do not account for the environmental impact of tourism directly. While there is currently no specific tax on tourists in Andalusia, these taxes exist in several European countries at the subnational government level, such as in the cities of Amsterdam and Lisbon, as well as in the Spanish Autonomous Communities of Catalonia and the Balearic Islands (Responsible Travel, 2022[7]). Typically, tourist taxes are a fixed charge per night or are charged as a percentage of the price of the accommodation. Whilst several tourist taxes use revenues to relieve some of the (environmental) pressures caused by tourism, there are few cases in Europe where tourist taxes incorporate environmental criteria explicitly. The tourist tax in the Balearic Islands is a notable example of a tax that incorporates environmental considerations.

Andalusia could consider reducing the preferential tax treatment of the tourism sector in existing levies and, if needed, explore additional options to internalise external environmental costs and fund infrastructure. Reform to existing taxes on use or consumption, such as the water improvement levy, is one option to account for the external costs. This would allow the tax treatment to align more closely with actual costs incurred; for example, tourists using more water would face a higher tax liability. However, as Andalusia does not have direct control over certain taxes, the scope for action at the regional level will be limited in some policy areas. Where reform of existing taxes would not be legally or politically feasible or where existing taxes would not be sufficient to cover external costs, Andalusia could explore alternate options to internalise these costs such as a tourist tax. In addition, a tourist charge could account for the costs of constructing and maintaining additional infrastructure capacity, which is only used during tourism peaks and could incorporate environmental impacts of tourist activities, which cannot be reflected by reforming existing taxes. However, care should be taken when using proxy measures that do not reflect actual external costs; for example, higher rates on more luxurious lodging reflect the greater environmental impact, but do not incentivise tourists to reduce this impact.

References

Comité de personas expertas (2022), *Libro Blanco Sobre la Reforma Tributaria.* [1]

European Commission (2022), *Commission proposes new Euro 7 standards to reduce pollutant emissions from vehicles and improve air quality [press release]*, https://ec.europa.eu/commission/presscorner/detail/en/IP_22_6495. [8]

European Commission (2021), *Green taxation and other economic instruments - Internalising environmental costs to make the polluter pay.* [5]

European Commission (2017), "REFIT evaluation of Regulation (EC) No 166/2006 concerning the establishment of a European Pollutant Release and Transfer Register (E-PRTR)", https://eur-lex.europa.eu/legal-content/EN/TXT/PDF/?uri=CELEX:52017SC0710&from=en. [4]

Labandeira, X. (2022), *Taxation and Ecological Transition during Climate and Energy Crises: the Main Conclusions of the 2022 Spanish White Book on Tax Reform*, Real Instituto Elcano WP 09-2022. [2]

OECD (2022), *Air pollution effects* (indicator), https://doi.org/10.1787/573e3faf-en (accessed on 29 November 2022). [3]

OECD (2020), *Financing Water Supply, Sanitation and Flood Protection: Challenges in EU Member States and Policy Options*, OECD Studies on Water, OECD Publishing, Paris, https://doi.org/10.1787/6893cdac-en. [6]

Responsible Travel (2022), *Tourist taxes map*, https://www.responsibletravel.com/copy/tourist-taxes-map (accessed on 29 November 2022). [7]

Notes

[1] DG REFORM project 21ES30 ("Technical support for an integral reform of the environmental tax legal framework of the Autonomous Region of Andalusia) under the conditions set in the DG REFORM/OECD Framework Delegation Agreement Reform/Im2021/006.

[2] The main GHGs are carbon dioxide (CO_2), methane (CH_4), nitrous oxide (N_2O) and F-gases.

[3] The main air pollutants are sulphur oxides (SOx) and nitrogen oxides (NOx) (generally expressed as quantities of SO_2 and NO_2), carbon monoxide (CO), ammonia (NH_3), volatile organic compounds excluding methane (NMVOC), particulate matter (PM).

[4] External costs in road transport do not only relate to private driving i.e. passenger cars that are registered in the region (which are at the centre of the analysis) but can come from drive-through traffic, tourist and company cars, as well as non-passenger cars such as heavy and light duty vehicles, motorcycles, etc. These driving and vehicle types are not considered in detail in the analysis.

[5] The Euro emission standards for vehicles (set out in European Regulations 715/2007 and 595/2009) set limits on the emissions of air pollutants for vehicles. Recently, the European Commission presented a proposal of a new Euro 7 standard to further reduce air pollution from vehicles and to improve air quality (European Commission, 2022[8]).

[6] The White Book for Tax Reform in Spain also suggest such a reform proposal at the medium term, namely a tax on the actual use of vehicles that varies according to location, time and type of vehicle (Comité de personas expertas (2022[1])). Such a charge would replace most of the existing taxes in road transport (fuel, vehicles) and also those on congestion and infrastructure (should they be introduced). Such a charge would best be implemented gradually and considering potential distributional impacts – likely through the help of pilot evaluations (Labandeira (2022[2])).

Part I Governance in Andalusia

1 The multi-level governance framework and tax competences in Spain and Andalusia

1.1. Spain has a complex and asymmetric multi-level governance framework

Spain underwent a decentralisation process beginning in 1978 that led to the establishment of a quasi-federal country based on a three-tier system of subnational government. The Spanish multi-level governance framework is complex and characterised by strong political, administrative and fiscal asymmetries at the regional and local levels. This complexity is reflected in the allocation of responsibilities across the different levels of government as well as in the subnational finance system and tax competences, which differ from one region to another. Andalusia, the most populous and the second largest autonomous community by size, is part of this complex system, with its eight provinces and 785 municipalities. Andalusia has extensive responsibilities to develop and implement policy measures related to climate change and the environment.

1.1.1. Spain is a quasi-federal country

Although the 1978 Spanish Constitution established the country as a unitary parliamentary monarchy, Spain is also referred to as "the State of Autonomies" and described as a quasi-federation that has features of both a federal and a unitary country.

Spain has a three-tier system of subnational government whose autonomy is constitutionally recognised (article 137), composed of 17 autonomous communities, 50 provinces, 8 131 municipalities and two Autonomous Cities (Ceuta and Melilla) in 2022. Autonomous communities have a large degree of autonomy, including the exclusive ability to decide on the organisation of municipalities and provinces within their territory, which is often a prerogative of "state governments" in federal countries. However, unlike federal countries, the functions and finances of lower levels of government are determined within the framework of the national law and not by regional constitution or law, like in unitary countries (OECD, 2022[1]).

Like in many federations, regional governments are represented in the Parliament, in particular in the Senate which is meant to be the house of "territorial representation" (art. 69). Out of the Senate's 266 members, 58 are appointed by the autonomous communities' regional assemblies (see below). Each autonomous community appoints one senator and an additional senator for every million inhabitants within its territory (art. 69.5), according to a proportional system reflecting the composition of the regional assembly. The Senate has the authority to seize responsibilities from the autonomous communities if the Community is in breach of the Constitution (art. 155) (Gobierno de Espana, 1978[2]).

1.1.2. Spain has an asymmetric system of subnational governance

In addition to being a quasi-federation, Spain also has an asymmetric system of subnational governance, meaning that not all autonomous communities have the same statute of autonomy, resulting in differentiated responsibilities and fiscal systems between communities as well as asymmetries at the local government levels (Allain-Dupré, Chatry and Moisio, 2020[3]; Garcia-Milà and T. McGuire, 2007[4]).

Regional level

At the regional level, the decentralisation process that followed the 1978 constitution paved the way for the development of the autonomous communities, which were subsequently created through complex procedures from 1979 to 1983. The responsibility transfer process was carried out via a two-speed system with seven "fast-track" autonomous communities (*vía rápida*) that received a broad range of devolved responsibilities immediately and ten "slow track" communities, which received these responsibilities later. Andalusia was part of the "fast track" autonomous communities. Constituted as an autonomous community in February 1980, the first statute of Andalusia was approved in 1981 by the Spanish national government. As of 2003, the "slow track" autonomous communities have assumed all the same responsibilities as the "fast track" ones.

Despite this harmonisation between the fast and slow track autonomous communities, there are some remaining asymmetries between them. First, 15 of the 17 autonomous communities fall under a "common regime", while the Basque Country and Navarra fall under a "foral regime", which provides them with special financial responsibilities and more fiscal autonomy than the other autonomous communities from the common regime. Within the common regime, the Canary Islands has however a specific economic and tax system, especially as an EU outermost region. Second, while the autonomous communities are governed by the Constitution, they also have their own organic law, the Statute of Autonomy through which the central government may transfer or delegate some of its responsibilities to the autonomous communities. The law shall provide the appropriate transfer of financial means to the autonomous communities and the type of control that the central government retains regarding the responsibilities (art.150) (Gobierno de Espana, 1978[2]). As a result, autonomous communities each have their specific statute, allowing for some distinctive features. As the responsibilities of the autonomous communities may vary without the need to change the Constitution, provided that their transfer is adopted within this constitutional framework, this differentiation can increase over time. Since the mid-2000s, several statutes have been reformed on a case-by-case basis, for example, in Catalonia and Valencia in 2006, and in Andalusia (Box 1.1), Aragon, and the Balearic Islands in 2007.

As a general rule, regional governing bodies are composed of a regional assembly, the President of the regional government (*presidente*) and a government council. The regional assembly is the deliberative body of the autonomous communities and has devolved legislative powers. Its members are elected by direct universal suffrage for a four-year term. The President of the regional government is elected from among the regional assembly members for a four-year mandate (absolute majority of voting members). The government council is composed of the regional president and various regional ministers in charge of different offices (Consejerías). Andalusia has some specific institutional characteristics (Box 1.1).

Vertical coordination between the central government and the autonomous communities is made, on a voluntary basis, through the Conference of Presidents (*Conferencia de Presidentes*), created in 2004. Chaired by the Prime Minister, it includes the presidents of the 17 regional governments and the two autonomous cities and the central government. Vertical coordination also takes place through sectoral conferences such as the Council of Fiscal and Financial Policy (*Consejo de politica fiscal y financiera,* CPFF) in economic, fiscal and financial matters.

Horizontal cooperation is facilitated through the Conference of the Governments of the Autonomous Communities, which facilitates identifying shared positions of autonomous communities in negotiations with the central government as well as through the Federation of Spanish Municipalities and Provinces at local level.

Local level

At the local level, the Spanish Constitution guarantees the full legal personality and autonomy of municipalities and provinces (art. 140 and 141) (Gobierno de Espana, 1978[2]). As referred to in art. 2.1 of the law 7/1985, "for the effectiveness of the autonomy constitutionally guaranteed to local entities, the legislation of the central government and that of the Autonomous Communities, regulating the different sectors of public action, in accordance with the constitutional distribution of powers, must ensure that municipalities, provinces and islands have the right to intervene in all matters directly affecting their interests, attributing to them the appropriate powers in accordance with the characteristics of the public activity in question and the management capacity of the local entity, in accordance with the principles of decentralisation, proximity, effectiveness and efficiency, and strictly subject to the regulations on budgetary stability and financial sustainability" (Gobierno de Espana, 1985[5]).

Generally, the provinces' deliberative body is the provincial deputation (*diputación provincial*), which is composed of members elected by and from the municipal councillors of the province, following municipal elections. The deputation elects a president (*presidente de la provincia*) from among its members. The Balearic and Canary Islands are organised as "island councils" instead of provincial governments. At the municipal level, the deliberative body is the local council (*pleno*), whose members are elected every four years by direct universal suffrage. The council is chaired by a mayor (alcalde), elected from amongst the local council members, who is the head of the executive body.

The two Autonomous Cities of Ceuta and Melilla, located in North Africa, are municipalities with more responsibilities, close to those of the autonomous communities. They each hold a special individual Statute of Autonomy, approved in 1985, which establishes their institutional system (i.e. an assembly, a President and a governing council), their responsibilities and their economic and financial structure (The Congress of Local and Regional Authorities, 2013[6]).

Andalusia counts 8 provinces and the largest number of municipalities in Spain: 785 municipalities i.e. 9% of all Spanish municipalities. In 2019, more than 66% of Andalusian municipalities have fewer than 5,000 inhabitants, and in them live just over 10% of the population, occupying approximately 51% of the Andalusian territory (Junta de Andalucia, 2020[7]). There are also some characteristics specific to Andalusia, related to local responsibilities and funding as well as institutional relations between the regional and local governments (Box 1.1).

Vertical coordination between the central government and local governments takes place with the National Commission for local Administration (*Comisiòn Nacional de Administracion Local*), which was created in 1985. Autonomous communities have their own fora for coordinating with local governments under their jurisdiction, including in Andalusia (Box 1.1).

Horizontal coordination is facilitated by the Federation of Spanish Municipalities and Provinces. Inter-municipal cooperation happens through mancomunidades and comarcas which carry out joint projects or provide common services, for example in the environmental sector (water, waste). There are around 1 000 inter-municipal cooperation entities in Spain, including around 115 in Andalusia (to check and update). The law 27/2013 also promotes the integration or coordination of municipal services (e.g. education, social services, healthcare) through financial incentives.

Box 1.1. Institutional organisation of Andalusia at the regional and local levels

Andalusia was recognised as an autonomous community on February 28, 1980 and its Statute of Autonomy (*Estatuto de Atuonomia de Andalucia*) was approved by the Spanish government in 1981. A new Statute of Autonomy for Andalusia was approved in March 2007 by the Spanish parliament and by referendum to deepen self-government and the decentralised possibilities enabled by the Spanish Constitution (Junta de Andalucia, 2007[8]). The last regional election in Andalusia was held in June 2022.

The Andalusia Statute established the regional government of Andalusia (*Junta de Andalucia*), composed of a regional assembly (*Parlamento de Andalucía*), the President of the regional government (*Presidente de la Junta de Andalucía*) and a government council. The Statute also established a defender of the Andalusian people (i.e. ombudsperson), a consultative council, a regional chamber of accounts, an audiovisual council of Andalusia and a regional economic and social council.

The main functions of the Parliament of Andalusia are to enact, amend or repeal laws and to appoint and remove the President of the Regional Government. The President of the Regional Government of Andalusia is the executive chief of the Autonomous Community and the representative of the State in daily affairs. The Government Council of Andalusia is in charge of carrying out the executive and administrative functions. The current Council of government is composed of 13 ministries (Junta de Andalucia, 2022[9]).

The Statute recognises Seville as the capital city and the eight provinces that compose the territory of Andalusia (Huelva, Seville, Cordoba, Jaen, Cadiz, Malaga, Granada and Almeria).

The 2007 Statute of Autonomy of Andalusia provides full guarantee and protection of local autonomy. Local autonomy is grounded in art. 92.1 of the Statute, which recognises municipalities' own responsibilities. Art. 192.1 grants the participation of local governments in the tax system of the autonomous community, through the implementation of a municipal fund of an unconditional nature. The Statute also recognises the full capacity of local self-organisation and the principle of subsidiarity.

The institutional relations between the regional government of Andalusia and local governments are defined in art. 98 of the Statute (Junta de Andalucia, 2007[10]), which was followed by the law 5/2010 on local autonomy in Andalusia *(Autonomia Local de Andalucia)*. As per art. 57 of the law, the Andalusian Council of Local Government *(Consejo Andaluz de Gobiernos Locales)* was created as the representative body of the municipalities and Provinces before the regional government of Andalusia in order to guarantee the respect of local responsibilities. Through this body, local governments are involved in all parliamentary proceedings and legislation affecting local responsibilities in Andalusia (Junta de Andalucia, 2010[11]). The body adopts its own rules for procedure and organisation. It is composed of local governments' representatives and five locally elected officials proposed by the association of municipalities and provinces. The president shall be elected by an absolute majority of the council. Additionally, the Andalusian Council of Local Consultation *(Consejo Andaluz de Concertacion Local)* was established as a joint consultative body, gathering representatives from the regional government of Andalusia, municipalities and Provinces (The Congress of Local and Regional Authorities, 2013[6]).

1.1.3. Allocation of responsibilities across the EU, national, regional and local levels of government

As a member of the European Union, Spain shares some responsibilities with the European Union. According the Treaty on the Functioning of the European Union (TFEU), there are three types of EU "competences"[1]: exclusive, shared, and supporting (i.e. competence to carry out actions to support, coordinate or supplement the actions of the Member States) (European Union, 2012[12]).

Environment is a shared responsibility, for which both the EU and Member States are able to legislate and adopt legally binding acts. Other areas of shared responsibilities that may be related to environment and climate change are social and territorial cohesion, agriculture and fisheries, transport, trans-European networks, and energy.

The EU is obligated to exercise its responsibilities according to the principle of proportionality (i.e. the content and scope of EU action may not go beyond what is necessary) and the principle of subsidiarity (i.e. the EU may act only if the action of Members States is insufficient to achieve an objective for non-exclusive responsibilities).

In Spain, many responsibilities are also shared between the national and subnational levels of government. A major decentralisation process took place with the adoption of the 1978 Constitution and the subsequent laws. In the 2000's, two major areas of responsibility were transferred from the central government to the autonomous communities (education in 2000 and healthcare in 2002). Reforms of autonomous communities' statutes were also carried out on a case-by-case, transferring other areas of responsibility.

As per the Spanish Constitution, the autonomous communities may assume responsibilities that do not fall under the central government's jurisdiction. As a general rule, 23 areas are listed in the Constitution as responsibilities not expressly attributed to the central state and therefore devolved to autonomous communities.

In addition, there are also shared responsibilities between the central government and the regional governments. In particular, they are responsible for the development and implementation of the central government's basic legislation on economic activity, education, universities, public health, social protection, municipal and provincial supervision and environment as well as for the execution of the central government's legislation on labour, administration of justice, and intellectual and industrial property. Andalusia, as other autonomous communities, has extensive responsibilities regarding policy measures in the environment and climate sphere, as environmental protection is a regional responsibility. The region has also responsibilities in areas related to the green transition, such as transport, economic development, agriculture and forestry, water management, regional planning and housing (see Table 1.1).

As indicated above, the exact allocation of responsibilities is determined by each Community's Statute of Autonomy. Conflicts on the overlap of responsibilities between the central and regional governments are settled by the Constitutional Court.

At the local level, the organic law 7/1985 sets the framework of the local government system (*Ley reguladora de las bases del régimen local - LBRL*) and defines the basis of local responsibilities (Gobierno de Espana, 1985[5]). According to art. 2.1, the allocation of responsibilities in Spain shall respect the principle of subsidiarity, meaning that public responsibilities shall be exercised by authorities which are the closest to citizens (The Congress of Local and Regional Authorities, 2021[13]; Gobierno de Espana, 1985[5]). The organic law adopted in 2013 (Law 27/2013 on the Rationalisation and Sustainability of Local Administration – LRSAL) aimed at clarifying responsibilities between municipalities and provinces and preventing duplication (Gobierno de Espana, 2013[14]; OECD-UCLG, 2019[15]).

Provincial responsibilities are generally defined as ensuring the coordination and provision of municipal services, as well as investment projects of supra-municipal interest. They are in charge of the overall coordination of local government with the autonomous community and the central government, and guaranteeing compliance with solidarity and budget-balance principles among the municipalities they are comprised of. They must provide technical, legal, and economic assistance to small municipalities (fewer than 5 000 inhabitants).

Municipal responsibilities vary between mandatory "core responsibilities" and optional tasks clarified by the law LRSAL, according to their population size. All municipalities are responsible for local services including local public utilities, public lighting, road maintenance and municipal police. Larger municipalities (more than 20 000 inhabitants) have additional responsibilities such as social service allowances, civil protection, public transport and environmental protection.

The devolution of powers to municipalities may differ substantially from one autonomous community to the next. Besides the responsibilities allocated by the law, local governments may also adopt their own rules in accordance with national and regional legislation.

Table 1.1. Responsibilities across the levels of government according to the Constitution and the Statute of autonomy of Andalusia

Categories	Central government	Andalusia	Provinces	Local governments (depending on the size of the municipality)
General public services (administration)	Regulation guaranteeing equality of all Spanish in the exercise of their rights; Nationality, immigration, emigration, foreign policy and asylum law; International relations; Post Office services; Basic legislation of public administration; Statistics for general purposes; Authorisation for referendums; Municipal and provincial supervision (shared with the autonomous communities)	**Exclusive** responsibilities: Organisation and structure of regional government institutions; Electoral rules and procedures in Andalusia; Management of assets of public domain and patrimonial of Andalusia **Shared responsibilities:** Legal regime and statutory regime of regional staff; Common administrative procedures; Administrative contracts and concessions	Internal administration; Coordination of municipalities with the autonomous communities and the central government; Technical, legal, and economic assistance to municipalities with less than 5.000 inhabitants; Provision of public services of supra-municipal character	Internal administration
Public order and safety	Defence and security; Justice administration; Commercial, criminal and penitentiary legislation; Procedural legislation; Civil legislation; Intellectual and industrial property; Production and sale of arms and explosives; Public safety	**Exclusive responsibilities:** Supervision and protection of regional facilities; Coordination with local police forces; Establishment of Andalusia's public security policies under the terms in art. 149 of the Constitution; Creation, organisation and command of an Andalusian Police; Civil protection		Municipal police; Civil protection; Firefighting services (municipalities with more than 20 000 inhab.)
Economic affairs and transports	Customs and tariff regulations; Foreign trade; Monetary system; General finances and central government's debt; National ports, airports, control of air traffic, weather service; Railway and transports of supra-regional interest; Maritime fisheries; Merchant navy and shipping registry	**Exclusive responsibilities:** Agriculture, livestock and rural development; Maritime and recreational fishing, aquaculture; Transport (see Part II, Table 2.6); Commercial activity; Cooperatives and social economy entities; Promotion of competition for economic activities in Andalusia; Promotion and planning of economic activity in Andalusia; Industry, except for the responsibilities of central government; Consumer rights; Regional tourism **Shared responsibilities:** Planning of the fishing sector, as well as for fishing ports	Cooperation in the promotion of economic and social development and in planning of the provincial territory; Implementation of capital expenditure projects outside the municipal territorial boundaries (including secondary road networks, some hospitals etc.)	Local public road maintenance (all municipalities); Collective urban transportation (municipalities with more than 50 000 inhab.); Markets
Environment protection	Legislation, regulation and concession of hydraulic resources when the waters flow through more than one autonomous community; Basic legislation on environmental protection; Organisation of mining and energy	**Exclusive responsibilities:** Environment (see Part II, Table 2.4) Energy (see Part II, Table 2.4) Water (see Part III, Table 5.1) **Shared responsibilities:** Environment (Part II, Table 2.4) Energy (see Part II, Table 2.4) Water (see Part III, Table 5.1)		Waste collection; Cleaning; Drinking water supply systems; Sewage (all municipalities); Public park; Waste treatment (municipalities with more than 5 000 inhabitants.); Urban environmental protection (municipalities with more than 50 000 inhab.)

Housing and community amenities	Public works of general interest or of supra-regional interest	**Exclusive responsibilities:** Housing; Public works or regional interest; Town, land and costal planning **Shared responsibilities:** Right of reversion in urban expropriations		Urban policies; Water supply; Public lightning; Cemetery and funeral services (all municipalities)
Health	External health measures; Bases and coordination of health matters; Legislation on pharmaceutical products	**Exclusive responsibilities:** Organisation, internal functioning, evaluation, inspection and control of health centres; Research for therapeutic purposes **Shared responsibilities:** Internal health; Protection and promotion of public health in all areas; Implementation of the central government's legislation on pharmaceutical products; Planning and coordination in health with the central government		Participation in the management of first healthcare
Culture and recreation	Basic legislation on the organisation of press, radio, television and social communication; Promotion of Spanish cultural and artistic heritage and national monuments	**Exclusive responsibilities:** Museums, libraries and music conservatories of regional interest; Handicraft activities; Artistic and cultural activities in Andalusia; Cultural heritage; Promotion of the regional language; Planning, coordination and promotion of sports and leisure activities; Organisation of recreational activities		Public library (municipalities with more than 5 000 inhab.); Sport facilities (municipalities with more than 20 000 inhab.)
Education	Promotion and coordination of scientific and technical research; Regulation on academic degrees and professional qualifications; Education (shared); Universities (shared)	**Exclusive responsibilities:** Early childhood education; Programming and coordination of the Andalusian University system, creation of public Universities and authorisation of private Universities; approval of the Statutes of public Universities; funding of Universities; remuneration for staff; Organisation, control, monitoring and accreditation of research centres in Andalusia **Shared responsibilities:** Establishment of curricula and issuance of academic and professional qualifications in non university education; All matters other than those referred above regarding Universities; Coordination of the research centres of Andalusia		
Social protection	Legislation and financial system of the Social security; social assistance (shared)	**Exclusive responsibilities:** Social services; Volunteer work; Social protection of minors; Social protection of family and child		Social service allowances (municipalities with more than 20 000 inhab.)

Source: Author's own elaboration based on (Gobierno de Espana, 1978[2]; Junta de Andalucia, 2007[8]; OECD-UCLG, 2022[16]).

1.2. Subnational government finance in Spain and tax competences across levels of government

1.2.1. Subnational government finance in Spain

Provisions on fiscal matters relating to subnational governments are detailed in Articles 156, 157 and 158 of the Constitution, in law 22/2009 on the financing of autonomous communities and the Basic Law on Local Government 7/1985, revised in 2013 by the LRSAL, as well as in the successive Budgetary Stability Acts adopted in 2001, 2006, 2009 and 2012. Most fiscal powers are concentrated in the autonomous communities, to the detriment of local governments (OECD-UCLG, 2022[17]).

The two autonomous communities of the foral system (Basque Country and Navarra) have an almost complete spending and revenue autonomy. They benefit from all taxes, except import duties, payroll taxes, VAT and excise duties, under the condition that the overall effective tax burden does not fall below that of the rest of Spain. Besides, within the common regime, the Canary Islands has a specific economic and tax system due to historical and geographic reasons and its status as an EU "outermost region". The particularities of the "foral" territories and of the Canary Islands are mentioned in the Additional Provisions of the Spanish Constitution.

The rest of the Autonomous Communities have a more homogeneous financing system. Under the common regime, as per article 157 of the Constitution, the autonomous communities receive revenue from different sources: taxes (own-source and shared), user charges and fees (see Box 1.2 for definitions), inter-government transfers including grants from the central government and the inter-regional clearing fund[1], property and private law income and revenues from financial operations (Gobierno de Espana, 1978[2]). This constitutional framework was completed by the organic law 8/1980 on the Financing of the autonomous communities (LOFCA) and by the Statutes of Autonomy.

Box 1.2. OECD definition of taxes, user charges and fees

It is not always straightforward to distinguish between user charges and fees that are treated as taxes and those that are not, since the strength of the connection between a charge or fee and the service provided may largely vary, as well as the amount of the charge or fee and the cost of service provision. The OECD Interpretative Guide to Revenue Statistics provides a definition of taxes and user charges and fees, which is used as a reference in this report. The Guide also provides examples of borderline cases where user charges and fees could be considered as a tax.

Taxes: the term "taxes" is defined as compulsory unrequited payments to the general government or to a supranational authority. Taxes are unrequited in the sense that benefits provided by government to taxpayers are not normally in proportion to their payments. The term does not include fines, penalties and compulsory loans paid to government.

Fees, user charges and licence fees: where the recipient of a service pays a fee clearly related to the cost of providing the service, the levy may be regarded as requited and would not be considered as a tax. The main charges and fees include court fees, driving licence fees, harbour fees, passport fees, radio and television licence fees where public authorities provide the services, etc.

In the following cases, however, a levy could be considered as 'unrequited':

- where the charge largely exceeds the cost of providing the service;
- where the payer of the levy is not the receiver of the benefit (e.g., a fee collected from slaughterhouses to finance a service which is provided to farmers);
- where government is not providing a specific service in return for the levy which it receives even though a licence may be issued to the payer (e.g., where the government grants a hunting, fishing or shooting licence which is not accompanied by the right to use a specific area of government land);
- where benefits are received only by those paying the levy but the benefits received by each individual are not necessarily in proportion to his payments (e.g., a milk marketing levy paid by dairy farmers and used to promote the consumption of milk).

For the purpose of this report, the term "levy" refers to both taxes and user charges and fees.

Source: (OECD, 2021[18])

Recent fiscal decentralisation reforms modified the subnational financing structure, resulting in a significant increase in tax revenue as a percentage of total subnational government revenue. In particular, the organic law 22/2009 on the financing of the autonomous communities has introduced major changes including on subnational taxation (increase in the regional shares of shared taxes), a reform of the equalisation system and a change in intergovernmental transfers. Despite the reform, inter-governmental transfers remain the primary source of regional government revenue (Box 1.3).

As per the Constitution and laws, local governments have the capacity to regulate their own finances, which includes the power to establish their own taxes, to benefit from spending autonomy and to receive revenue from an unconditional nature from higher levels of government. In Andalusia, the 2007 Statute of Autonomy also grants local governments with the principles of autonomy, fiscal responsibility, equity and solidarity in Andalusia. It also stipulates that local governments shall have sufficient resources for the provision of local services.

Box 1.3. Subnational government finance in 2020: key data

Spain has undergone thorough decentralisation in recent decades, shifting from a highly centralised system before 1978 to a highly decentralised one.

Today, Spain is one of the most decentralised countries in the OECD, with subnational governments responsible for almost half of total public spending in 2020 (i.e. 47.3%), amounting to 24.8% of GDP. This lies above the OECD average (respectively 36.6% and 17.1%) and the average for OECD federal countries (respectively 43.5% amounting to 20.6% of GDP)[2]. The regional level represented almost three-quarters of total subnational government expenditure, while the local level accounted for the remaining (see the below note for the scope of fiscal data).

Spanish subnational governments are responsible for almost all public spending in health, education, environmental protection, and housing and community amenities. The autonomous communities, in particular, play a crucial role in infrastructure investment, research and development, and economic development policies. Subnational government direct investment represented 67.1% of total public investment in 2020, above the OECD average (54.6%) and the OECD average for federal countries (61.5%).

On the revenue side, tax revenue accounted for almost 40% of total Spanish subnational government revenues (37.5%) in 2020. This lies below the OECD average (42.4%) and the OECD average for federal countries (45.8%) (Figure 1.1). Tax revenue of subnational government amounted to 9.3% of GDP and 40.8% of total tax revenue in Spain, similar to the OECD average for federal countries in terms of GDP but below the average in terms of total tax revenue (44.5%). By contrast, the share of grants and subsidies in Spanish subnational government revenue remained quite high compared to other OECD federations (respectively 55.1% vs. 35.4%).

Figure 1.1. Subnational government revenue breakdown in all OECD countries, OECD federal countries and Spain in 2020

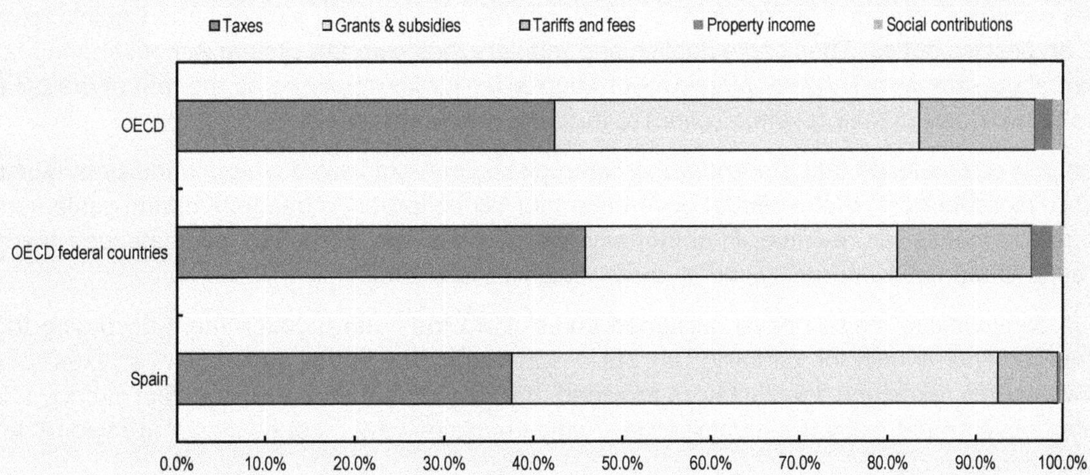

Source: Author's own elaboration based on OECD National Accounts

StatLink https://stat.link/lnxgt8

At the regional level, tax revenue represented 35.5% of regional revenue in 2020, while grants and subsidies represented 60.2% of revenue and tariffs and fees 6% (OECD-UCLG, 2022[17]). At the local level, tax revenue represented 48.8% of local government revenue in 2020, while grants and subsidies accounted for 41.2% and tariffs and fees 9% (OECD-UCLG, 2022[17]).

Note: The scope of fiscal data for Spanish subnational governments encompasses: (i) at the "regional level": autonomous communities, regional administrative agencies, regional universities and regional corporations that are non-market producers; (ii) at the "local level", local authorities (municipal, provincial and islands), associations and groupings of municipalities, autonomous cities (Ceuto and Melila) and bodies reporting to them (e.g. public organisations, corporations and foundations).
Source: (OECD-UCLG, 2022[17]).

1.2.2. Tax competences across levels of government

The EU has limited competences on tax policy, which remain in the hands of Member States. Tax proposals at the EU level typically require unanimity voting at the Council of the EU. The EU shall ensure harmonisation of legislation concerning turnover taxes, excise duties and other forms of indirect taxations under art. 110 to 113 (European Union, 2012[12]).

The primary power to raise taxes in Spain is provided to the central government by the 1978 Spanish Constitution (art. 133). However, the autonomous communities and local governments benefit from "assigned taxes" (shared taxes) and may also establish and raise taxes in accordance with the Constitution and legislation (Gobierno de Espana, 1978[2]).

Parts of the Spanish tax system is currently under review. The "White Book for the reform of the tax system and its adaptation to the reality of the 21st century" (White Book for Tax Reform in Spain, Comité de personas expertas (2022[19]) published on March 2022 proposed a diagnosis of the tax system, including on environmental taxation.[3]

Autonomous communities' taxes

Under the foral regime, the Basque Country and Navarra have higher fiscal autonomy compared to other autonomous communities. They benefit from a full autonomy on all taxes, except import duties, payroll taxes, VAT and excise duties. They can establish and regulate their own tax system without sharing taxes with the central government. They should however keep a similar tax pressure as the rest of the country and have to provide part of their revenue (*cupo*) to the central government.

For autonomous communities that are under the common regime, autonomous communities may act as "delegates" or "collaborators" of the central government for tax collection, management and settlement of the central government's tax revenue, in conformity with the law (art. 156). Tax competences are also defined in each Statute of Autonomy, such as the Andalusia's one (Box 1.4).

Autonomous communities may (i) have "assigned taxes" from the central government according to tax sharing arrangements (wholly or partially), (ii) put a surcharge on central government's taxes, or (iii) establish own-source taxes and special levies provided that they do not levy a taxable fact already levied by the central government or by the municipalities, which explains why most own-source regional taxes are on environmental facts.

Newly created taxes established by the central government, which were originally levied by the autonomous communities and represent a reduction in the autonomous community's revenue, requires compensatory measures in favour of the autonomous communities.

Likewise, newly created taxes established by the autonomous communities, which were originally levied by the municipalities and entail a decrease in the municipality's revenue, requires compensatory measures in favour of the municipalities (art. 6) (Gobierno de Espana, 2009[20]).

Under the common regime, the "assigned taxes" imply that the central government is responsible for the establishment and regulation of these taxes, while the revenue is wholly or partially shared and distributed to the autonomous communities (art. 10) (Gobierno de Espana, 2009[20]). Autonomous communities have some leeway on assigned taxes (ceilings on rates, tax exonerations and exemptions, etc.). For example, in the context of the personal income tax (PIT), they are able to increase or decrease tax exemptions on the regional share (e.g. max. of 10% greater or less than the national level; art. 69 ley 35/2006) and can also have discretion regarding the number of tax brackets, although they must have a progressive rate scale (art. 46 ley 22/2009).

Taxes assigned to the autonomous communities include the following:

- assignment of 50% of the personal income tax (PIT) receipts (*impuestos sobre la renta de las personas fisicas; IRPF*) (instead of 33% before the 2009 tax reform);
- assignment of 50% of the value added tax (VAT) receipts (*impuestos sobre el valor añadido*);
- assignment of 58% of the receipts from excise taxes (*impuestos sobre consumos especificos*) on beer, wine and fermented beverages, intermediate products, alcohol and derived beverages, **hydrocarbons** and tobacco products;
- assignment of the full receipts from the **electricity tax** and certain means of transport (**vehicle registration tax**), the wealth tax, inheritance and gift/donation tax, tax on capital transfers and documented legal acts (stamp duty), gambling tax and **vehicle excise tax.**

Autonomous communities may establish surcharges on these "assigned taxes", as well as on non-assigned taxes at the national level that tax the income or assets of persons with residence within their territories, provided that the surcharges does not imply a reduction in the central government's revenue, nor distort their nature or structure (art. 12) (Gobierno de Espana, 2009[20]).

Own-source taxes are created by the autonomous communities and shall

- be on assets located or income or expenditure originated within their territory, and
- not harm the free movement of persons, goods and capital services, nor affect the location of residence of persons or the location of companies and capital within the country (art. 9) (Gobierno de Espana, 2009[20]).

The autonomous communities may also establish fees on the use of their public domain, on the provision of a public service or the performance of an activity that affects the taxable person, with an expected return that do not exceed the cost of these services or activities. When the central government or municipalities transfer goods of public domain to the autonomous communities, the fees levied on the services or activities related to these specific goods are also transferred to the communities (art. 7) (Gobierno de Espana, 2009[20]).

In Andalusia, the PIT represented 37.9% of tax revenue in 2020, the VAT 37.5%, excise taxes 14.8% and other taxes the remaining 9.8%, below the averages of autonomous communities (respectively 43.7%, 33.1% and 12.6%, excluding the Basque Country, Navarra and the Canary Islands that have specific financing systems). (Figure 1.2) The autonomous communities under the common regime do not receive corporate income tax (CIT) receipts.

Figure 1.2. Revenues from PIT, VAT and excise taxes as a share of total tax revenue in autonomous communities in 2020

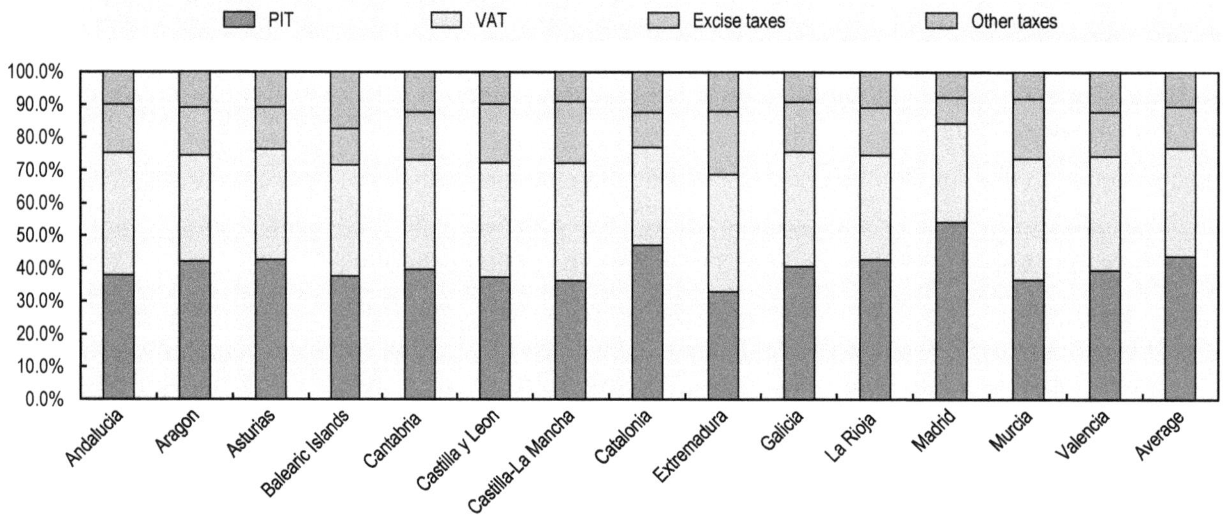

Note: The Basque Country, Navarra and the Canary Islands are not presented in the above figure as they have a specific financing system.
Source: Author's own elaboration based on (Ministerio de Hacienda y Funcion Publica, 2022[21]).

StatLink https://stat.link/35wju9

The Superior Council of Tax Coordination, made up of representatives of the central government's tax administration and the autonomous communities, is responsible for coordinating the management of "assigned taxes", together with the Territorial Councils for Tax Coordination and Management that operate within each autonomous community's territory (OECD-UCLG, 2022[17]).

Three quarters (75%) of tax revenue from the autonomous communities goes to the Guarantee Fund for Basic Public Services *(Fondo de Garantía de Servicios Públicos Fundamentales)* (the rest being central government's contribution), which is an equalisation fund to ensure that each autonomous community receives the same revenue according to its need to finance essential public services. The redistribution is based on an adjusted population criterion.

Box 1.4. Tax competences of the Autonomous Community of Andalusia

Under the 2007 Statute of Autonomy, art. 180 provides that Andalusia is responsible for the establishment and regulation of its own taxes, as well as their management, settlement, collection, inspection and review (Junta de Andalucia, 2007[10]). Andalusia established several own taxes, user charges and fees, of which several that are related to the topic of the present report: the tax on underused land, the tax on gas emissions into the atmosphere, the tax on discharges into coastal waters, the tax on deposit of radioactive waste, the tax on the deposit of hazardous waste, the charge for the improvement of hydraulic infrastructures of interest of the autonomous community, the tax on customer deposits with credit institutions and the tax on single-use plastic bags.

Regarding "assigned taxes" that have been fully transferred to Andalusia by the central government, the autonomous community exercises the regulatory powers and, by delegation of the central government, the management, settlement, collection, inspection and review of these taxes, in accordance with the law that establishes the scope and conditions of the assignment.

For the other "assigned taxes" that have not been fully transferred to the autonomous community, the management, settlement, collection, inspection and review are entitled to the central government, without prejudice to the collaboration that may be established between the central government and the community (art. 180) (Junta de Andalucia, 2007[10]).

A Tax Agency has been created for the purpose of managing the above-mentioned tasks associated to own and assigned taxes in Andalusia (*Agencia Tributaria*). The Tax Agency may also collaborate with other administrations (art. 181) (Junta de Andalucia, 2007[10]).

Provincial and municipal taxes

The Spanish Constitution grants provinces and municipalities the autonomy to manage their respective interests. Each Statute of Autonomy also grants local governments with principles of fiscal autonomy, fiscal responsibility, equity and solidarity, like in Andalusia (Box 1.5).

Provinces have the power to levy a surtax on the local business tax and are also entitled to some shared tax revenue (PIT, VAT and CIT). They do not have own-sources taxes.

Municipalities can finance their responsibilities through their own taxes and assigned taxes from the autonomous communities and the central government (art.142) (Gobierno de Espana, 1978[2]).

Across all communities, municipal own-source tax revenue represents the main source of revenue, which includes a property tax (IBI), a vehicle tax (IVTM), a local business tax and two optional taxes (a tax on the increase in the value of urban land "IIVTNU – plus valia" and a tax on construction, facilities and infrastructure). Given their extended scope of responsibilities, larger municipalities (more than 75 000 inhabitants) have a special status and benefit from additional assigned taxes (PIT, VAT and excise taxes).

Although this is rarely used, municipalities can also raise environmental-related taxes (OECD-UCLG, 2022[17]), such as the circulation tax or fees for the provision of water supply, sewage and wastewater treatment services. Unlike regions, municipalities are not allowed to create new taxes. Local taxes should be listed in a national law.

Box 1.5. Tax competences of local governments in Andalusia

According to the 2007 Statute of Autonomy, Andalusian local governments are responsible for the management, collection and inspection of their taxes, without prejudice to the delegation or collaboration they may establish with other levels of government (art. 191) (Junta de Andalucia, 2007[10]).

Art. 192 of the Statute defines the collaboration between local governments and the autonomous community. Local governments shall participate in the tax system of the autonomous community through the implementation of a municipal levelling fund of an unconditional nature.

They may also delegate to the autonomous community the management, settlement, collection and inspection of their own taxes or through other forms of collaboration.

Regarding the taxes they share or unconditional subsidies they receive from the central government, local governments shall receive them through the autonomous community, which then redistribute them according to criteria established by the law. The article also specifies that any allocation of responsibilities shall be accompanied by appropriate compensation (Junta de Andalucia, 2007[10]).

The White Book for Tax Reform in Spain

Parts of the Spanish tax system is currently under review. The Secretary of State for Finance has commissioned a Committee of 17 external experts to work on a "White Book for the reform of the tax system and its adaptation to the reality of the 21st century" (*Libro Blanco para la reforma del sistema tributario y su adaptación a la realidad del siglo XXI*). Published on March 2022, the White Book elaborates a diagnosis of the tax system as a whole and includes specific analyses in the domain of environmental taxation, corporate taxation, property taxation, the digital economy and the promotion of innovation (Comité de personas expertas, 2022[19]). The present analysis considers the proposals and considerations in the area of environmental taxation put forward in the White Book where relevant.

Chapter 2 of the White Book is fully dedicated to environmental tax reform, which is at the centre of the current report. In total, the Committee has formulated 19 proposals and included analyses in the following domains: electricity, transport, waste and water. The proposals and recommendations of the Committee have a technical basis (environmental, socioeconomic and legal), an analysis of significant experiences in other countries and existing academic evidence, and are specified in recommendations that are judged viable from an administrative and management point of view. When the information is available, the Committee's proposals are accompanied by quantitative simulations of their environmental, distributive and revenue impacts. According to a recent report, the IMF estimates that a harmonisation of environmental taxes in Spain with EU average would represent an additional 0.7 to 0.9 point of GDP (IMF, 2022[22]).

Among the proposed measures, the Committee considers that transport and energy are two of the priority action sectors for reviewing current taxation. In addition, it emphasises the importance of improving the existing water and waste taxation design, as these are sectors where the challenges are of great relevance for Spanish society. The proposals for environmental tax reform are presented in the Table 1.2 below and further detailed in the respective sections of the report.

Table 1.2. Environmental tax reform proposals put forth by the White Book

Environmental areas	Proposals
Electricity	1. Elimination of the tax on the value of electricity production 2. Introduction of measures to improve the design and effectiveness of regional taxes with effects on the electric sector 3. Modifications in the electricity tax to promote electrification and energy efficiency
Transport	4. Taxation of aviation, maritime and agricultural fuels 5. Equalisation of the taxation of diesel and automotive gasoline 6. General increase in taxation of hydrocarbons 7. Modification of the registration tax to promote a sustainable vehicle fleet 8. Configuration of the circulation tax to penalise the most polluting technologies 9. Creation of a municipal tax on congestion in certain cities 10. Consideration of tax mechanisms for the use of certain road infrastructures 11. Creation of a tax on airline tickets
Waste	12. Intensification and extension of the taxes of the law on waste and contaminated soils 13. Reformulation of municipal waste taxation to link it to generation payment systems* 14. Creation of a tax on gravel extraction 15. Creation of a tax on nitrogenous facilities 16. Extend and harmonise taxation on certain emissions from large industrial and livestock facilities
Water	17. Introduction of coordination and cooperation measures to improve the design and effectiveness of regional taxes on environmental damage to water 18. Reform of taxes associated with coverage of hydraulic infrastructure costs 19. Creation of a tax on the extraction of water resources

Note : The Spanish Agency for Fiscal Responsibility (AIREF) is currently working on a spending review of the regional and local cycle for waste treatment (Autoridad Independiente de Responsabilidad Fiscal, 2022[23]).
Source: (Comité de personas expertas, 2022[19])

References

Allain-Dupré, Chatry and Moisio (2020), *Asymmetric decentralisation: trends, challenges and policy implications*, OECD Regional development Papers, https://doi.org/10.1787/0898887a-en. [3]

Autoridad Independiente de Responsabilidad Fiscal (2022), *Plan de actuaciones de la AIReF*. [23]

Bird, R. (ed.) (2007), *Garcia-Milà, T. and T. McGuire (2007), "Fiscal Decentralisation in Spain: An Asymmetric Transition to Democracy"*, Edward Elgar Publish. [4]

Comité de personas expertas (2022), *Libro Blanco Sobre la Reforma Tributaria*. [19]

European Commission (2022), *Commission proposes new Euro 7 standards to reduce pollutant emissions from vehicles and improve air quality [press release]*, https://ec.europa.eu/commission/presscorner/detail/en/IP_22_6495. [24]

European Commission (2021), *Green taxation and other economic instruments - Internalising environmental costs to make the polluter pay*. [29]

European Commission (2017), "REFIT evaluation of Regulation (EC) No 166/2006 concerning the establishment of a European Pollutant Release and Transfer Register (E-PRTR)", https://eur-lex.europa.eu/legal-content/EN/TXT/PDF/?uri=CELEX:52017SC0710&from=en. [27]

European Union (2012), *Consolidated version of the Treaty on the Functioning of the European Union*. [12]

Gobierno de Espana (2013), *Law 27/2013, Of 27 December, Rationalization And Sustainability Of The Local Administration*. [14]

Gobierno de Espana (2009), *Ley Orgánica 3/2009, de 18 de diciembre, de modificación de la Ley Orgánica 8/1980, de 22 de septiembre, de Financiación de las Comunidades Autónomas*. [20]

Gobierno de Espana (1985), *Ley 7/1985, de 2 de abril, Reguladora de las Bases del Régimen Local*. [5]

Gobierno de Espana (1978), *The Spanish Constitution*. [2]

IMF (2022), *2021 Article IV Consultation with Spain*. [22]

Junta de Andalucia (2022), *Decreto del Presidente 10/2022, de 25 de julio, sobre reestructuración de Consejerías*. [9]

Junta de Andalucia (2020), *New Trends in the Structuring of Andalusia*, https://www.centrodeestudiosandaluces.es/. [7]

Junta de Andalucia (2010), *Ley 5/2010, de 11 de junio, de autonomía local de Andalucía*. [11]

Junta de Andalucia (2007), *Ley Orgánica 2/2007, de 19 de marzo, de reforma del Estatuto de Autonomía para Andalucía*. [10]

Junta de Andalucia (2007), *Organic law 2/2007 dated 19 March 2007 on Reform of the Statute of Autonomy for Andalusia*. [8]

Labandeira, X. (2022), *Taxation and Ecological Transition during Climate and Energy Crises: the Main Conclusions of the 2022 Spanish White Book on Tax Reform*, Real Instituto Elcano WP 09-2022, https://www.realinstitutoelcano.org/en/work-document/taxation-and-ecological-transition-during-climate-and-energy-crises/. [30]

Ministerio de Hacienda y Funcion Publica (2022), *Autonomous Community Funding*. [21]

OECD (2022), *2022 Synthesis Report World Observatory on Subnational Government Finance and Investment*, OECD Publishing, Paris, https://doi.org/10.1787/b80a8cdb-en. [1]

OECD (2022), *Air pollution effects* (indicator), https://doi.org/10.1787/573e3faf-en (accessed on 29 November 2022). [26]

OECD (2021), *Interpretative Guide to Revenue Statistics*. [18]

OECD (2020), *Financing Water Supply, Sanitation and Flood Protection: Challenges in EU Member States and Policy Options*, OECD Studies on Water, OECD Publishing, Paris, https://doi.org/10.1787/6893cdac-en. [28]

OECD-UCLG (2022), *2022 Report of the World Observatory on Subnational Government Finance and Investment - Country Profiles*. [16]

OECD-UCLG (2022), *2022 Report of the World Observatory on Subnational Government Finance and Investment – Key Findings*, https://doi.org/10.1787/b80a8cdb-en. [17]

OECD-UCLG (2019), *Report of the World Observatory on Subnational Government Finance and Investment – Country Profiles*, https://www.sng-wofi.org/country-profiles/Fiche%20SPAIN.pdf. [15]

Responsible Travel (2022), *Tourist taxes map*, https://www.responsibletravel.com/copy/tourist-taxes-map (accessed on 29 November 2022). [25]

The Congress of Local and Regional Authorities (2021), *Monitoring of the application of the European Charter of Local Self-Government in Spain*. [13]

The Congress of Local and Regional Authorities (2013), *Local and regional democracy in Spain*. [6]

Notes

[1] Although the Treaty uses the term of "competence", this report uses the term "responsibility" to refer to a "legal competence" and uses the term "tax competence", in accordance with this project's Detailed Project Description (DPD), in order to clearly distinguish between "legal competences" and "tax competences".

[1] The inter-regional clearing fund is a transfer from the central government to the autonomous communities and Provinces for investment expenditure, with the aim to correct economic imbalances between the autonomous communities and implement the principle of solidarity (art. 158).

[2] OECD federal and quasi-federal countries include Australia, Austria, Belgium, Canada, Germany, Mexico, Spain, Switzerland and United States.

[3] A non-official English summary of these suggestions were recently published (Labandeira, 2022[30]).

Part II GHG emissions and air pollution

2 Legal stocktake: GHG emissions and air pollution

In Andalusia, greenhouse gas (GHG) emissions increased from 42 015 kt CO_2 eq. in 1990 to 54 416 kt CO_2 eq. in 2019 (Instituto de Estadistica y Cartografia de Andalucia, 2020[1]). Hydrocarbons still represent the largest share in the energy mix of Andalusia compared to Spain and EU average, especially driven by the transport sector (Junta de Andalucia, 2020[2]). In 2020, transport represented 41.3% of CO_2 emissions (41.2% in 2019), the largest emitters, followed by generation (25.6% vs 31.1% in 2019) and industry (12.8% vs 11.8% in 2019) (Table 2.1) (Agencia Andaluza de la Energia, 2020[3]).[1] Andalusia has a high potential for renewable energy as there is a high availability of renewable energy sources, which is judged capable of meeting the energy demand of the autonomous community (Junta de Andalucia, 2020[2]). The gross electricity production of renewables as compared to final electricity consumption sharply grew from 7.2% in 2005 to 44.6% in 2020 (Instituto de Estadistica y Cartografia de Andalucia, 2020[1]). Regarding the transport sector, the number of vehicles per 100 inhabitants increased from around 64 in 2010 to 71 in 2020, of which 0.57 per thousand vehicles were electric, hybrid or used biofuels in 2020 (Instituto de Estadistica y Cartografia de Andalucia, 2020[1]).

Table 2.1. CO_2 emissions by sector in Andalusia from 2010 to 2020 (million tonnes of CO_2)

	2010	2011	2012	2013	2014	2015	2016	2017	2018	2019	2020
Generation	16 494.9	17 191.1	18 234.9	14 820.2	14 361.2	18 093.0	14 333.8	17 129.3	16 723.7	11 584.3	7 166.0
Transport	14 547.7	13 094.2	12 234.2	12 361.3	12 707.9	13 242.9	13 305.7	14 537.7	14 975.5	15 361.8	11 551.6
Industry	4 353.3	4 347.2	3 859.9	3 711.4	3 809.9	3 705.9	4 069.8	4 697.9	4 320.5	4 398.0	3 578.3
Primary sector	2 839.0	2 717.1	2 477.0	2 316.3	2 230.4	2 221.1	2 269.3	2 281.9	2 310.7	2 373.3	2 383.5
Services	313.3	431.4	313.3	262.9	271.9	390.3	410.1	378.5	395.2	497.1	463.0
Residence	1 557.1	1 408.0	1 462.6	1 391.9	1 273.8	1 331.6	1 257.5	1 179.4	1 199.2	1 162.7	1 009.3
Energy	2 152.5	2 069.4	2 103.4	2 125.9	2 323.8	2 179.8	2 079.0	2 074.6	2 170.7	1 903.0	1 816.5
Total	**42 257.7**	**41 258.4**	**40 685.3**	**36 989.9**	**36 979.0**	**41 164.7**	**37 725.1**	**42 279.3**	**42 095.5**	**37 280.2**	**27 968.2**

Note: the energy sector includes emissions associated with the consumption of fossil fuels for the development of the activities of extraction, production, transformation and distribution of energy.
Source: Author's own elaboration based on Agencia Andaluza de la Energia (2020[3]).

StatLink https://stat.link/6cipxu

In addition to carbon emissions, small particulate matter (PM2.5) remains one of the largest cause of human mortality induced by air pollution (OECD, 2021[4]). Air pollution also amplifies infectious disease, such as COVID-19, and affects children the most. In most autonomous communities in Spain, at least 25% of the population was exposed to PM2.5 above the World Health Organisation (WHO) threshold in 2019 (OECD, 2021[4]).

Policies to reduce GHG emissions and air pollution are therefore a key priority in Andalusia. This chapter proposes possible opportunities for reform to Andalusia's existing environmental tax system governing

GHG emissions and air pollution for both stationary[2] and non-stationary sources.[3] While the discussion on stationary sources will focus on industrial and electricity generation facilities, non-stationary sources will focus on personal vehicles. The proposed possibilities are derived from two sequential analyses. First, an analysis of the legal and policy framework governing GHG emissions and air pollution at the EU, national, and regional government levels; second, an analysis on the distribution of responsibilities in policy areas relevant to reducing GHG emissions and air pollution, between the different levels of government (EU, national, regional and local). The key possibilities will be assessed against environmental tax policy principles in the economic analysis.

2.1. Legal Framework on Greenhouse Gas Emissions and Air Pollution

This section outlines the legal and policy instruments governing GHG emissions and air pollution at the EU, national, and regional levels. In doing so, it provides context on the policies, targets, and strategies in place for these two environmental domains which serves as the basis for the subsequent section on the responsibilities across levels of government in these two domains.

2.1.1. At the EU level

The EU Green Deal: a legally binding objective of climate neutrality by 2050

The EU established the EU Green Deal as the core of its policies to fight against climate change (Box 2.1). The objective of the EU Green Deal is to achieve climate neutrality in Europe by 2050, which means net zero GHG emissions for EU countries as a whole. This objective implies the involvement of all sectors of the economy, including industry, energy and transport, and a global response, in line with the UN Framework Convention of Climate Change (UNFCCC) and the Paris Agreement[4].

Box 2.1. The EU Green Deal

The EU Green Deal comprises an ambitious package of measures to achieve carbon neutrality by 2050, including cutting GHG emissions, investing in green technologies and protecting Europe's natural environment. The measures comprise climate mitigation actions, of which the European Climate Law, the 2030 Climate Target Plan and the European Climate Pact. They also comprise climate adaptation measures, under the new EU Adaptation Strategy, to confront the adverse impacts of climate change (Figure 2.1). A third of the EUR 1.8 trillion from the NextGenerationEU Recovery Plan and the EU 2021-2027 budget will finance the European Green Deal.

Figure 2.1. Climate mitigation and adaptation measures under the EU Green Deal

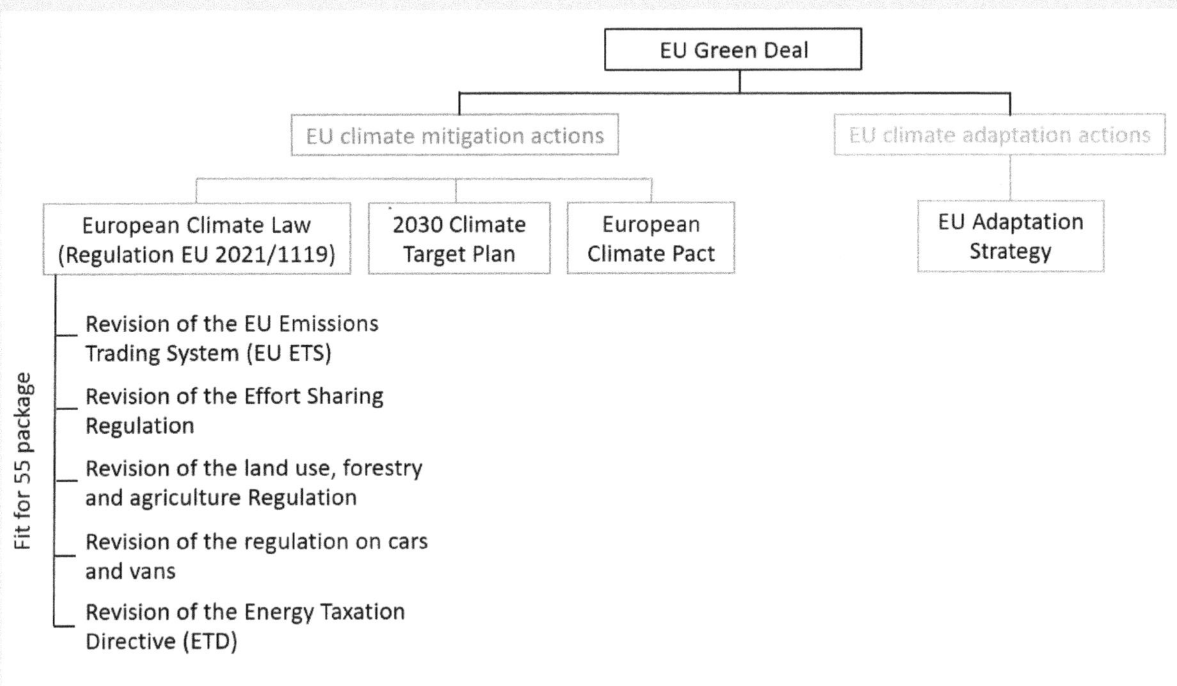

Note: Climate mitigation measures are depicted in blue; climate adaptation measures are depicted in orange. The "Fit for 55" package is a series of legislative proposals to revise current EU legislation aiming to mitigate climate change. The list of proposals is not exhaustive but focuses on the relevant domains for the report.
Source: Author's own elaboration.

To make the objective of carbon neutrality by 2050 legally binding, the European Commission enshrined it in the EU law with the implementation of the European Climate Law on 29 July 2021 (Regulation EU 2021/1119) (European Commission, 2021[5]). Based on a comprehensive impact assessment, the law also sets an intermediary target of reducing net GHG emissions by at least 55% by 2030 compared to 1990 levels. It also includes a process for establishing an EU-wide climate target by 2040, based on an indicative GHG budget for 2030-2050, and a commitment for negative GHG emissions after 2050. In addition, the law provides the creation of a European Scientific Advisory Board on Climate Change for independent scientific advice and a commitment to prepare sector-specific roadmaps towards carbon neutrality. The law also grants coherence across EU policies on carbon neutrality.

The emission objectives of 2030 and 2050 set within the European Climate law are immediately applicable to all Member States, of which Spain and its autonomous communities, since EU regulation is a legally binding instrument that overrules national laws.[5] The law comprises measures for Member States to keep track record of their progress based on national climate and energy plans (see below) and to adjust actions accordingly. Progress is reviewed every five years. A series of legislative proposals ("Fit for 55") have been implemented on 14 July 2021 in order to deliver the above-mentioned emission targets of 2030 and 2050, as set within the 2030 Climate Target Plan and in the European Climate Law. Several EU climate legislations have been revised accordingly, including the EU Emissions Trading System (EU ETS) (Box 2.2), the Effort Sharing Regulation, transport and land use laws.

Box 2.2. The European Union Emissions Trading System (EU ETS)

Established in 2005 through Directive 2003/87/EC of the European Parliament and the Council of the EU, the system is currently active in EU Member States and in Switzerland. The system works on the "cap and trade principle", meaning that a cap is set on the total amount of certain GHGs that are emitted by the installations covered by the system. The cap decreases over time as the total amount of emissions fall. Within the cap, installations can trade their emission allowances amongst each other. At the end of the year, an installation must surrender enough emissions to cover the entirety of its emissions, otherwise fines apply. The EU ETS works in trading phases and is now within its fourth phase (2021-2030). It covers the following sectors and gases:

- **Carbon dioxide (CO_2)** from electricity and heat generation, energy-intensive industry sectors (e.g. oil refineries, steel works, iron, aluminium, metals, cement, lime, glass, ceramics, pulp, paper, cardboard, acids, and bulk organic chemicals) and commercial aviation within the European Economic Area.
- **Nitrous oxide (N_2O)** from nitric, adipic and glyoxylic acids and glyoxal production.
- **Perfluorocarbons (PFCs)** from the production of aluminium.

The participation of companies in these sectors is mandatory. However, in some sectors, only installations above a certain size are included. Small installations may be excluded if central governments implement fiscal or other measures cutting their emissions by an equivalent amount. The EU ETS has been revised several times to remain in line with the EU climate policy objectives.

Source: (European Commission, 2003[6]).

EU legislation in the area of GHG emissions

The EU ETS is the centrepiece of the EU climate policy package. It has been proposed to be revised to become more ambitious under the "Fit for 55" package. Accordingly, the proposed emission reduction target for the sectors covered by the system will be increased from 43% to 61% by 2030 compared to 2005 levels (European Commission, 2021[7]). The distribution of freely allocated emissions allowances has also been proposed to be revised with the aim to further incentivise the development of low-carbon technologies:

- the Market Stability Reserve (i.e. the system that addresses the excess of allowances since 2009) will be strengthened; and
- the emission trading system will be extended to new sectors (i.e. maritime activities for ships above 5 000 gross tonnage travelling within the EU or at berth in EU ports) and a separate trading system will be established from 2026 to cover emissions from fuels in road transports and buildings (cost paid by fuel suppliers rather than households and car drivers). In addition, flights between EU

outermost regions and international flights between EU outermost regions and the European Economic Area will be included in the EU ETS. The number of free allowances allocated to aircraft operators will be also reduced gradually to reach full auctioning by 2027.

The European Commission also proposed the introduction of a Carbon Border Adjustment Mechanism (CBAM) as part of its "Fit for 55" package (European Commission, 2021[8]). This new mechanism aims to price the carbon content of imports on specific products in order to address competitive concerns of industries that pay a high carbon price and to avoid "carbon leakage". The Commission proposed that CBAM starts in 2026, following a transitional period from 2023-2025 with reporting requirements on EU importers.

As part of the "Fit for 55 package", the European Commission also proposed to revise the Effort Sharing Regulation (EU) 2018/842. The Regulation establishes binding annual GHG emission targets for Member States between 2021 and 2030 for the sectors that are not covered by the EU ETS, which accounts for around two-thirds of total domestic EU emissions (European Commission, 2021[9]). The European Commission proposed to increase the EU emission reduction target from 30% to 40% by 2030 compared to 2005 levels to be in line with the objectives of the EU Green Deal. To ensure fairness among Member States, targets are based on GDP per capita. They also reflect cost-efficiency for countries with GDP per capita above the average to avoid too highly costs for them. The European Commission's proposal increases GHG emission reduction target for Spain from 26% to 37.7% in 2030 compared to 2005 levels (European Commission, 2021[9]).

The legislative package also includes a proposed revision of the Energy Taxation Directive (ETD) (No 2003/96/EC) to align it with the objectives set within the EU Green Deal (European Commission, 2021[10]). The ETD has been implemented in 2003 and laid down minimum excise duty rates for the taxation of energy products used as motor and heating fuels and electricity. Member States are free to set their own rates within the minimum limits set by the law. The Commission has suggested to introduce a new structure of minimum tax rates based on the energy content and environmental performance of the fuels and electricity. It also proposed to broaden the tax base by including more products (e.g. mineralogical processes) and by removing some of the exemptions (e.g. kerosene used as fuel in aviation and heavy oil used in the maritime sector) and reductions. Nevertheless, certain reductions of the rates will remain possible (e.g. electricity or advanced energy products from renewables, primary sector such as farming).

The land use and transport laws has also been proposed to be revised as part of the EU "Fit for 55" legislative package. The LULUCF Regulation (EU) 2018/841 provides that all sectors, including land use, shall contribute to the EU's 2030 emission reduction target (European Commission, 2018[11]). Accordingly, GHG emissions from land use, agriculture or forestry must be balanced by at least an equivalent level of CO_2 removal from the atmosphere between 2021 and 2030. The Commission has proposed to increase the carbon removal to -310 million of tonnes CO_2 by 2030 and to introduce the objective of carbon neutrality in land use, agriculture and forestry by 2035 in the EU (European Commission, 2021[12]).

Regarding transport, the Regulation (EU) 2019/631 sets EU fleet-wide CO_2 emission performance standards for new passenger cars and new light commercial vehicles (vans) registered in the EU (European Commission, 2021[13]). The Commission has proposed to increase the targets as follows: (i) by 55% for cars and 50% for vans from 1 January 2030; and (ii) fully for both cars and vans from 1 January 2035. Targets are set annually for manufacturers. Additionally, the Alternative Fuels Infrastructure Regulation will aim to ensure the availability of the recharging and refuelling infrastructure for zero-emission vehicles (European Commission, 2021[14]).

Overall, the Regulation (EU) 2018/1999 on the governance of the Energy Union and climate action aims to ensure that the EU's Energy Union Strategy on energy, decarbonisation, research, innovation and competitiveness is implemented in a coherent and co-ordinated manner (European Commission, 2018[15]). From 2021 to 2030, the law requires Member States to produce integrated national energy and climate plans (NECPs), which includes consultation processes, regional co-operation, progress reports, policies and requirements for national and EU inventory systems for GHG emissions. The plans shall address energy efficiency, renewables, GHG emissions reduction, interconnections, research and innovation.

EU legislation in the area of air pollution

In addition to GHG emissions, the Commission adopted the Directive (EU) 2016/2284 on 17 December 2016 on the reduction of national emissions of certain atmospheric pollutants, amending the Directive 2003/35/EC and repealing the Directive 2001/81/EC on national emission ceilings (NEC directive). The Directive extends the period of the NEC Directive from 2020 to 2030. The directive does not only focus on GHG emission reduction but aims to improve national air quality by establishing national emission reduction targets for several pollutants, of which sulphur dioxide (SO2), nitrogen oxides (NOx), non-methane volatile organic compounds (NMVOC), ammonia (NH3) and fine particulate matter (PM2,5) (European Commission, 2016[16]). As a Directive, Member States have to transpose it into their national legislation and to achieve the objectives specified. For Spain, the Directive provides the thresholds listed in its Annex II (Table 2.2).

Table 2.2. Emission reduction commitments for Spain under the Directive (EU) 2016/2284 on the reduction of national emissions of certain atmospheric pollutants (compared to 2005 levels)

	SO_2 reduction	NOx reduction	NMVOC reduction	NH_3 reduction	PM2,5 reduction
From 2020 to 2029	67%	41%	22%	3%	15%
From 2030	88%	62%	39%	16%	50%

Note: Reductions are based on 2005 levels.
Source: Author's own elaboration based on European Commission (2016[16]).

StatLink https://stat.link/by3df5

2.1.2. At the national level

The Spanish Climate Change and Energy Transition Law: a general legislative framework to address climate change in Spain

Spain established its first climate change law on 21 May 2021 with the implementation of the Climate Change and Energy Transition (Law 7/2021). The law provides a general legislative framework to address climate change with the objective to achieve carbon neutrality by 2050 in Spain (Gobierno de Espana, 2021[17]), in line with EU regulations (of which the European Climate Law) and the Paris agreement. The law also establishes intermediary GHG emission reduction targets of 23% by 2030 compared to 1990s levels.

The law comprises both climate mitigation and adaptation measures. It focuses on several environmental domains, ranging from renewable energies and energy efficiency (including electricity generation), fuel transition and low-emission transports (road, maritime and ports). Article 3 specifies an objective of at least 42% renewables in final energy consumption and at least 74% renewable electricity generation by 2030, as well as a reduction of primary energy consumption by at least 39.5% compared to the baseline under EU legislation. These objectives may be increased periodically as of 2023.

In line with the Regulation (EU) 2018/1999 on the European climate change governance mechanism (see above), the Spanish climate law provides that the central government, the autonomous communities and local governments must achieve the objectives set by the law through co-operation and collaboration among them (art. 1). It also stipulates that climate plans from the autonomous communities must be submitted to the central government to secure policy coordination and compliance with the existing responsibilities distribution (Gobierno de Espana, 2021[17]). Ahead of the COP26 in November 2021, only three autonomous communities had climate change laws in place (Balearic Islands, Catalonia, and Andalusia) and seven others had started to elaborate theirs.

To foster coherence and coordination among Spanish climate policies, the law has established the Integrated National Energy and Climate Plan (PNIEC), which is the national strategic planning tool on energy and climate policies, reflecting Spain's contribution to the achievement of the objectives set by EU regulations (art. 4). The plan covers the period 2021-2030. It contains the objectives and quantitative contributions at the national and sectoral levels for the reduction of GHG emissions and removals by sinks, renewable energies and energy efficiency for all sectors of the economy, as well as the policies and measures to achieve these objectives.

Spanish legislation on GHG emissions

Regarding GHG emissions, the law 1/2005 on the trading of GHG emission rights scheme, adopted on 9 March 2005, transposed the Directive 2003/87/EC (see above) into national legislation and introduced the EU ETS at the national level for GHG emissions (Gobireno de Espana, 2005[18]). In 2021, the Spanish Climate Change and Energy Transition law also prohibited new explorations for hydrocarbon research and for exploitation concessions (art. 9), while it promoted the use of renewable gases (e.g. biogas, biomethane, hydrogen and other renewable fuels) (art. 12) (Gobierno de Espana, 2021[17]). On electricity, the Spanish Electric Sector law (law 54/1997, later consolidated in law 24/2013) provides the basic regulation of electricity in Spain (Gobierno de Espana, 1997[19]; Gobierno de Espana, 2013[20]). It aims to ensure efficient electricity supply, economic and financial sustainability of the electric system and effective competition. The law has been complemented by the Spanish Climate Change law, which promotes the use of reversible hydroelectric power plants, as well as the use of electricity generation for urban water supply and sanitation systems (art. 7) (see below on water).

Article 14 of the Spanish Climate Change and Energy Transition law also provides that the central government, the autonomous communities and local governments, within the scope of their responsibilities, must achieve the objective of a fleet of passenger cars and light commercial vehicles with zero GHG emission by 2050, in line with EU regulations (Gobierno de Espana, 2021[17]). To this end, the PNIEC sets targets for the share of zero or low-emission vehicles in the car fleet to reach by 2030. In addition, the sales of new passenger cars and light commercial vehicles emitting GHG emissions, not intended for commercial purposes, will be prohibited by 2040, in accordance with EU regulations. Large municipalities with more than 50 000 inhabitants shall also adopt sustainable urban mobility plans by 2023 with the aim to reduce emissions from mobility. This includes the development of low-emission zones[6] by 2023, the improvement of the public transport network (e.g. multimodal integration) and its electrification, the promotion of private electric transport (including charging points), the integration of last-mile electrification plans with low-emission zones and the establishment of criteria to improve air quality around schools, health or sensitive areas, in accordance with air quality regulation (see below). Article 15 provides for the installation of electric charging points in facilities that supply fuels.

The law also includes the objective of zero emission by 2050 for ships, vessels and naval devices in ports and the development of sustainable logistic chains with origin or destination in ports (art. 16) (Gobierno de Espana, 2021[17]). After agreement with the autonomous communities in their domain of responsibilities, the central government shall support the supply of electric or alternative sources on docked ships and rail transport from/to ports and establish objectives to reduce energy consumption in ports based on their activity.

In order to help the autonomous communities and cities to promote electric mobility, in the realm of the EU Recovery and Resilience Facility,[7] the central government launched the Moves III Plan on 13 April 2021 through the Royal Decree 266/2021 (Gobierno de Espana, 2021[21]). The Plan is coordinated by the national Institute for the Diversification and Saving of Energy (IDAE) and managed by the autonomous communities and cities. It is endowed with initial EUR 400 million, which may be extended to EUR 800 million, provided that there is adequate budget execution and budget availability. The funds are distributed from IDAE to the autonomous communities and cities, based on population criteria, to encourage the purchase of electric vehicles and to finance the deployment of charging facilities for these vehicles. The autonomous communities and cities are responsible for the calls within their territory to subsidy citizens for the purchase of efficient vehicles (e.g. electric, extended-range or plug-in hybrid cars). Subsidies may amount up to EUR 7 000 for passenger cars and EUR 9 000 for light commercial cars. This program provides continuity to the previous Efficient Vehicle Incentive Program and the Renewal Plans, which provided financial support to drivers exchanging their cars for the ones respecting tighter environmental and social criteria.

In addition, *es.movilidad*, the Safe, Sustainable and Connected Mobility Strategy 2030 was approved by the Spanish Council of Ministers on 10 December 2021 (Ministerio de Transportes Movilidad y Agenda Urbana, 2021[22]). The strategy will serve as a guide to the actions of the Spanish Ministry of Transport, Mobility and Urban Agenda (MITMA) in the areas of mobility, infrastructure and transport for the next 10 years. It is made up of 150 measures, structured around 40 lines of action and 9 strategic axes, of which (i) mobility for all, (ii) new investment policy, (iii) safe mobility, (iv) low emission mobility, (v) smart mobility, (vi) smart intermodal logistics chains, (vii) connectivity, (viii) social and labour aspects, and (ix) evolution of the MITMA. The fourth axis aims to develop sustainable energy sources for transport (e.g. electrification, hydrogen) and low-emission technologies, to decrease the age of the vehicle fleet and to support the sustainability of transport facilities (e.g. terminals). The Strategy will also promote administrative co-operation and co-ordination, as well as public participation (e.g. the Open Mobility Dialogue in 2020, territorial workshops, surveys). The Strategy is based on the Sustainable Mobility Law, whose the bill has been approved by the Council of Ministers on 13 December 2022 (Ministerio de Transportes Movilidad y Agenda Urbana, 2022[23]), and is financed under the Recovery, Transformation and Resilience Plan.

Spanish legislation on air pollution

Spain also targets other atmospheric pollutants to fight against air pollution. The Spanish National Program on Atmospheric Pollution Control, which transposed the Directive EU 2016/2284 (see above) into national legislation through Royal Decree 818/2018, sets sectoral measures to achieve the emissions reduction targets of several pollutants (Table 2.2). The Program gives continuity to the previous Spanish Air Plans with the objective to improve national air quality.

2.1.3. At the regional level

The urgency of combatting climate change and reaching environmental protection objectives has led the region to take action since the early 2000s. In 2002, the region passed the Andalusian Climate Change Strategy, the first Spanish autonomous region to develop a strategy of measures and actions. In 2007, it published its first Climate Action Plan (PAAC) (Junta de Andalucia, 2007[24]). In 2018, the region passed a law entitled "Measures against climate change and for the transition towards a new Andalusian energy model", calling for the creation of a new PAAC to act as a general strategy/planning instrument for climate change action in the region in the short, medium, and long-term. The law also called for the creation of municipal climate action plans (PMCC). Following this law, Andalusia implemented a new PAAC in 2021 with three new programs to fight against climate change.

The 2002 Andalusian Strategy on Climate Change and 2007 Andalusian Climate Action Plan

Andalusia established the Andalusian Strategy on Climate Change on 3 September 2002. The initiative was the first in Spain. The document presented a set of measures to reduce regional emissions of GHGs. It provided the ground for the 2007-2012 PAAC, which was structured around three Action Programs.

- Mitigation program (approved on 5 June 2007): the program aimed to reduce GHG emissions of 19% by 2012 compared to 2004 and from 8 tonnes of CO_2 per capita to 6.5 tonnes per capita in Andalusia, as well as to promote carbon sequestration by enhancing the carbon sink capacity of ecosystems (Junta de Andalucia, 2007[25]). The objectives were met by 2012, with 21% reduction of GHG emissions and 6.1 tonnes of CO_2 per capita (Junta de Andalucia, 2015[26]). The development of environmental taxation to reduce GHG effect was underlined as part of the measures of the program (M.138 under "Development of new intervention tools").
- Adaptation program (approved on 13 August 2010): the program aimed to reduce the vulnerability of the autonomous community to climate change by increasing its adaptation capacity through planning instruments (Junta de Andalucia, 2010[27]). The program comprised four subprograms, of which measures for immediate actions, sectoral analysis of the effects of climate change, sectoral measures of adaptation, and continuous development of knowledge and governance.
- Communication program (approved on 21 January 2012): the program aimed to promote knowledge, raise awareness and increase participation of citizens in climate action (Junta de Andalucia, 2012[28]).

The Plan is still at the core of Andalusia's climate policy and the climate component of the Andalusian Sustainable Development Strategy 2030 *(Estrategia Andaluza de Desarrollo Sostenible 2030; EADS)*, adopted on 5 June 2018 (Junta de Andalucia, 2018[29]).

The 2018 Andalusian Law on Measures against Climate Change and the Transition towards a New Energy Model

On 8 October 2018, the Andalusian Parliament approved the Andalusian Law on Measures against Climate Change and the Transition Towards a New Energy Model in Andalusia (Andalusian law 8/2018), which aimed to reduce GHG emissions, to limit fossil fuel consumption and to increase cities' adaptation to climate change (Junta de Andalucia, 2018[30]). The law also provided for the creation of an Interdepartmental Climate Change Commission, as a transversal commission responsible for climate planning, and of the Andalusian Climate Change Office, the administrative unit in charge of managing mitigation, adaptation and communication policies.

In line with the Regulation (EU) 2018/1999 and the Spanish Climate Change and Energy Transition law, chapter II of the law provided for the co-ordination of Andalusia with the other levels of government on climate policies. Municipalities must elaborate municipal plans against climate change, within the scope of their responsibilities laid down in law 5/2010 (law on local autonomy in Andalusia) and within the framework of the PAAC (art. 15) (Junta de Andalucia, 2010[31]; Junta de Andalucia, 2018[30]). Municipal plans shall include an analysis and evaluation of GHG emissions within their territory, objectives and strategies for mitigation and adaptation to climate change, actions to reduce emissions, actions to promote research and innovation, actions for awareness on climate change, actions for progressive replacement of fossil fuels with renewable energies, actions to rehabilitate municipal buildings, actions to optimise public lighting, actions to promote energy transition within urban mobility plans, and temporary planning of the actions. Municipal plans shall be reviewed along with the revision of the PAAC to align objectives. They shall approve a report on the degree of compliance with their plans every two years. Provinces may provide support to municipalities for the preparation of their plans, within their scope of responsibilities.

The Autonomous Community of Andalusia shall approve economic resources allocated to the plans of the municipalities within its territory, as per art. 25 of the law 5/2010[8] (Junta de Andalucia, 2010[31]). Art. 16 of the Andalusian law 8/2018 provides that the autonomous community must collaborate with the central government, within the scope of its responsibilities, to promote mitigation, adaptation and communication measures established in the PAAC through specific instruments (Junta de Andalucia, 2018[30]).

The 2021 Andalusian Climate Action Plan

The Andalusian law 8/2018 prepared the ground for the new PAAC, approved on 13 October 2021 and published by the Andalusian Decree 234/2021 (Junta de Andalucia, 2021[32]). The 2021 PAAC is the general strategic planning instrument to fight against climate change in Andalusia. It aims to integrate climate change into regional and local planning and align them with the central government's plans, the European Green Deal and the Paris Agreement, to achieve the Sustainable Development Goals set by the 2030 Agenda of the United Nations. The 2021 PAAC comprises six strategic objectives, 12 sectoral objectives and more than 137 lines of actions, structured under three programs:

- the Mitigation and Energy Transition Program,
- the Adaptation to Climate Change Program, and
- the Communication and Participation Program (Figure 2.2).

The main objective of the Plan, under its Mitigation and Energy Transition Program, is to achieve a reduction of 39% of GHG emissions in Andalusia by 2030 compared to 2005 levels through emission reductions in strategic sectors listed in (Figure 2.2).

Figure 2.2. Objectives under the Programs of the PAAC 2021-2030 in Andalusia

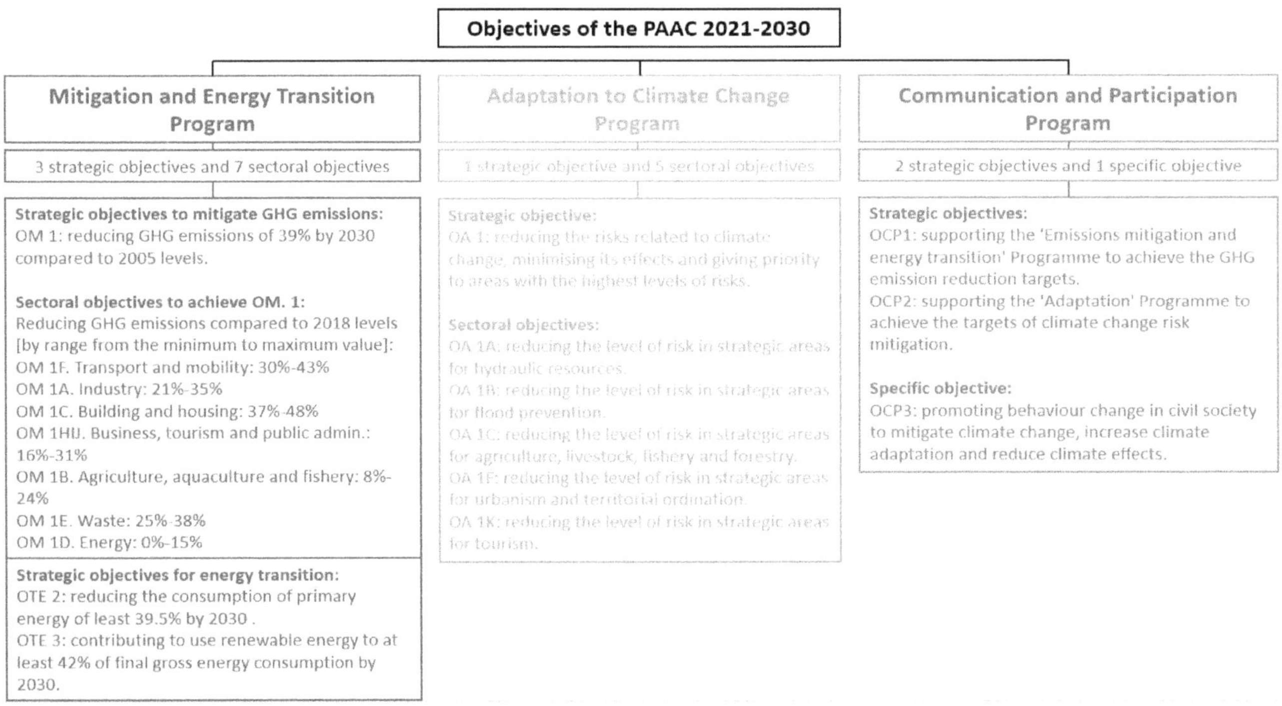

Source: Author's own elaboration based on Junta de Andalucía (2021[32]).

Andalusia has also established the Andalusian Energy Strategy (EEA) 2020, which is in line with the objectives set in the PAAC for 2030 and 2050. The EEA Strategy follows the previous energy plans of the autonomous community, of which the Andalusian Plan for Sustainable Energy 2007-2013 *(Plan Andaluz de Sostenibilidad Energetica 2007-2013; PASENER)*. It aims to foster renewable energy generation projects, to increase buildings' energy efficiency, to optimise energy consumption, promote bioeconomy, to decarbonise transport, to prepare the workforce to adapt to technological changes and to enhance private investment in renewable energy projects (Junta de Andalucia, 2020[33]).

On June 7, 2022, in line with the Andalusian law 8/2018 and the PAAC of the region, the government council of Andalusia approved the Andalusia Energy Strategy 2030 (Junta de Andalucia, 2022[34]). This strategy contains six objectives and 12 strategic lines, each with action programs, to support the use of renewable energy and the development of sustainable energy networks. It aims to (i) supporting the decarbonisation of energy consumption, (ii) reducing energy consumption, (iii) reducing the dependence on petroleum derivatives in transport; (iv) having the necessary infrastructure to harness renewable resources and provide quality supply, (v) improving the effectiveness and efficiency of the administration as a facilitator of the transition and decarbonise its energy consumption, and (vi) strengthening the Andalusian energy business and industrial base.

Regarding transportation, Andalusia has also approved the Transport and Mobility Plan of Andalusia for 2021-2030 (PITMA 2030) on 2 February 2021 (replacing 2021-2027) (Junta de Andalucia, 2021[35]). The Plan contains several objectives with indicators to be achieved by 2030, such as (i) improving research and innovation for mobility and transport infrastructure (e.g. share of expenditure on innovation over turnover of 1.65 by 2030 in vehicle manufacturing companies compared to 0.46 in 2018), (ii) supporting mobility services through digitalisation (e.g. 1 500 000 users of new digital mobility services and applications by 2030), (iii) promoting energy efficiency, climate change mitigation and adaptation (e.g. 7.6% of electricity in transport sector energy consumption by 2030 compared to 0.4% in 2019), (iv) developing a good and sustainable network of transport infrastructure to meet demand for mobility (e.g. only 20% of roads with very deteriorated or somewhat deteriorated pavement by 2030 compared to 39.6% in 2019), (v) achieving sustainable regional mobility (e.g. 10% of freight transport over land transport by 2030 compared to 0.77% in 2018), and (vi) developing sustainable multimodal urban and metropolitan mobility (5.5 kt eq. CO2 emitted by road traffic in all Andalusian cities with more than 10 000 inhabitants by 2030 compared to 9.1 kt eq. CO2 in 2018) (Junta de Andalucia, 2022[36]). In line with the PITMA and the PAAC, which are the main regional instruments for designing policies in mobility, transport and climate change, the Governing Council of Andalusia has also approved the formulation of the Andalusian Strategy for Sustainable Mobility and Transport 2030 on 12 January 2021 (Junta de Andalucia, 2021[37]). This Strategy will guide the region to achieve its GHG emissions reduction target that ranges between 30% and 43% by 2030 compared to 2008 levels for transport and mobility sector.

The Andalusian Strategy for Air Quality

Regarding air pollution, Andalusia adopted the law 7/2007 on the integrated management of atmospheric quality and its development to encompass the three dimensions of sustainable development (i.e. environmental, social and economic). In accordance with the Spanish Air Plans, of which the Spanish National Program on Atmospheric Pollution Control (Royal Decree 818/2018) and the Directive (EU) 2016/2284, Andalusia also established its Andalusian Strategy for Air Quality *(Estrategia Andaluza de Calidad del Aire; EACA)*, approved on 22 September 2020 (Junta de Andalucia, 2020[38]). This Strategy is a framework instrument to facilitate the preparation of air quality improvement plans by local governments in Andalusia. It is based on a comprehensive assessment of air quality at the local level between 2017 and 2019. All atmospheric pollutants that are required to be assessed are included in the analysis and compared with national legislation, EU regulations and the Air Quality Guide of the WHO. The main pollutants emitted by sectors are provided in the Table 2.3. The objectives of the future air quality improvement plans are to remain below the limit values set by these regulations (Junta de Andalucia, 2020[2]). Currently no specific reduction targets for these pollutants exist at the regional level (Table 2.2 shows targets at the national level).

Table 2.3. The main atmospheric pollutants by sector under the Andalusian Strategy for Air Quality

Sectors	Main atmospheric pollutants									
	PM10	NO_2/NOx	SO_2	CO_2	Pb	As	Cd	Ni	Benceno	Benzo(a)pireno
Transport	x	x		x	x					
Construction, demolition	x									
Maritime transport	x	x	x							
Airport	x	x								
Agriculture sector	x	x								
Industry		x	x	x		x	x	x	x	x
Residence, business and administration	x									

Source: Author's own elaboration based on (Junta de Andalucia, 2020[2]).

2.2. Responsibilities Related to Greenhouse Gas Emissions and Air Pollution across Levels of Government

This section details the responsibilities related to GHG emissions and air pollution across the levels of government in Spain and maps where Andalusia has the power to set, manage, or implement taxes in these environmental domains.

In line with the analysis carried out in Activity 1.3 of the project, emissions are separated into stationary source emissions and non-stationary source emissions (see definitions above). For stationary sources, the analysis will focus on industry and electricity, as they represent a key source of GHG emission and air pollution. For non-stationary sources, the analysis concentrates on personal vehicles (i.e. vehicles for individuals with no commercial purpose), that are large emitters of pollutants into the atmosphere. Stationary sources: industry and electricity

This section maps the responsibilities of each level of government regarding emissions from stationary sources in the industrial and electricity sectors. These are identified by examining the environment, climate change, and the energy sector responsibilities of each level of government.

2.2.1. At the EU level

The EU responsibilities related to environment and energy are both shared responsibilities between the EU and Member States (art. 4) (European Union, 2012[39]). Regarding environment, art. 11 and art. 191 to 193 of the TFEU set that the EU has responsibilities in all domains of environmental policy, such as air and water pollution, waste management and climate change, to promote sustainable development (European Union, 2012[39]). Its scope for action is, however, limited by the principle of subsidiarity (see Part I) and the requirement of unanimity in the Council for fiscal matters, urban and country planning, land use, quantitative water resource management, choice of energy source and structure of energy supply (art. 192) (European Union, 2012[39]). Under art. 191, the preparation of EU environmental policies shall be based on scientific and technical data, the environmental conditions of the regions in the EU, the benefits and costs of action and the economic and social development of the EU as a whole. They shall pursue the following objectives: (i) preserving and protecting the environment, (ii) protecting human health and (iii) promoting the prudent use of natural resources (art. 191). The fight against climate change is also set as an explicit objective of the EU environmental policy (art.191) (European Union, 2012[39]). In addition, Member States may adopt more stringent protective measures on environment that those set in EU environmental policies, provided that they are compatible with the Treaty and are notified to the Commission (art. 193).

Regarding energy, art. 194 of the TFEU stipulates that EU energy policy shall aim to: (i) ensure the functioning of the Energy Union, (ii) ensure the energy supply in the Union, (iii) promote energy efficiency and the development of renewable energy and (iv) promote the interconnection of energy networks (European Union, 2012[39]). The measures to achieve these objectives shall be adopted after consultation with the Economic and Social Committee and the Committee of the Regions (art. 194). As stipulated under art. 192 of the TFEU, EU energy policy shall not affect Member State's right to determine their energy sources and the general structure of their energy supply. It is also limited by the requirement of unanimity in the Council regarding any fiscal matters (art. 194) (European Union, 2012[39]).

2.2.2. At the national, regional and local levels

The distribution of responsibilities across the levels of governments in Spain, including environment and energy, are described in the Constitution, the Statute of Autonomy of Andalusia and the LBRL. The responsibilities related to environment and climate change and to energy are described in Table 2.4.

The Spanish Constitution provides that the autonomous communities may assume responsibilities related to the management of environment protection (art. 148) (Gobierno de Espana, 1978[40]). Accordingly, the Statute of Andalusia requires that the autonomous community shall adopt measures and strategies to mitigate climate change, which includes the rational use of energy resources (Junta de Andalucia, 2007[41]). In addition, art. 149 of the Constitution stipulates that the autonomous communities may establish additional rules on environmental protection. This has been reflected in art. 57 of the Statute and in art. 49, which grants shared responsibilities to Andalusia over the facilities of production, distribution and transport of energy, when this transport remains within its territory (Junta de Andalucia, 2007[41]). The responsibilities of municipalities are set within the LBRL (Junta de Andalucia, 2010[31]), while provinces are responsible for ensuring the co-ordination and provision of municipal services (Table 2.4).

Table 2.4. Distribution of environmental and energy sector responsibilities across levels of government in Spain

Matters	Central government	Andalusia	Provinces	Municipalities
Environment	**Exclusive responsibilities**: Basic legislation on environmental protection, without prejudice to the responsibilities of autonomous communities to establish additional protection norms (149.1.23).	**Exclusive responsibilities:** Forestry, exploitation, utilisation and forest services; marshes and lagoons, and aquatic ecosystems; pastures and special treatment of mountain areas; delimitation, regulation, planning and comprehensive management of protected natural spaces; environmental prevention (57.1 AS). **Shared responsibilities:** Establishment and regulation of environmental planning instruments and the procedure for processing and approving these instruments; the establishment and regulation of environmental sustainability and research measures; the regulation of natural resources; the regulation of the atmospheric environment and the	**Responsibilities:** Securing co-ordination and provision of municipal services.	**Responsibilities:** Urban environmental protection (25.2.bLBLR). Cooperation with other public administrations to promote, defend, and protect the environment and public health (92 AS). Municipalities with more than 50.000 inhabitants are obliged to provide urban environmental services (26.1.dLBLR).

		different types of contamination of the same; the regulation of the system of authorisations and monitoring of GHG emissions; the establishment and regulation of ecological taxation measures; and the prevention, restoration and repair of damage to the environment, as well as the corresponding sanctioning regime over (57.3 AS).		
Energy	**Exclusive responsibilities:** Bases of the organisation of mining and energy (149.1.25) The authorisation of electrical facilities when their use affects another autonomous community or the transport of energy exceeds its territorial scope (149.1.22). Establish the basic regulation of activities aimed at supplying electricity (3.1 ESL). Establish the remuneration system (3.3 ESL). Regulate the organisation and operation of the market (3.9 ESL).	**Shared responsibilities:** Facilities for the production, distribution and transport of energy, when this transport runs entirely through the territory of Andalusia and its use does not affect another territory (49.1 AS). Promotion and management of renewable energies and energy efficiency (49.1 AS). Energy and mines, without prejudice to the provisions of article 149.1.25 of the Spanish Constitution (49.2 AS). Regulation of energy production, storage and transportation activities, as well as their authorisation and inspection and control, establishing, where appropriate, the quality standards for supply services (49.2 AS).	**Responsibilities:** Securing co-ordination and provision of municipal services.	**Responsibilities:** Public lighting (LBLR). Municipalities can approve ordinances for the use of renewable energy in buildings and facilities (Court ruling 2339/2015).

Note: ELS: Electricity Sector Legislation; AS: Andalusian Statute; LBLR: Regulatory Law of the Bases of the Local Regime.
Source: Author's own elaboration based on (Gobierno de Espana, 1978[40]; Junta de Andalucia, 2007[41]; Junta de Andalucia, 2010[31]).

Nevertheless, Table 2.4 shows that the responsibilities of the autonomous communities and local governments in energy remain limited. They are mainly centralised, with the central government being responsible for establishing regulations and economic instruments on electricity production. The Constitutional Court Decision 87/2019 on the Catalonian Climate Change Act also underlines the limited room for the autonomous communities to implement energy sector reforms (Box 2.3).

Box 2.3. The Constitutional Court Decision 87/2019 in Catalonia

In 2019, the Constitutional Court Decision 87/2019 declared unconstitutional 15 articles of the Climate Change Act in Catalonia (law 16/2017), some of which related to energy (Generalitat de Catalunya, 2017[42]). The decision outlined that "autonomous communities cannot establish quantitative, measurable, and time-bound emission reduction, renewable energy and energy efficiency objectives", on the ground that it is contrary to basic national legislation and in breach of the responsibilities of the central government. On 16 November 2019, Catalonia approved the Decree-law 16/2019 on urgent measures to deal with the climate emergency and the promotion of renewable energies and, on 30 December 2019, the law 9/2019, amending the law 16/2017, on climate change was published (Generalitat de Catalunya, 2019[43]). The Decree-law 33/2020 on urgent measures in the field of the tax on GHG emissions from mechanical traction vehicles and the tax on stays in tourist establishments (Generalitat de Catalunya, 2020[44]) and the Decree-law 24/2021 on the acceleration of the development of renewable energy (Generalitat de Catalunya, 2021[45]) also amend the law 16/2017.

Although the targeted legislation was the Catalonian Climate Change Act, there are similarities between the Catalonian Climate Change Act and the Andalusian Climate and Energy Transition law (Table 2.5). However, no ruling on the Andalusian Climate and Energy Transition law has been identified at the time of writing.

Table 2.5. Comparison between Catalonian and Andalusian climate legislation

Catalonian Act excerpt ruled unconstitutional	Justification	Andalusian Law excerpt
"The goal of reducing greenhouse gas emissions for the year 2030 is 40% compared to the base year (1990), 65% for 2040 and 100% for 2050."	"... Violate the basic legislation of the State on polluting emissions, in the same terms as art. 7.3 also appealed (and already declared unconstitutional and null in the preceding legal basis 8). These sections directly indicate binding, specific, measurable and term objectives for the reduction of polluting emissions that are irreconcilable with the possibility and the right to emit greenhouse gases recognised by the State in the Law mentioned above 1/2005 and in Law 34/2007. and Royal Legislative Decree 1/2016, also cited, with no other condition than requesting the mandatory authorisation, paying for it, complying with the corresponding formal obligations and not exceeding the emission limit values and the air quality objectives, as we have already explained more above. Consequently, they must be declared unconstitutional and null".	"The objective for Andalusia for the year 2030 is to reduce at least 18% of diffuse greenhouse gas emissions per inhabitant concerning 2005."
"1. The measures adopted in the field of energy must be aimed at the energy transition towards a one hundred per cent renewable, denuclearised and decarbonised model, neutral in greenhouse gas emissions, which reduces the vulnerability of the Catalan energy system and guarantees the right of access to energy as a common good, and	"Pursuing "the energy transition towards a one hundred per cent renewable, denuclearised, decarbonised model [and] neutral in greenhouse gas emissions" (art. 19, first and second paragraphs) is nothing more than a constitutionally legitimate programmatic guideline, protected by the arts. 45 CE and 27 and 46 EAC (right to an adequate environment and sustainable development) that by itself does not violate central government responsibilities. However, this art. 19 goes beyond this guideline and imposes specific, detailed, term, measurable and therefore binding objectives, such as those already mentioned to close nuclear power plants	"1. The measures adopted as a development of this law or Law 2/2007, of 27 March, on the promotion of renewable energies and energy-saving and efficiency in Andalusia, must be aimed at the energy transition towards a renewable and decarbonised energy model, neutral in greenhouse gas emissions, that reduces the vulnerability of the Andalusian energy system and guarantees the right of access to energy as a common good. In

specifically they must be aimed at:	in 2027 "ensuring the preservation of direct jobs", reduce consumption of energy by 2% per year "to reach at least 27% in 2030" and the consumption of fossil fuels at 50% in 2030 and zero in 2050."	particular, they should focus on: a) Promote energy saving and efficiency policies, to reduce the trend consumption of primary energy in the year 2030, by at least 30%, excluding non-energy uses. b) Promote renewable energies and an energy model in which the consumption of fossil fuels tends to be zero so that by 2030 renewable energies can contribute at least 35% of gross final energy consumption."

Source: Author's own elaboration based on Constitutional Court Ruling 87/2019.

2.2.3. Non-stationary sources: personal vehicles

The analysis for non-stationary source emissions focuses on transport and personal vehicles in particular. Findings from the analysis on environment and energy above may also apply to emissions from and energy used in vehicles.

At the EU level

Transport is a shared responsibility between the EU and Member States according to art. 4 of the TFEU. The EU responsibilities on transport by rail, road and inland waterway are regulated by Title VI of the TFEU, which encompasses art. 90 to art. 100. According to art. 90, the EU has the power to establish a common transport policy. This policy shall include common rules applicable to international transport or transport from one or several Member States, the conditions under which non-EU resident carriers may operate transport services within a Member State, measures to improve transport safety and any other appropriate provisions (art. 91). Regarding transport within the EU, discrimination based on different rates and conditions for the carriage of the same goods over the same transports shall be prohibited (art. 95). In addition, charges or fees related to the crossing of frontiers shall not exceed a reasonable level compared to the costs (art. 97).

At the national, regional and local levels

The Constitution, the Statute of Andalusia and the LBRL provide a more balanced distribution of transport responsibilities across levels of government (Table 2.6). The Constitution grants the central government exclusive responsibility over maritime, air, railway and road transport that passes through the territory of multiple autonomous communities (art. 149). By contrast, it allocates responsibilities to the autonomous communities regarding these forms of transports as long as they fall exclusively within the territory of the autonomous community and they do not pursue any commercial activity (art. 148) (Gobierno de Espana, 1978[40]). This is reflected in art. 64 of the Statute (Junta de Andalucia, 2007[41]). Local level responsibilities relies on urban and rural public roads, as well as on the provision of public transport services for large municipalities (Junta de Andalucia, 2010[31]).

Table 2.6. Distribution of responsibilities in transport across levels of government in Spain

Matter	Central government	Andalusia	Provinces	Municipalities
Transport	**Exclusive responsibilities:** Merchant marine, ports and airports of general interest and control, air transport traffic (149.1.20). Railways and land transport passing through the territory of more than one autonomous community; traffic and circulation of motor vehicles (149.1.20).	**Exclusive responsibilities:** Regional railway and road networks lying exclusively within their territories, regional and recreational ports and airports with no commercial activities (148.1.5); Andalusian road network, made up of railways, highways and roads; maritime and river transport of people and merchandise that takes place entirely within the waters of Andalusia; ports and airports and other transport infrastructure in the territory of Andalusia (64.1 AS).	**Responsibilities:** Securing co-ordination and provision of municipal services.	**Responsibilities:** Provision of public transport; conservation of urban and rural public roads; management of mobility and accessibility of people and vehicles on urban roads (92.1 AS). Vehicle parking and mobility (25.2.g). Municipalities with more than 50 000 inhabitants are obliged to provide public transport services (26.1.d LBLR).

Note: AS: Andalusian Statute; LBLR: Regulatory Law of the Bases of the Local Regime.
Source: Author's own elaboration based on (Gobierno de Espana, 1978[40]; Junta de Andalucia, 2007[41]; Junta de Andalucia, 2010[31]).

2.3. Current Levies Related to Greenhouse Gas Emissions and Air Pollution across Levels of Government in Spain

This section provides the existing levies on GHG emissions and on air pollution that apply to emitting sources in Andalusia. As for the previous section, emissions are separated between stationary sources and non-stationary sources.

2.3.1. Stationary sources: industry and electricity

The different levies applicable to stationary sources of GHG emissions and air pollution in Spain and Andalusia, with a focus on industry and electricity, are presented in the Table 2.7. As an EU Member State, Spanish GHG emissions are managed under the EU ETS (Box 2.2).

Table 2.7. Current levies applicable to stationary sources of GHG emissions and air pollution in Spain and Andalusia

Current levies	Competence
EU ETS	EU
Tax on fluorinated GHGs	National
Hydrocarbon tax	National
Value Added Tax	National
Hydroelectric development fee	National
Electricity production tax	National
Electricity tax	National
Tax on gas emissions into the atmosphere	Regional
Charge for administrative services in industrial, energy and mining matters	Regional

Source: Author's own elaboration.

At the national level, there is the **hydrocarbon tax**, which is regulated under the law 38/1992 on special taxes (Gobierno de Espana, 1992[46]), amended in 2019 to harmonise regional tax rates into a national hydrocarbon tax system. The law has been established in the context of the EU ETD, which set minimum tax rates, and is currently under revision (see above). The tax is levied on hydrocarbons (e.g. petrol, diesel, natural gas, oil, and biofuels) that are used as fuel. Some exemptions apply, of which (i) natural gas used for purposes other than fuel, (ii) fuel supply in air and sea navigation, (iii) rail transport, (iv) construction and maintenance of vessels and aircrafts, and (v) pilot projects of less polluting products. Since 2018, natural gas and biogas used to produce electricity and heat or self-consumption are tax-exempted. Total and partial refunds also apply depending on the fuel use. The tax represented 8.9% of Andalusia's tax revenue in 2020, above the average of the autonomous communities (7.2%, excluding the Basque country, Navarra and the Canary Islands) (Ministerio de Hacienda y Funcion Publica, 2022[47]). The regions of Catalonia and Andalusia receive the largest receipts from the hydrocarbon tax in absolute terms, although it does not represent the highest shares in tax revenue among the autonomous communities (Figure 2.3).

Figure 2.3. The hydrocarbon tax as a share of total tax revenue in autonomous communities in 2020

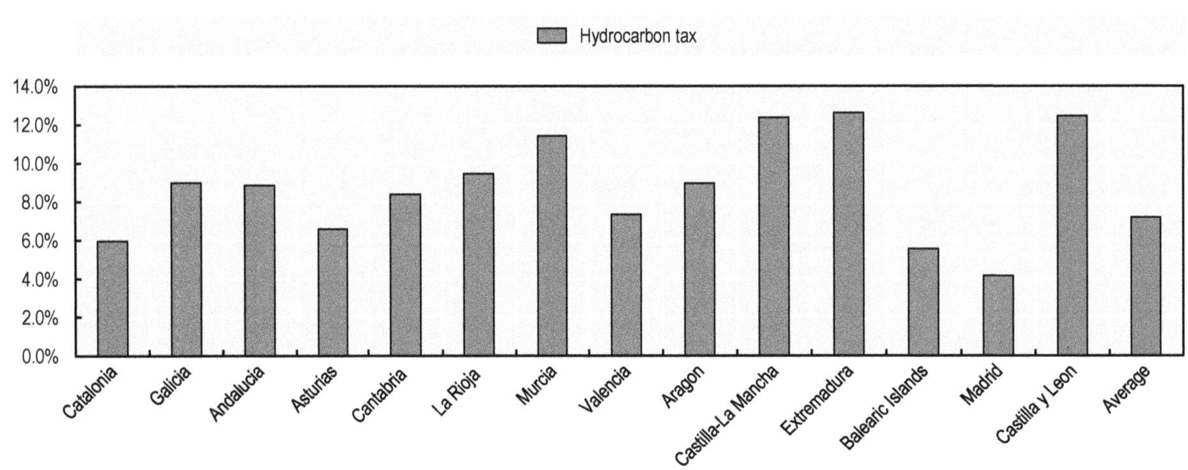

Note: The Basque country, Navarra and the Canary Islands are not presented in the above figure as they have a specific financing system.
Source: Author's own elaboration based on *Ministerio de Hacienda y Funcion Publica* (2022).

StatLink https://stat.link/6g2drk

The **tax on fluorinated GHGs** is regulated under the law 16/2013 of 29 October (Gobierno de Espana, 2012[48]). It is an indirect tax levied on the consumption of certain fluorinated gases used as refrigerants or solvents based on their global warming potential. It includes hydrofluocarbons (HFCs), perfluorocarbons (PFCs), sulphur hexafluoride (SF6), regenerated and recycled gases, etc. The taxable matter is the first sale or delivery of fluorinated gases after production, import or acquisition. The tax base is the weight of the products, measured in kilograms. Several exemptions apply to the first sale or delivery when they aim to be: (i) resale only (i.e. no use of fluorinated gases in the production process), (ii) exported, (iii) used to chemical transformations that alter its composition, (iv) incorporated for the first time into new equipment and devices, (v) used to produce, import or acquire medical aerosols for inhalation, (vi) imported or acquired in new equipment and devices. A 90% exemption applies when the first sale or delivery is used to fire extinguishing equipment. Reductions are possible for waste management, destruction, recycling or reclamation of waste in accordance with sectoral waste legislation (Gobierno de Espana, 2012[48]).

Regarding electricity, the majority of levies are established and regulated by the central government. They apply to electricity generation, transmission and consumption. They include:

- The **Value Added Tax** (Royal Decree 1624/1992): national tax on consumption of goods and services (Gobierno de Espana, 1992[49]; Gobierno de Espana, 1992[46]). General VAT is levied at 21% on most products and services in Spain. Since June 2021, under the Royal Decree-law 12/2021, electricity falls under the category of "products and services eligible for reduced VAT" (Gobierno de Espagna, 2021[50]). A reduced rate of 10% applies since then for consumers with contracted power of less than 10kW. The Decree-law 6/2022, adopted in response to the economic and social consequences of the war in Ukraine, set that reduced VAT shall be maintained as long as the price in the market is higher than EUR 45 per MWh and extended until 30 June 2022 (Gobierno de Espana, 2022[51]). With effect from 27 June 2022, the rate of VAT on electricity consumption was reduced to 5% due to rising inflation under the Royal Decree-law 11/2022. Recently, this reduced tax rate was extended to 31 December 2023 under the Royal Decree-law 20/2022 (Gobierno de Espana, 2022[52]).
- The **hydroelectric development fee** (art. 112 to Royal Decree 1/2001): national fee on the public hydraulic domain for hydroelectric development purposes (Gobierno de Espana, 2001[53]). The taxable matter is the use of the dams of the reservoirs or the channels built with funds from public administrations for the purposes of hydroelectric exploitation. The tax base is determined by the competent river basin authority (see Box 6.2) based on the economic value of the hydroelectric energy produced.
- The **electricity production tax** (art. 1 to 11 Law 15/2012): national tax levied on the production of electricity and its incorporation in the electrical system. The tax applies to economic capacity of electricity producers whose facilities give rise to significant investments in the transmission and distribution. It is calculated based on the gross revenue generated during the production of electricity and its inclusion in the electrical system received by the taxpayer. The rate is at 7%. Under the Royal Decree-law 12/2021, this tax has been suspended temporarily in 2021 (Gobierno de Espagna, 2021[50]). The suspension was extended to 31 December 2022 under the Royal Decree-law 11/2022 (Gobierno de Espana, 2022[51]) and recently to 31 December 2023 under the Royal Decree-law 20/2022 (Gobierno de Espana, 2022[52]).
- The **electricity tax** (art. 89 to 104 Law 38/1992): special national ad valorem tax levied on the supply of electricity to a person for its own consumption at a rate of 5.113% of the total consumption and power charge, which is the applicable tax base for VAT. Since the Royal Decree-law 12/2021, the tax rate has been reduced to 0.5% (Gobierno de Espagna, 2021[50]), which has been extended to 31 December 2022 (Gobierno de Espana, 2022[51]). The reduced tax rate was extended to 31 December 2023 under the Royal Decree-law 20/2022 (Gobierno de Espana,

2022[52]). The electricity tax represented 1.3% of Andalusia's tax revenue in 2020, in line with the average of the autonomous communities (1.2%, excluding the Basque country, Navarra and the Canary Islands) (Ministerio de Hacienda y Funcion Publica, 2022[47]).

At the regional level, the main tax on GHG emissions and air pollution is the **tax on gas emissions into the atmosphere** in Andalusia. The tax is regulated by the law 18/2003 (art. 21 to 28) (Junta de Andalucia, 2003[54]). It is levied on industries that emit carbon dioxide (CO_2), nitrogen oxides (NOx) and sulphur oxides (SOx) above a certain threshold during their production processes (Box 2.4). The industries subject to this tax are listed in Annex 1 of Law 16/2002. They include energy production, oil processing, steelmaking and chemical industries. Emissions from landfills, facilities for intensive rearing of animals and those from the combustion of biomass and biofuel are exempt. Deductions apply to industries investing in emission reduction. The tax represented a negligible amount of revenue in Andalusia in 2020. Similar taxes on air pollution or gas emissions exist in other autonomous communities (Table 2.8).

Table 2.8. Taxes on air pollution and gas emissions applicable in other autonomous communities

	Andalusia	Cataluna	Galicia	Murcia	Valencia	Aragon	Castilla-La-Mancha
Instrument	Tax on gas emissions into the atmosphere *(Impuesto sobre las emisiones de gases a la atmosfera)*	Tax on industrial emission of gases and particles into the atmosphere *(Impuesto sobre la emisión de gases y partículas a la atmósfera producida por la industria)*	Air pollution tax *(Impuesto sobre la contaminación atmosférica)*	Tax on emission of pollutants into the atmosphere *(Impuesto sobre emisiones de gases contaminantes a la atmósfera)*	Tax on activities that affect the environment *(Impuesto sobre actividades que inciden en el medio ambiente)*	Environmental tax on the emission of pollutants into the atmosphere *(Impuesto mediomabiental sobre la emisión de contaminantes a la atmósfera)*	Tax on certain activities that affect the environment *(Impuesto sobre determinadas actividades que inciden en el medio ambiente)*
Tax base	Annual CO_2, NOx and SOx emissions expressed in polluting units above a certain threshold (3 polluting units)	Annual NOx, SO_2, particulate matter and total organic carbon emissions above a certain threshold (SO_2: 150 tonnes; NOx: 100 tonnes; Particulate matter: 50 tonnes; Total organic carbon: 150)	Annual NOx and SOx emissions	Annual SO_2, NOx, volatile organic compounds and ammonia emissions expressed in polluting units above a certain threshold (3 polluting units)	Annual NOx (expressed in NO_2 eq.) and SO_2 emissions above a certain threshold (150 metric tonnes)	Annual SOx, NOx and CO_2 emissions above a certain threshold (SOx: 150 tonnes; NOx:100 tonnes; CO_2: 100,000 tonnes)	Annual SOx and NOx emissions
Tax rate	Rate in EUR per polluting unit: ≤ 10: 5,000 10 and ≤ 20:8,000 20 and ≤ 30:10,000 30 and ≤ 50:12,000 50: 14,000 The polluting units are obtained by dividing the total amount of each substance emitted by a reference value	Rate in EUR per pollutant tonne: SOX: 45 NOX: 75 Particulate matter: 60 Total organic carbon: 45	Rate in EUR per tonne: ≤ 100:0 > 100 and ≤ 1 000:36 > 1,000 and ≤ 3,000:50 > 3,000 and ≤ 7,000:70 > 7,000 and ≤ 15,000:95 > 15,000 and ≤ 40,000:120 > 40,000 and ≤ 80,000:150 > 80:200	Rate in EUR per polluting unit: ≤ 10: 5,000 > 10 and ≤ 20:8,000 > 20 and ≤ 30:10,000 > 30 and ≤ 50:12,000 > 50: 14,000 The polluting units are obtained by dividing the total amount of each substance emitted by a reference value.	Rate in EUR per tonne: ≤ 1,000: 9 > 1,000 and ≤ 3,000:12 > 3,000 and ≤ 7,000:18 > 7,000 and ≤ 15,000:24 > 15,000 and ≤ 40,000:30 > 40,000 and ≤ 80,000:38 > 80,000:50 Sum of NOX emissions in tonnes of NO2 multiplied by 1.5 and SO2 emissions in tonnes.	Rate in EUR per pollutant tonne: SOX: 50 NOX: 50 CO_2: 0.2	Rate in EUR per tonne: ≤ 500: 0 > 501 and ≤ 5,000:34 > 5,000 and ≤ 10,000:60 > 10,000 and ≤ 15,000:80 > 15,000:100 Sum of NOX and SO2 emissions weighted by 1 and 1.5, respectively.

Source: Author's own elaboration based on (Junta de Andalucia, 2003[54]; Generalidad de Cataluna, 2014[55]; Comunidad Autonoma de Galicia, 1995[56]; Comunidad Autónoma de la Región de Murcia, 2005[57]; Comunidad Autonoma Valenciana, 2012[58]; Comunidad Autónoma de Aragón, 2007[59]; Comunidad Autónoma de Castilla-La Mancha, 2005[60]).

The tax on gas emissions into the atmosphere is complemented by the **charge for administrative services in industrial, energy and mining matters** in Andalusia, which is regulated under the law 10/2021 (art. 42 to 46). It is levied on the provision of services and performance of administrative activities concerning the planning of industrial, energy and mining activities.

Box 2.4. The tax on gas emissions into the atmosphere in Andalusia

Key features of the tax on gas emissions into the atmosphere in Andalusia are:

Tax base: the pollutant load of different pollutants emitted from the same industrial plant. The pollutant load is determined by the sum of the polluting units of all the substances emitted from the same industrial facility. The polluting units are obtained by dividing the total amount of each substance emitted yearly divided by a reference value.

A direct estimation of the tax base can only be carried out when the industrial installations have the respective monitoring system in place and when the percentage of monitored data meets the requirements laid down in the rules.

In the case where industrial installations have no monitoring devices or when the percentage of monitored data does not meet the requirements of the standard, the tax base is calculated as the sum of the quantities of substances emitted, by applying specific coefficients depending on the industrial activity carried out in each installation according to an equation.

Indirect estimation is possible in the cases provided for in Law 58/2003 General Tax Law.

Tax rate: This tax is progressive and consists of five brackets whose taxable base varies between 5 000 and 14 000 euros per polluting unit. Thus, the more gases are emitted into the atmosphere, the more expensive the polluting unit becomes.

Source: Author's own elaboration based on (Junta de Andalucia, 2003[54]).

2.3.2. Non-stationary sources: personal vehicles

The different levies applicable to personal vehicles in Spain and Andalusia are presented in the Table 2.9. The EU ETS does not currently cover fuels from transport, but an extension to transport is being discussed (see above). Regarding air pollution, the Directive (EU) 2016/2284 on 17 December 2016 on the reduction of national emissions of certain atmospheric pollutants also establishes national emission reduction targets for several pollutants, of which sulphur dioxide (SO2), nitrogen oxides (NOx), non-methane volatile organic compounds (NMVOC), ammonia (NH3) and fine particulate matter (PM2,5), as described above.

Table 2.9. Current levies applicable to personal vehicles in Spain and Andalusia

Current levies	Competence
Hydrocarbon tax	National
Vehicle registration tax	National
Road tolls (in discussion)	National
PIT deductions for the purchase of electric vehicles	Regional
Circulation tax	Municipal

Source: Author's own elaboration.

At the national level, there is the **hydrocarbon tax** levied on fuels amongst those fuels used in transport (described above) and the **vehicle registration tax**. The latter is regulated by the law 38/1992 (art. 65 to 74) (Gobierno de Espana, 1992[61]). It is a national tax on the first registration of motorised vehicles (including cars, boats and airplanes). The tax base is the vehicle market price and the tax rate progressively increases as a function of the vehicles CO_2 emissions. Exemptions and reductions apply. Exemptions include, for instance, two-seat vehicles used exclusively for industrial, commercial, agricultural, clinic or scientific use, three-wheel motorbikes, vehicles for disabled people, vehicles used by the public administration. The tax revenue has been assigned to the autonomous communities since 2002. It represented a small share of tax revenue in Andalusia in 2020 (0.3%), close to the average of the autonomous communities (0.4%) (Ministerio de Hacienda y Funcion Publica, 2022[47]).

As part of the Spanish recovery plan under the European Recovery and Resilience Facility, Spain also announced in 2021 that **road tolls** would be implemented in all national roads starting in 2024. This proposal is currently under discussion.

There is no levy implemented specifically on personal vehicles at the regional level in Andalusia. However, in order to incentive the purchase of electric vehicles, the autonomous communities have the competency to establish **tax deductions from the PIT**, which is a partially assigned tax (see Part I, Section 1). For example, the Autonomous Community of La Rioja established it under the law 10/2017 (art. 32) of 27 October (Comunidad Autónoma de La Rioja, 2017[62]). Under this law, deduction of 15% of the PIT applies for the acquisition of new electric vehicles, provided that it corresponds to the conditions set by the law. Among these conditions: (i) the vehicle must not be for professional or business activities, (ii) its amount shall not exceed EUR 50 000euros, and (iii) they must belong to the categories defined by the Directive 2007/46/EC (i.e. M1 passenger cars, N1 vans or light trucks, mopeds L1e, L2e tricycles, L6e light quadricycles, heavy quadricycles L7e, L3e motorcycles, category L5e and pedal-assisted bicycles with electric motor). For the passenger cars and vans or light trucks, electric vehicles must be: (i) powered by internal combustion engines that can use approved alternative fossil fuels such as LPG/Autogas, Compressed Natural Gas (CNG), Liquefied Natural Gas (LNG) or bi-fuel gasoline-gas., (ii) pure electric, or (iii) extended range electric vehicles, propelled entirely by electric motors.

At the municipal level, there is a **circulation tax**, regulated by the central government under the Royal Decree 2/2004 (art. 92 to 99) (Junta de Andalucia, 2003[54]). The tax is paid annually for the right of circulating on public roads. Amounts to be paid depend on multiple criteria, such as the vehicle category, the horsepower and the number of seats. There is no explicit environmental aspects considered. Exemptions and reductions apply. Exemptions include vehicles for disabled and personal vehicles used by the public administration.

2.4. Possibilities for Improvement on Taxation Related to Greenhouse Gas Emissions and Air Pollution in Andalusia

This section identifies some opportunities to reform the environmental taxation on GHG emissions and air pollution in Andalusia. It also includes opportunities at the national and municipal levels, which may improve environmental outcomes in Andalusia. Opportunities are based on the legal framework, the responsibilities mapping and the existing levies as discussed in the previous section. A selection of these opportunities will be further analysed in Activity 1.3 of the report, with an emphasis on their alignment with good environmental tax policy principles: notably opportunities in the area of the current tax on gas emissions into the atmosphere, vehicle taxation and distance-based charging. Case studies on the use of such instruments in other countries and Spanish regions will also be included. Where relevant this discussion looks at related aspects such as distributional consequences and health. As for the previous sections, emissions are separated between stationary sources, focusing on industry and electricity, and non-stationary sources, focusing on personal vehicles.

2.4.1. Stationary sources: industry and electricity

GHG emissions and air pollutants are already taxed in Andalusia under the tax on gas emissions into the atmosphere (Junta de Andalucia, 2003[54]). The main possibilities pertain to amending the current tax to broaden its scope, either by expanding the taxable matter or the tax base. The tax may also need to be updated with the current regulatory framework at the national level. In addition, the establishment of a national tax on emissions into the atmosphere is currently being discussed, which, in case of implementation, would repeal the Andalusian tax on gas emissions into the atmosphere due to double-taxation issue and would lead to compensation from the central government to Andalusia. The White Book for Tax Reform in Spain also made a recommendation to harmonise this tax across the autonomous communities (Box 2.5).

Possibility 1 (regional): improving the current tax on gas emissions into the atmosphere

Currently, the taxable matter is restricted to emissions of CO_2, NO_X and SO_X. There is an opportunity to extend the taxable matters to other kinds of emissions, such as particulate emissions (e.g. Catalonia) or ammonia (NH_3) and organic compounds (e.g. Murcia). To this end, it would be sufficient to amend article 23 of the current Andalusian law.

Today, the tax only applies to industrial activities included in Annex I of law 16/2002. There is an opportunity to expand the tax base to emissions from other facilities and productive activities. The tax base could include other industrial activities polluting the atmosphere, such as those comprised in groups A and B defined in Royal Decrees 100/2011 and 115/2017 or other productive activities, such as waste management and poultry. Activities in groups A and B includes (i) electricity generation for distribution through public grid, combustion in non-industrial sectors and industrial processes with combustion, classified by their level of thermal megawatts, (ii) industrial processes without combustion, classified by industrial process (e.g. CO fluid-furnace catalytic cracking for group A, storage of petroleum products in refineries for group B), (iii) use of solvents and other products, classified by process (e.g. vehicle coating for group A), (iv) waste treatment and disposal, classified by process (e.g. production of liquid fuels from plastic waste for group A), (v) crops with fertilisers (except animal manure), classified by production capacity (e.g. above 85 000 chickens for group B) (Gobierno de Espana, 2011[63]).

The expansion of the taxable matters and the extension of the industrial activities subject to the tax would support the national government's efforts to address the requests of increasing green taxes and reducing subsidies to actions that harm the environment made by the European Commission to the Spanish federal government.

Additionally, the calculation of the tax value is complex and may benefit from a simplification, e.g. following the Catalonian model in which the tax base is made up of the mass emissions of each of the polluting substances into the atmosphere emitted by the same facility. Another possibility is to adapt the wording of article 30 of law 18/2003 to replace the reference to art. 50 with art. 54 of the General Tax law (law 58/2003) (Gobierno de Espana, 2003[64]).

The current tax rate is low. There may be an opportunity to increase the current tax rate using the Catalonian or Aragonese models as references following the suggestion to harmonise taxes across autonomous communities (see Box 2.5).

Art. 25 and 36 of the law (Junta de Andalucia, 2003[54]) are not adapted to refer to the current Spanish General Tax law (Gobierno de Espana, 2003[64]). The suggestion is to adapt the wording of art. 25 to the current General Tax law by modifying the wording of Section 1 to replace the reference to art. 33 of the mentioned General Tax law with a reference to art. 35.4 of the current General Tax law, which attributes the status of taxpayers to entities without legal personality. Along the same lines, the wording of section 2 of art. 25 should be modified to refer to art. 35.7 of the current General Tax law. Similarly, the wording of section 3 of art. 36 should be modified to replace the reference to art. 58.2.c. of the old General Tax law with a reference to art. 58.2.a. of the current General Tax law, which configures default interest as a component of the tax debt because letter "c' refers to the surcharges of the executive period.

Box 2.5. Recommendations from the White Book on the tax on gas emissions into the atmosphere in Spain

Similar taxes on gas emissions into the atmosphere exist in other autonomous communities for both emissions from industry and large farms (Table 2.8), with different tax bases, tax rates and deductions. To harmonise these taxes, the Committee of experts from the White Book for Tax Reform in Spain (Table 1.2) recommends to maintain existing regional taxes and to introduce a national tax setting a minimum tax base (i.e. NOX, CH_4, NH_3, COVDM and N20 industrial emissions and CH_4, NH_3 and N_{20} emitted by intensive farming) and tax rates, as well as to fully assign the collection and regulatory responsibilities to the autonomous communities.

Source: (Comité de personas expertas, 2022[65]).

More specifically on electricity, the room for manoeuvre for the autonomous communities is limited due to the number of existing levies and the highly centralised responsibilities for this sector. The possibilities rely on creating regional taxes levied on the externalities associated with the electricity production, storage, transformation, and transmission activities. The White Book for Tax Reform in Spain also made several recommendations on taxes related to the electricity sector (Box 2.6).

Possibility 2 (regional): establishing a wind fee

To some extent wind power plants can deteriorate the visual and environmental conditions of the location where they are built. Similarly, the development of water reservoirs for electricity production do put pressure on ecosystems and environmental conditions.

For example, some experts suggest the creation of a fee on the adverse visual and environmental conditions and impacts on the natural environment associated with installing wind turbines to produce electricity. Similar fees already exist in Galicia, Aragon, and Castilla y Leon.

The wind fee is however subject to significant litigation at the European Court of Justice (European Court of Justice, 2017[66]) since it does tax zero-carbon energy sources (as opposed to putting a tax on fossil fuel power plants) on the ground of an environmental consideration. Professor Francisco D. Adame Martinez underlined that such a fee, if "well designed", can be justified as environmental policy. However, it is paradoxical to tax clean energy for environmental purpose while not taxing carbon-intensive energy sources. The purpose of such a tax could therefore mainly be to raise revenue (Adame Martinez, 2021[67]). Also the White Book for Tax Reform in Spain underlines the limitations of such fees (Box 2.6).

This opportunity will therefore not be considered in the following analysis.

Possibility 3 (regional): creating a tax on electricity production facilities that affect the environment

The infrastructure associated with electricity production and transmission negatively impacts the natural environment. An opportunity could be to establish a tax on the worsening environmental conditions associated with electricity production and transportation infrastructure (e.g., transmission lines), attributing exemptions and deductions to renewable energy facilities to not hamper their use. Similar taxes already exist in Catalonia and Extremadura.

The issue related to this tax is that, from an GHG emissions perspective, it is not levied on the direct factor of damage (i.e. GHG emissions) but on the adverse impact that electricity production and transportation may have on other environmental outcomes. In addition, the implementation of such a tax based on GHG emissions in Andalusia would lead to double taxation since thermal energy production industries are already taxed on gas emissions.

This opportunity will therefore not be considered in the following analysis.

Box 2.6. Recommendations from the White Book on taxes related to electricity

The White Book for Tax Reform in Spain establishes several recommendations related to taxation on electricity (Table 1.2).

1. Elimination of the electricity production tax

According to the Committee, this national tax was created to reduce the tariff deficit of the electricity sector (as per law 17/2012 of 27 December) (Gobierno de Espana, 2012[68]) rather than for environmental purposes. Therefore the configuration of the tax does not differentiate the environmental effects of the different electricity generation technologies. It may even harm the energy transition by making electrification more difficult, since it raises the relative prices of electricity, and by constraining technological change in electricity generation (Comité de personas expertas, 2022[65]).

2. Improvement of the design and effectiveness of regional taxes with effects on the electricity sector

Fees are levied on certain generation facilities (e.g. hydroelectric development fee, wind fee) and electricity distribution in several autonomous communities (e.g. Aragón, Asturias, Castilla y León, Castilla-La Mancha, Catalonia, Extremadura, Galicia, La Rioja and the Valencian Community). The Committee considers they do **not** meet their environmental objectives, which are (i) to establish incentives to reduce the environmental impacts of these facilities, (ii) to be visible and (iii) to promote appropriate synergies with other policy instruments. Regarding the first objective, most of these regional taxes do not include actions to protect biodiversity or to promote the repowering of wind installations. Regarding the second objective, most of the taxes have low visibility due to both their designs, which are not inclined to introduce behavioural changes, and the low level of the fees. In addition, these facilities usually require a detailed environmental impact analysis prior to their authorisation, which generates additional doubts about the role of corrective environmental taxation in these cases. Finally, these taxes may also hinder the necessary electrification process (Comité de personas expertas, 2022[65]).

3. Modification of the electricity tax

The Committee of experts recommends to modify the tax base of this national tax so that it is based on the physical quantity of electricity consumed and not on the applicable base for VAT, as at present, which would provide a more direct incentive for energy saving and efficiency. The Committee also suggests to adjust the current tax rate to the minimum established by the revised Energy Taxation Directive (currently under discussion EUR 0.54/MWh) (see above).

Source: (Comité de personas expertas, 2022[65]).

2.4.2. Non-stationary sources: personal vehicles

There is currently no direct taxation on non-stationary source emissions in Andalusia. The main opportunities identified include the establishment of new taxes, such as on commercial aviation, maritime transportation, and mechanical traction vehicles. Additional opportunities comprise creating a congestion tax for polluting vehicles circulating in central urban areas, and pushing for reforms of the national registration tax. The White Book for Tax Reform in Spain also provides recommendations on taxation related to transport in Spain (Box 2.7). The forthcoming economic analysis will focus on assessing possibilities in the road transport sector.

Possibility 1 (regional): creating a tax on emissions from commercial aviation.

Commercial aviation benefits from a favourable tax regime despite being an important source of emissions. The suggestion is to establish a tax on emissions from commercial aviation, based on the Catalonian tax on the emission of NO_X into the atmosphere produced by commercial aviation (Catalonian Law 12/2014) (Generalitat de Catalunya, 2014[69]). The Catalonian tax charges emissions of NO_X from aircraft on commercial passenger flights at airports belonging to municipalities. They are declared special environmental protection zones during the landing and take-off (LTO) cycle, which includes the taxing phases of entry to the airport, taxing out of the airport, take-off and landing. Other EU Member States adopted similar taxes (e.g. Germany and France). This is also recommended in the White Book for Tax Reform in Spain (Box 2.7).

Possibility 2 (regional): creating a tax on emissions from maritime transportation.

Maritime transportation benefits from a favourable tax regime despite being an important source of emissions. The suggestion is to create a tax on the emissions of NOX and SOX from ships that dock in Andalusian ports, based on the Catalonian tax on maritime emissions from large ships (Catalonian Law 16/2017) (Generalitat de Catalunya, 2017[70]).[9] The 87/2019 Spanish Constitutional Court Decision has concluded that the autonomous communities have the competency to establish taxes on emissions during the manoeuvre of ships docked at ports located within the territory of the autonomous communities. This is also recommended in the White Book for Tax Reform in Spain (Box 2.7).

Possibility 3 (regional): creating a tax on mechanical traction vehicles (CO_2 and NO_X) emissions.

Mechanical traction vehicles (i.e. motor vehicles) are not charged for their emissions. There is a possibility to establish a tax on vehicle emissions, similar to the Catalonian tax on vehicle emissions (Catalonian Law 16/2017) (Generalitat de Catalunya, 2017[70]). The Spanish Constitutional Court has declared on its judgment of June 87/2019 the constitutionality of the Catalonian tax on mechanical traction vehicles emissions (Annex Table 4.A.2). The absence of environmental criteria (e.g. emissions) in the local circulation tax in Catalonia has probably led the autonomous community to create a specific tax on vehicles' emissions. If the circulation tax was amended to include environmental criteria (e.g. Germany, Denmark) as suggested in the White Book for Tax Reform in Spain (Box 2.7), the amended national tax would replace all regional taxes (e.g., the Catalonian tax on mechanical traction vehicles emissions), and the Spanish government would have to compensate each autonomous community as foreseen in article 6.2 of law 8/1980 on the Financing of the autonomous communities (*Ley Organica de Financiacion de las Comunidades Autonomas; LOFCA*).

Possibility 4 (national, regional or municipal): creating a congestion charge for polluting vehicles circulating in central urban areas.

Urban central areas often suffer from high pollution levels due to the large fleet of vehicles in circulation and the high population exposure in these areas. An opportunity could be to establish a charge on vehicle congestion (i.e. to charge vehicles for circulating in particular areas of cities). The development of this charge could be aligned with the creation of the Low Emission Zones in cities with more than 50 000 inhabitants, as described in the Spanish Climate Change and Energy Transition Law (Gobierno de Espana, 2021[17]) and the "Guidelines for the creation of Low Emission Zones" prepared by the Spanish Ministry for Ecological Transition and Demographic Challenge (Ministerio para la Transicion Ecologica y el Reto Demografico, 2021[71]). This document entitles municipalities with the responsibility to define Low Emissions Zones. Cities like London, Oslo, Milan, Singapore and Gothenburg already have implemented such congestion charges.

As described in Section 2, local governments have considerable responsibilities related to transport, including mobility management, and are therefore responsible for implementing Low Emission Zones. The document "Guidelines for the Creation of Low Emission Zones" suggests to establish the congestion charge as a measure complementary to the Low Emission Zones (Ministerio para la Transicion Ecologica y el Reto Demografico, 2021[71]). The tax would therefore most likely fall under municipal competency. However, the autonomous communities still have the possibility to substitute the existing local taxes (regulated at the national level) for regional ones as long as they financially compensate municipalities for revenue decreases (Gobierno de Espana, 2009[72]). The White Book for Tax Reform in Spain also provides recommendations on the congestion charge at the municipal level (Box 2.7).

In addition, it may interesting to discuss the introduction of other distance-related charges, as recommended in the White Book for Tax Reform in Spain (Box 2.7).

Possibility 5 (national): updating the registration tax.

Many vehicle types currently in circulation are exempt from the registration tax (art. 66) (law 38/1992) (Gobierno de Espana, 1992[46]). The tax rates are based on nine headings, which correspond to categories of transport according to their emission of CO_2. However, the rates are not regularly updated despite technological advances. In addition, the current tax does not account for other externalities created by vehicles (e.g. accidents, local emissions).

There is a possibility to update the tax rates more regularly to strengthen the link with CO_2 emissions of new vehicles and to include other pollutants emissions into tax rate calculation. This argument is also underlined in the recommendations of the White Book for Tax Reform in Spain (Box 2.7).

Box 2.7. Recommendations from the White Book on taxes related to transport

The White Book for Tax Reform in Spain provides several recommendations related to taxation on transport (Table 1.2). Overall, the Committee considers appropriate the introduction of a tax on the use of vehicles to replace most of the existing taxes on road transport. The tax shall be distributed among the different levels of government according to their respective competences and the territorial nature of the costs covered.

1. Taxation on aviation, maritime and agricultural fuels [national]

The Committee suggests creating a tax on aviation, maritime and agricultural fuels, in line with the introduction of these energy products in the suggested revision of the EU ETD in the context of the EU "Fit for 55" legislative package (see above). The main reasons are: (i) the favourable tax treatment of these sectors do not correspond to the externalities they generated, (ii) the necessary contribution from these sectors to mitigate climate change, (iii) the incentive for development and investment in less polluting technologies. Given the importance of these sectors for the Spanish economy, the Committee however recommend a gradual introduction of the taxes.

2. Equalisation of the taxation of diesel and automotive gasoline [national]

In line with the proposed revision of the EU ETD, the Committee suggests to equalise the taxation on diesel with the one on automotive gasoline, considering that the new suggested minimum for gasoline at the EU level is below the current one in Spain: EUR 0.359 (EU level) vs. EUR 0.474 per litre (Spain), while it is the contrary for diesel: EUR 0.391 (EU level) vs. EUR 0.329 per litre (Spain). The current refund for professional uses in case of non-residential automotive diesel is suggested to be maintained (EUR 0.049 per litre).

3. General increase in taxation of hydrocarbons [national]

The Committee recommends to increase the hydrocarbons tax rates, especially on natural gas and automotive fuels to align with the proposed revision of the EU ETD. This would lead to significant decrease in polluting emissions and a strong increase in revenue. However, this recommendation may imply some distributional and competitive impacts than the previous recommendations. The suggestion is thus to follow gradual implementation.

4. Modification of the registration tax to promote a sustainable vehicle fleet [national]

The Committee of Experts highlights that the current registration tax design does not update the tax rate according to technological change in the sector, which reduces the incentives to develop a more sustainable vehicle fleet. The Committee suggests to increase the tax rates to encourage the purchase of low-emission vehicles and to introduce a surtax on vehicles above a certain weight to reduce the number of large vehicles (that are greater emitters and use more material resources). However, the Committee suggests that the surtax has a general application, regardless whether it is based on propulsion technologies or not, as this does not impact the external costs of heavy vehicles.

5. Configuration of the circulation tax to penalise the most polluting technologies [local]

The Committee of Experts recommends to modify the circulation tax to incorporate an environmental criterion and thus promote a more sustainable vehicle fleet in Spain. The tax shall be based on environmental damage indicators rather than on horse power, as at present.

6. Creation of a municipal charge on congestion in certain cities [local]

The Committee of Experts recommends the creation of municipal congestion charges in certain cities. It is suggested that the tax rate varies according to the volume of traffic, depending on the location and time of the day, to address local congestion, population density and pollution problems in the most efficient manner.

7. Consideration of tax mechanisms for the use of certain road infrastructures

The Committee recommends the introduction of taxes for the use of certain transport infrastructures. They may take the form of distance-related payments via electronic devices, which exist is many EU Member States for both light and heavy-duty vehicles. Payments for use are efficient and transparent tax mechanisms.

8. Creation of a tax on airline tickets [national]

In addition to create a tax on aviation fuels and thus correct the favourable tax treatment of this sector, the Committee recommends to introduce additional taxation to incorporate environmental costs in airline tickets. This tax aims to change behaviours through the internalisation of air transport costs and to promote the development of more sustainable technologies. This tax shall only apply to flights that generate direct greenhouse gas emissions.

Source: (Comité de personas expertas, 2022[65])

References

Adame Martinez (2021), *Análisis desde la perspectiva ambiental de la tributación de la Comunidad Autónoma de Andalucía en el entorno de las restantes comunidades autónomas y la Unión Europea*, Universidad de Sevilla. [67]

Agencia Andaluza de la Energia (2020), *Emisiones CO2 asociadas al consumo de combustibles fósiles por sectores*. [3]

Comité de personas expertas (2022), *Libro Blanco Sobre la Reforma Tributaria*. [65]

Comunidad Autónoma de Aragón (2007), *Decreto Legislativo 1/2007, de 18 de septiembre, del Gobierno de Aragón, por el que se aprueba el Texto Refundido de la Legislación sobre los impuestos medioambientales de la Comunidad Autónoma de Aragón*. [59]

Comunidad Autónoma de Castilla-La Mancha (2005), *Ley 16/2005, de 29 de diciembre, del Impuesto sobre determinadas actividades que inciden en el medio ambiente y del tipo autonómico del Impuesto sobre las Ventas Minoristas de Determinados Hidrocarburos*. [60]

Comunidad Autonoma de Galicia (1995), *Ley 12/1995, de 29 de diciembre, del Impuesto sobre la Contaminación Atmosférica*. [56]

Comunidad Autónoma de la Región de Murcia (2005), *Ley 9/2005, de 29 de diciembre, de Medidas tributarias en materia de tributos cedidos y tributos propios para el año 2006*. [57]

Comunidad Autónoma de La Rioja (2017), *Ley 10/2017, de 27 de octubre, por la que se consolidan las disposiciones legales de la Comunidad Autónoma de La Rioja en materia de impuestos propios y tributos cedidos*. [62]

Comunidad Autonoma Valenciana (2012), *Ley 10/2012, de 21 de diciembre, de Medidas Fiscales, de Gestión Administrativa y Financiera, y de Organización de la Generalitat*. [58]

European Commission (2021), *Proposal for a Directive of the European Parliament and of the Council amending Directive 2003/87/EC establishing a system for greenhouse gas emission allowance trading within the Union, Decision (EU) 2015/1814 concerning the establishment and operation of a market stability reserve for the Union greenhouse gas emission trading scheme and Regulation (EU) 2015/757*. [7]

European Commission (2021), *Proposal for a Regulation of the European Parliament and of the Council amending Regulation (EU) 2018/842 on binding annual greenhouse gas emission reductions by Member States from 2021 to 2030 contributing to climate action to meet commitments under the Paris Agreement*. [9]

European Commission (2021), *Proposal for a Regulation of the European Parliament and of the Council amending Regulation (EU) 2019/631 as regards strengthening the CO2 emission performance standards for new passenger cars and new light commercial vehicles in line with the Union's increased climate ambition*. [14]

European Commission (2021), *Proposal for a Regulation of the European Parliament and of the Council amending Regulations (EU) 2018/841 as regards the scope, simplifying the compliance rules, setting out the targets of the Member States for 2030 and committing to the collective achievement of climate neutrality by 2035 in the land use, forestry and agriculture sector, and (EU) 2018/1999 as regards improvement in monitoring, reporting, tracking of progress and review.* [12]

European Commission (2021), *Proposal for a Regulation of the european Parliament and of the council establishing a carbon border adjustment mechanism.* [8]

European Commission (2021), *Regulation (EU) 2019/631 of the European Parliament and of the Council of 17 April 2019 setting CO2 emission performance standards for new passenger cars and for new light commercial vehicles, and repealing Regulations (EC) No 443/2009 and (EU) No 510/2011.* [13]

European Commission (2021), *Regulation (EU) 2021/1119 of the European Parliament and of the Council ('European Climate Law').* [5]

European Commission (2021), *Revision of the Energy Taxation Directive (ETD).* [10]

European Commission (2018), *Regulation (EU) 2018/1999 of the European Parliament and of the Council of 11 December 2018 on the Governance of the Energy Union and Climate Action, amending Regulations (EC) No 663/2009 and (EC) No 715/2009 of the European Parliament and of the Council, Directives 94/22/EC, 98/70/EC, 2009/31/EC, 2009/73/EC, 2010/31/EU, 2012/27/EU and 2013/30/EU of the European Parliament and of the Council, Council Directives 2009/119/EC and (EU) 2015/652.* [15]

European Commission (2018), *The Regulation on the inclusion of greenhouse gas emissions and removals from land use, land use change and forestry.* [11]

European Commission (2016), *Directive 2016/2284 of the European Parliament and of the Council of 14 December 2016 on the reduction of national emissions of certain atmospheric pollutants, amending Directive 2003/35/EC and repealing Directive 2001/81/EC.* [16]

European Commission (2003), *Directive 2003/87/EC of the European Parliament and of the Council of 13 October 2003 establishing a system for greenhouse gas emission allowance trading within the Union and amending Council Directive 96/61/EC.* [6]

European Court of Justice (2017), *Joined Cases C-215/16, C-216/16, C-220/16 and C-221/16.* [66]

European Union (2012), *Consolidated version of the Treaty on the Functioning of the European Union.* [39]

Generalidad de Cataluna (2014), *Ley 12/2014, de 10 de octubre, del impuesto sobre la emisión de óxidos de nitrógeno a la atmósfera producida por la aviación comercial, del impuesto sobre la emisión de gases y partículas a la atmósfera producida por la industria y del impuesto sobre la producción de energía eléctrica de origen nuclear.* [55]

Generalitat de Catalunya (2021), *Decret Llei 24/2021, de 26 d'octubre, d'acceleració del desplegament de les energies renovables distribuïdes i participades.* [45]

Generalitat de Catalunya (2020), *Decret Llei 33/2020, de 18 de febrer, pel qual s'estableixen la composició i el règim de funcionament de la Comissió Interdepartamental del Canvi Climàtic i sobre el Fons Climàtic i la Comissió del Fons Climàtic.* [44]

Generalitat de Catalunya (2019), *Decreto-ley 9/2019, de 21 de mayo, de medidas urgentes en materia de contención de rentas en los contratos de arrendamiento de vivienda y de modificación del libro quinto del Código civil de Cataluña en el ámbito de la prenda*. [43]

Generalitat de Catalunya (2017), *Ley 16/2017, de 1 de agosto, del cambio climático*. [70]

Generalitat de Catalunya (2017), *LLEI 16/2017, de l'1 d'agost, del canvi climàtic*. [42]

Generalitat de Catalunya (2014), *Decret 12/2014, de 21 de gener, pel qual s'atorga a l'Institut d'Estudis Aranesi el caràcter d'acadèmia i d'autoritat lingüística de l'occità, llengua pròpia a Era Val d'Aran i oficial a Catalunya*. [69]

Gobierno de Espagna (2021), *Real Decreto-ley 12/2021, de 24 de junio, por el que se adoptan medidas urgentes en el ámbito de la fiscalidad energética y en materia de generación de energía, y sobre gestión del canon de regulación y de la tarifa de utilización del agua*. [50]

Gobierno de Espana (2022), *Real Decreto-ley 6/2022, de 29 de marzo, por el que se adoptan medidas urgentes en el marco del Plan Nacional de respuesta a las consecuencias económicas y sociales de la guerra en Ucrania*. [51]

Gobierno de Espana (2022), *Royal Decree-Law 20/2022 of 27 December on measures in response to the economic and social consequences of the war in Ukraine and in support of the reconstruction of the island of La Palma and other situations of vulnerability, provided in Articles 1 and 72 for changes in VAT and equivalence surcharge rates.*. [52]

Gobierno de Espana (2021), *Ley 7/2021, de 20 de mayo, de cambio climático y transición energética*. [17]

Gobierno de Espana (2021), *Real Decreto 266/2021, de 13 de abril, por el que se aprueba la concesión directa de ayudas a las comunidades autónomas y a las ciudades de Ceuta y Melilla para la ejecución de programas de incentivos ligados a la movilidad eléctrica (MOVES III) en el marco del Plan de Recuperación, Transformación y Resiliencia Europeo*. [21]

Gobierno de Espana (2013), *Ley 24/2013, de 26 de diciembre, del Sector Eléctrico*. [20]

Gobierno de Espana (2012), *Ley 15/2012, de 27 de diciembre, de medidas fiscales para la sostenibilidad energética*. [48]

Gobierno de Espana (2012), *Ley 17/2012, de 27 de diciembre, de Presupuestos Generales del Estado para el año 2013*. [68]

Gobierno de Espana (2011), *Real Decreto 100/2011, de 28 de enero, por el que se actualiza el catálogo de actividades potencialmente contaminadoras de la atmósfera y se establecen las disposiciones básicas para su aplicación*. [63]

Gobierno de Espana (2009), *Ley Orgánica 3/2009, de 18 de diciembre, de modificación de la Ley Orgánica 8/1980, de 22 de septiembre, de Financiación de las Comunidades Autónomas*. [72]

Gobierno de Espana (2003), *Ley 58/2003, de 17 de diciembre, General Tributaria*. [64]

Gobierno de Espana (2001), *Real Decreto Legislativo 1/2001, de 20 de julio, por el que se aprueba el texto refundido de la Ley de Aguas*. [53]

Gobierno de Espana (1997), *Ley 54/1997, de 27 de noviembre, del Sector Eléctrico*. [19]

Gobierno de Espana (1992), *Ley 38/1992, de 28 de diciembre, de Impuestos Especiales*. [46]

Gobierno de Espana (1992), *Ley 38/1992, de 28 de diciembre, de Impuestos Especiales*. [61]

Gobierno de Espana (1992), *Real Decreto 1624/1992, de 29 de diciembre*. [49]

Gobierno de Espana (1978), *The Spanish Constitution*. [40]

Gobireno de Espana (2005), *Ley 1/2005, de 9 de marzo, por la que se regula el régimen del comercio de derechos de emisión de gases de efecto invernadero*. [18]

Instituto de Estadistica y Cartografia de Andalucia (2020), "Environment indicators". [1]

Junta de Andalucia (2022), *Boletín número 112 de 14/06/2022*. [34]

Junta de Andalucia (2022), *Plan de Infraestructuras de Transporte y Movilidad de Andalucía (PITMA) 2030*. [36]

Junta de Andalucia (2021), *Acuerdo de 2 de febrero de 2021, del Consejo de Gobierno, por el que se modifica el Acuerdo de 21 de mayo de 2019, por el que se aprueba la formulación del Plan de Infraestructuras de Transporte y Movilidad de Andalucía, 2021-2027.*. [35]

Junta de Andalucia (2021), *Estrategia Andaluza de Movilidad Sostenible y Transporte 2030*. [37]

Junta de Andalucia (2021), *Plan Andaluz de Accion por el Clima (2021-2030)*. [32]

Junta de Andalucia (2020), *Acuerdo de 22 de septiembre de 2020, del Consejo de Gobierno, por el que se aprueba la Estrategia Andaluza de Calidad del Aire*. [38]

Junta de Andalucia (2020), *Estrategia Andaluza de Calidad del Aire*. [2]

Junta de Andalucia (2020), *Estrategia Energetica de Andalucia 2020*. [33]

Junta de Andalucia (2018), *Acuerdo de 5 de junio de 2018, del Consejo de Gobierno, por el que se aprueba la Estrategia Andaluza de Desarrollo Sostenible 2030*. [29]

Junta de Andalucia (2018), *Ley 8/2018, de 8 de octubre, de medidas frente al cambio climático y para la transición hacia un nuevo modelo energético en Andalucía*. [30]

Junta de Andalucia (2015), *Informe para la valoracion de medidas de mitigacion de emisiones de gases de efecto invernadero*. [26]

Junta de Andalucia (2012), *Plan Andaluz de Accion por el Clima: Programa de Communicacion*. [28]

Junta de Andalucia (2010), *Ley 5/2010, de 11 de junio, de autonomía local de Andalucía*. [31]

Junta de Andalucia (2010), *Plan Andaluz de Accion por el Clima: Programe de Adaptacion*. [27]

Junta de Andalucia (2007), *Ley 7/2007, de 9 de julio, de Gestión Integrada de la Calidad Ambiental y sus desarrollos*. [24]

Junta de Andalucia (2007), *Organic law 2/2007 dated 19 March 2007 on Reform of the Statute of Autonomy for Andalusia*. [41]

Junta de Andalucia (2007), *Plan Andaluz de Accion por el Clima: Programa de Mitigacion*. [25]

Junta de Andalucia (2003), "Ley 18/2003, de 29 de diciembre, por la que se aprueban medidas fiscales y administrativas". [54]

Ministerio de Hacienda y Funcion Publica (2022), *Autonomous Community Funding.* [47]

Ministerio de Transportes Movilidad y Agenda Urbana (2022), "Anteproyecto de Ley de Movilidad Sostenible". [23]

Ministerio de Transportes Movilidad y Agenda Urbana (2021), *Estrategia de movilidad.* [22]

Ministerio para la Transicion Ecologica y el Reto Demografico (2021), *Directrices para la creación de zonas de bajas emisiones (ZBE).* [71]

OECD (2021), *OECD Regional Outlook 2021: Addressing COVID-19 and Moving to Net Zero Greenhouse Gas Emissions*, OECD Publishing, Paris, https://doi.org/10.1787/17017efe-en. [4]

Notes

[1] The year 2020 is an outlier because of the COVID-19 pandemic.

[2] Stationary sources are fixed sources (e.g. a building, a power plant or any facility) that emit greenhouse gas emissions or air pollutants.

[3] Non-stationary sources are mobile sources (e.g. motor vehicles, airplanes or any other equipment that can move from one location to another) that emit greenhouse gas emissions or air pollutants.

[4] The Paris Agreement is a legally binding international treaty on climate change, adopted by 196 countries (of which EU Member States) during the COP 21 in Paris on 12 December 2015, which entered into force on 4 November 2016. The objective of the treaty is to limit global warming to below 2°C, preferably to 1.5°C, compared to pre-industrial levels, through the reduction of greenhouse gas emissions and carbon neutrality.

[5] By contrast, an EU Directive is a legislative act setting objectives that must be achieved by EU Member States and transpose into their national legislation within a defined time period.

[6] A low emission zone is an area delimited by public administration, within its territory, in which access, circulation and parking of vehicles are restricted to meet air quality standards and reduce greenhouse gas emissions, in accordance with the classification of vehicles based on their levels of emissions (Gobierno de Espana, 2021[17]).

[7] The EU Recovery and Resilience Facility is a key EU instrument to support Member States' recovery from the pandemic. It is made of EUR672.5 billion, of which up to EUR312.5 billion in grants and up to EUR360 billion in loans, to finance public investments and structural reforms with a focus on environment and digitalisation.

[8] Andalusia, the provinces and the municipalities have established the City 21 (*Ciudad 21*) programme, which became the Sustainable City (*Ciudad Sostenible*) program in 2011, as an instrument for achieving

sustainable development at the local level through the development of environmental analyses and the drafting and implementation of local action plans for sustainable development. The program bring together 291 municipalities and has been implemented through more than 600 urban development projects. It focuses on several areas of action (e.g. urban waste management, urban water cycle, energy use, air quality, protection of urban flora and fauna, sustainable urban mobility, environmental awareness and citizen participation) (Junta de Andalucia, 2010[31]).

[9] Although established in 2017, this tax has is not yet in force. In June 2022, the Plenary of the Parliament of Catalonia approved a motion by En Común Podem calling for the approval before the end of the year of the Tax on port emissions from cruise ships and large ships.

3 Assessment: stationary sources

This section examines how Andalusian, Spanish and EU-level pricing of emissions from stationary relate to climate change and air pollution. While climate change and air pollution are two separate environmental issues, they partly overlap. Climate change is mostly due to Greenhouse Gas (GHG) emissions and their impact are at the global level. Even if GHGs are emitted in a specific area, their concentration in the atmosphere will contribute to climate change across the globe. Air pollution, on the other hand, is mostly due to other pollutant emissions with generally local impacts.

After a brief exposition of GHGs and air pollutants, this section presents the taxes or similar instruments[1] that apply in Andalusia on stationary sources for these two types of emissions. The main part of the stationary sources analysis covers power plants and industry.[2] The buildings sector (residential and commercial heating) is also part of the stationary source category, but as it represents a somewhat smaller share of emissions and is not subject to any regional tax in Andalusia, it is not covered in this analysis. Activities in the agricultural sector (to be understood as livestock farming and cultivation) generate emissions that may fall into both categories: stationary and non-stationary sources (non-stationary emissions in that sector arise from the use of agricultural engines such as tractors). For ease of exposition and given that a large share of emissions in that sector are from stationary albeit diffuse sources, agriculture is analysed in this stationary source section. The focus on the agricultural, electricity and industry sectors enables an alignment with the Polluter Pays Principle, as these are the main sectors responsible for stationary source GHG and air pollutant emissions taken together.

3.1. Greenhouse gases and air pollutants

3.1.1. Greenhouse gases

There are seven main GHGs: carbon dioxide (CO_2), methane (CH_4), nitrous oxide (N_2O), chlorofluorocarbons (CFCs), hydrofluorocarbons (HFCs), perfluorocarbons (PFCs), sulphur hexafluoride (SF6) and nitrogen trifluoride (NF3). GHG emissions are directly responsible for climate change through global warming: by absorbing long-wave infrared radiation reflected by the earth's surface, they prevent part of the infrared radiation from being reflected back to space. This results in the absorbed energy being converted into heat.

The global warming impact of GHGs is generally independent of where the emissions occur, but it can change over the years. These changes are mainly measured for GHGs relative to one another: when GHG concentrations change, so does the relative energy absorption of one additional tonne of a given GHG. For example, the energy absorption of CH_4 and N_2O have increased over the years.

Some GHGs have a stronger global warming impact than others. This mainly depends on their radiative forcing and their lifetime.[3] The 100-year global warming potential (GWP_{100}) index takes CO_2 as the reference and indicates its relative radiative forcing (the amount of warming) over 100 years[4] following the release of one unit mass of GHG into the atmosphere. For example, according to the IPCC Fifth Assessment Report (AR5) (IPCC, 2014[1]), 1 tonne of N_2O causes 265 times more warming over 100 years than 1 tonne of CO_2, so that N_2O has a GWP_{100} of 265. CH_4 has a GWP_{100} of 28. GHG emissions can then be expressed in CO_2-equivalent (CO_2e), which is obtained by multiplying the unit mass of emissions of a GHG by its GWP_{100}.

In Andalusia, GHG emissions are principally from CO_2 and have steadily declined since 2007. Indeed, they have gone from about 75 $MtCO_2e$ in 2007 to about 54 $MtCO_2e$ in 2019. In 2019, CO_2 emissions represented 80% of GHG emissions in Andalusia (see Figure 3.1), close to the national share of 78% (Spanish Ministry for Ecological Transition, 2020[2]). CH_4 represent about 10% of emissions in CO_2-equivalent, N_2O 6% and F-gases, 3%. In total GHG emissions in Andalusia represent about 16% of the national total.

Figure 3.1. GHG emissions in Andalusia

2019, percentages based on CO_2e

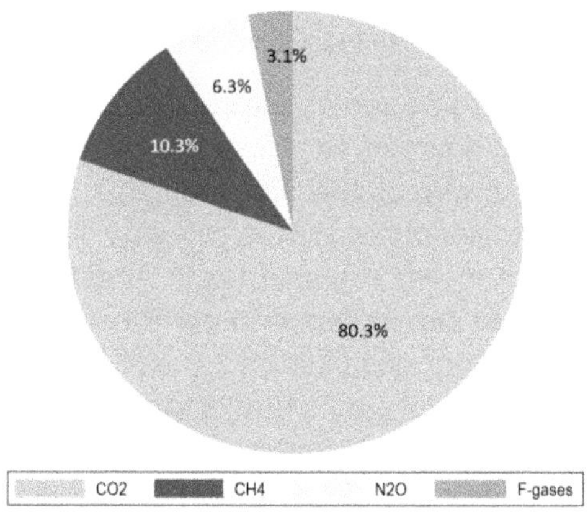

Note: The CO_2-equivalence was calculated using the IPCC AR5 GWP_{100} indicator.
Source: Consejería de Agricultura, Pesca y Desarrollo Rural de la Junta de Andalucía.

StatLink https://stat.link/kqe72v

The sources of GHG emissions vary across GHGs. The main sectors responsible for CO_2 emissions are the electricity (29%), industry (24%) and road transport (31%) sectors. The main sources of CH_4 emissions are the agriculture sector (56%), waste (29%) and biogenic activities (11%). N_2O emissions principally come from agriculture (68%) and biogenic activities (15%). F-gas emissions overwhelmingly stem from the industry sector (above 99.9%), and more specifically almost entirely from the use of refrigerants and propellants. This is reflected in the GHG emission breakdown by sector (Figure 3.2).

Figure 3.2. GHG emissions by sector

Industry, Electricity, Buildings, Agriculture, Transport sectors, 2019, percentages based on CO_2e

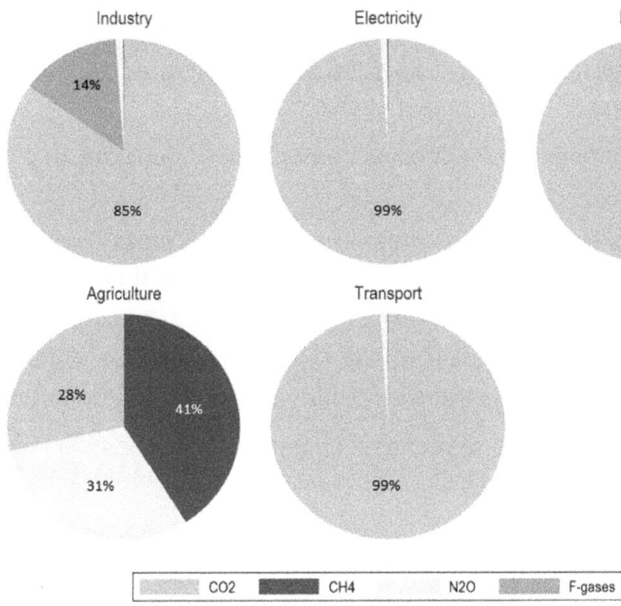

Note: The CO_2-equivalence was calculated using the IPCC AR5 GWP_{100} indicator.
Source: Consejería de Agricultura, Pesca y Desarrollo Rural de la Junta de Andalucía.

StatLink https://stat.link/v9u4pk

GHG emissions emanate both from fuel use and from other sources such as industrial process, cattle or waste. A specificity of CO_2 is that its emissions from fuel use are directly proportional to the amount of fuel used. Indeed, CO_2 emissions are constant per unit of fuel used.[5] Exact carbon emissions associated with the combustion of a given fuel may vary with local fuel characteristics but not the end-of pipe technology or combustion process chosen (U.S. EPA Center for Corporate Climate Leadership, 2016[3]). For example, on average the combustion of one litre of diesel generates around 2.76 kilograms of CO_2 be it combusted in a vehicle or by stationary machinery. CO_2 emissions from fuel use represent about 80% of worldwide CO_2 emissions.

The proportionality of CO_2 emissions from fuel use to the amount of fuel used makes fuel taxes a good policy instrument to reflect CO_2 emissions in consumer prices (and thus mimicking carbon taxes) or to relate tax levels to specific carbon benchmarks. This is reflected in the OECD effective carbon rates indicator (ECRs), which evaluates carbon pricing across countries, i.e. how CO_2 emissions from fossil fuel use are priced not only through carbon taxes and permit prices from emissions trading systems, but also through fuel excise taxes. Box 3.1 provides additional detail on these three components as well as on sectors, fuels and years covered by the OECD ECR. The ECR profile for the Andalusia industry and electricity sector is represented and analysed in Section 3.2.2.

Box 3.1. The OECD Effective Carbon Rates

The OECD Effective Carbon Rates (ECR) database (OECD, 2021[4]; OECD, 2019[5]) provides a breakdown of CO_2 emissions from energy use and corresponding effective carbon rates for 44 OECD and G20 countries by sector and fuel. Taken together, these 44 OECD and G20 countries represent 80% of worldwide CO_2 emissions from energy use. Effective carbon rates are the sum of explicit carbon taxes, emissions trading systems (ETSs) and fuel excise taxes.

More precisely, the three components of effective carbon rates, depicted in Figure 3.3, should be understood as follows:

Carbon taxes generally set a rate on fuel consumption based on its carbon content (e.g., on average, a EUR 30/tCO_2 tax on carbon emissions from diesel use would translate into a 7.99 eurocent per litre tax on diesel).

Fuel excise taxes typically set a rate per physical unit (e.g., litre, kilogram, cubic metre) or per unit of energy (e.g., gigajoule), which can then be translated into rates on the carbon content of these fuels.

The price of tradable emission permits, regardless of the permit allocation method, represent the opportunity cost of emitting an extra unit of CO_2.[1]

Figure 3.3. Components of Effective Carbon Rates

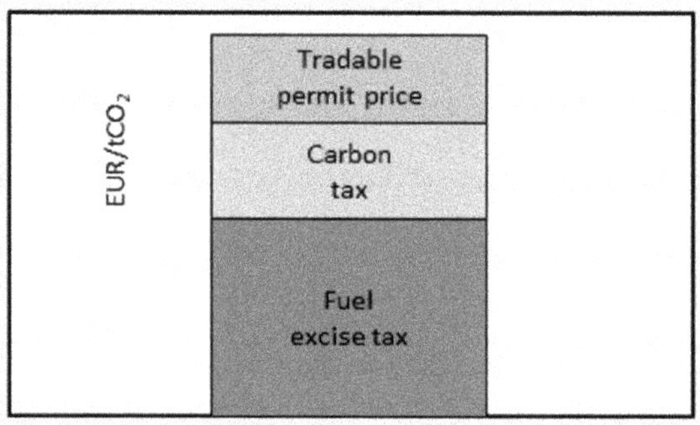

Source: Based on Figure 3.1 in OECD (2016[6]).

The database covers six sectors that together span all energy uses: agriculture and fisheries, buildings (i.e., residential and commercial heating), electricity, industry, off-road transport and road transport. More detail on sector definitions can be found in Annex Table 3.A.1.

Fuels are grouped into ten categories, which in turn can be grouped into two broad classes. Fossil fuels are composed of the categories *coal and other solid fossil fuels*, *diesel*, *fuel oil*, *gasoline*, *kerosene*, *liquefied petroleum gas* (LPG), *natural gas* and *other fossil fuels* (a category consisting in those fossil fuels that cannot be classified under the first seven categories in the list). Other combustible fuels are composed of biofuels and non-renewable waste. More detail on fuel categorisation can be found in Annex Table 3.A.2.

Note:
1. Thus, effective carbon rates are sometimes also referred to as effective marginal carbon rates. In the following, the discussion centres around those, but the sector-level discussion goes into more detail and highlights the share of free allocations in different sectors.
Source: OECD (2016[6]).

3.1.2. Air Pollutants

The main air pollutants are sulphur oxides (SOx) and nitrogen oxides (NOx) (generally expressed as quantities of SO_2 and NO_2), carbon monoxide (CO), ammonia (NH_3), volatile organic compounds excluding methane (NMVOC), particulate matter[6] (PM). These gases and particulate matter are directly responsible for air pollution.

Air pollution has effects on human health and on the environment. The World Health Organisation (WHO), for instance finds that 7 million premature deaths annually are linked to air pollution.[7] Even at the European Union (EU27) level, the European Environmental Agency estimates that, in 2019, approximately 307,000 premature deaths were attributable to $PM_{2.5}$, 40,400 premature deaths to NO_2 and 16,800 premature deaths to ground-level ozone. The OECD's *Air pollution effects* indicator, uses estimates of the "Value of a Statistical Life" (VSL) and computes the number of premature deaths attributable to ambient particulate matter (OECD, 2022[7]). It finds that in 2019, Exposure to $PM_{2.5}$ caused a mortality of 190 per 1 000 000 inhabitants in Spain. Additional details on the types of effects air pollutants might have are provided in Box 3.2. .

Box 3.2. Principal air pollutants and their impacts

Air pollutants may be harmful in and by themselves but also through their reaction with water, oxygen and other chemicals in the atmosphere, which can lead to the formation of other toxic substances.

For example, high concentrations of SO_2 in the air can lead to the formation of other sulfur oxides (SOx). NOx, SOx and NH_3 can react with other chemicals in the air to form particulate matter. Moreover, the reaction of NOx or CO with other chemicals in the atmosphere can result in the production of tropospheric ozone (O3). VOCs exacerbate the production of ozone in the lower atmosphere. The interaction of NO_2 and SO_2 with one another or with other substances, such as water can cause acid rains.

Environmental impacts

High concentrations of gaseous SOx can damage foliage and decrease plant growth. Particulate matter, either emitted directly or created through the reaction of other air pollutants with chemicals in the air can make the air hazy, hence reducing visibility as well as stain and damage stone and other materials. NOx in the atmosphere can contribute to nutrient pollution in coastal waters. This can result in algae growing faster than manageable for ecosystems. This damages water quality and decrease the oxygen that fish and other aquatic life need to survive. NH_3 is also harmful for the fish and aquatic life more generally. At high levels, ground-level O_3 damages vegetation, including crop yields.

Health impacts

Most air pollutants harm the human respiratory system, and some can cause further damages by increasing the risks of certain illnesses and conditions. For example, it has been found that longer exposures to elevated concentrations of NO_2 may contribute to the development of asthma. Breathing air with a high concentration of CO reduces the amount of oxygen that can be transported in the blood stream to critical organs like the heart and brain.

VOCs can be carcinogenic.

Through their impact on nutrient pollution and algal blooms, NOx emissions can cause sickness when humans come into contact with polluted water, consume tainted fish or shellfish, or drink contaminated water.

PM_{10} and $PM_{2.5}$ can get deep into the lungs and in some cases even into the bloodstream. $PM_{2.5}$ is the air pollutant that poses the greatest risk to health globally and affects more people than any other pollutant.

Climate impacts

O3 is a short-lived GHG, hence it also contributes to climate change. Its radiative forcing effect however, is mainly at regional level.

PM can influence climate "through both interactions that scatter or absorb radiation and through interactions with cloud microphysics and other cloud properties, or upon deposition on snow- or ice-covered surfaces thereby altering their albedo and contributing to climate feedback" (IPCC, 2019[8]).

Economic impacts

Evidence shows that beyond the health and environmental impacts, air pollution, and in particular particulate matter may also have detrimental effects on firms and more generally the economy through productivity of workers (Zivin and Neidell, 2018[9]; Dechezleprêtre, Rivers and Stadler, 2019[10]). For example, Leroutier and Ollivier (2022[11]) find that by negatively affecting workers' health and cognitive functions, $PM_{2.5}$ exposure impacts workers' absenteeism and firms' monthly sales. At the national level, this can have important consequence. For instance, they estimate if air pollution in France had been in line with the World Health Organization's guidelines, this would have saved at least 0.3% of GDP annually through avoided sales losses.

The economy can also be affected by air pollution through increase in public health expenditure and loss of crop yields. For example, Deryugina et al. (2019[12]) find that in the United States (US) $PM_{2.5}$ concentration increases lead to more emergency room visits, more hospitalizations, and higher inpatient spending. Mink (2022[13]) estimates that reducing NO_2 concentrations by 27% would results in an annual saving of EUR 5.2 billion in healthcare costs in France. Regarding crop yields, Lobell et al. (2022[14]) find that reducing NOx emissions by about half in Western Europe would improve yields by nearly 10% in the region. SOx and NH_3 may also be harmful to plants.[1]

Note:
1. https://www.ontario.ca/page/effects-air-pollution-agricultural-crops#:~:text=Agricultural%20crops%20can%20be%20injured,premature%20death%20of%20the%20plant.
Source: OECD (2020[15]; 2022[7]),https://www.epa.gov, IPCC (2019[8]) for environmental, health and climate impacts and Zivin and Neidell (2018[9]), Dechezleprêtre et al. (2019[10]), Leroutier and Ollivier (2022[11]), Deryugina et al. (2019[12]), Mink (2022[13]), Lobell et al. (2022[14]) for economic impacts.

The direct impact of air pollutants is often local, and their harmfulness generally depends on local conditions, such as population density, and local weather conditions (e.g. rainfall, wind regime or atmospheric stability). Age, standards of living and prevalence of certain pathologies can also impact the health effects on population. The impacts can also depend on pollutant densities in the air and can be more important beyond certain density levels. The United States, for example, have defined an Air Quality Index (AQI),[8] which depends on ground-level ozone, PM, CO, SOx and NOx emissions and ranges from "Good" to "Unhealthy" to "Hazardous" (it has a total of six categories). Individual threshold levels are also defined for each of these air pollutants (U.S. Environmental Protection Agency, 2018[16]). Air pollutants might also indirectly impact climate change.

In Andalusia, air pollutant emissions have followed a downward trend since 2003, with reductions of up to 80% for SO_2. NH_3 emissions went through a significant decrease up until 2011, but have gone up since, resulting in the lowest air pollutant decrease since 2004 of about 6%.

The main sources of air pollutants differ (Andalucia, 2021[17]). The main anthropogenic sources of SO_2 in 2019 are the industry sector (about 50%, with 22% from the petrochemicals industry, 11% from the metal industry and 11% from the non-metallic materials industry, 3% from the oil production industry and 2% from the chemical industry), maritime traffic (22%) and electricity production (20%). For NOx, these are road traffic (28%), agriculture (21%), maritime traffic (13%) and electricity production (11%). For CO emissions the main sources are agriculture (34%), domestic activities (23%), forest fires (13%) and road traffic (13%). Those for NH_3 are livestock (47%) and the rest of agriculture (46%). Those for NMVOC are the use of solvents. Finally, direct $PM_{2.5}$ emissions are mostly from domestic activities (38%), agriculture (30%), forest fires (9%) and road traffic (8%).[9]

Air pollutants, contrary to CO_2 are not necessarily proportional to fuel use. Indeed, their emissions intensity also depends on the end-of-pipe technology used and the combustion process (OECD, 2019[5]).

3.1.3. Interactions between GHGs and air pollutants

Air pollution and climate change interact and can influence each other.[10] For example, an increase in levels of GHGs leads to temperature changes that affect the chemical composition of the atmosphere, and can make air pollution impacts worse. On the contrary, certain air pollutant emissions may actually have negative radiative forcing, i.e., have a cooling effect on the climate – SOx emissions for example form light *reflecting* particles (Arneth et al., 2009[18]). Moreover, as explained in Box 3.2. , the interaction of air pollutants with other substances in the atmosphere can result in the production of other components, which do have a direct effect on climate change (e.g., black carbon, O_3).

In addition, the management of climate change and air pollution have consequences for each other. First, pricing GHG emissions can encourage a reduction in fuel use, which in turn not only reduces GHG emissions but also air pollutant emissions (and vice versa). Co-benefits of climate policy therefore include better health and environmental outcomes. However, this can also create trade-offs, due to the complex interactions between air pollutants and GHGs described above. This is the case, in particular if, at least in the short-term, reducing a pollutant's emissions leads to additional atmospheric warming rather than cooling. Second, trade-offs can also arise because of the consequences of pricing. This is particularly the case if fuels are replaced by more sustainable fuels such as biofuels and not by non-combustible renewables such as wind and solar. If sustainably sourced, the combustion of biofuels may result in lower GHG emissions over the life cycle because before being burnt, feedstocks have previously absorbed a similar amount of CO_2 from the atmosphere. However, it does all the same lead to higher PM in the air. The first issue can be dealt with, for example, by associating bioenergy expansion with effective implementation of post-combustion PM-control measures, such as filters and precipitators (Portugal-Pereira et al., 2018[19]). Similar issues can arise with carbon capture and storage technologies, which may induce larger amounts of primary energy requirements and hence higher air pollution overall.

3.2. Pricing emissions from stationary sources in Andalusia

This subsection deals with carbon and air pollutant taxes on stationary sources in the Andalusian context and focuses on the Andalusian tax on the emission of gases into the atmosphere (IEGA). A description of this tax is followed by its analysis in the more comprehensive context of carbon and air pollution pricing policies. This leads to the identification of how the current tax system compares with a system that covers emissions more comprehensively and more accurately according to sound environmental tax principles, including considerations for potential economic and behavioural impacts. Proposals for strategic reform options in the Andalusia context are made and best practice examples from other countries are presented throughout. First, this subsection provides a description of the Andalusian tax on the emission of gases into the atmosphere. Then, the first part of the analysis deals with GHG emissions, first with a focus on the industry and electricity sectors and CO_2 emissions from energy use: the Andalusian tax on the emission of gases into the atmosphere within the context of other national and European-level taxes dealing with carbon emissions. The second part of the analysis deals with air pollutant emissions. Finally, the agriculture sector is discussed, both from a GHG and air pollutant perspective.

CO_2 emissions from energy use in Andalusia

In Andalusia, the electricity and industry sectors[11] taken together represent more than 80% of stationary sources of CO_2 emissions from energy use. The rest of CO_2 emissions from energy use from stationary sources is principally from the buildings sector – or in other words residential and commercial heating. Stationary sources are responsible for 64% of overall CO_2 emissions from energy use – with the road and off-road transport sectors being responsible for the rest.

3.2.1. The Andalusian Tax on the Emission of Gases into the Atmosphere or the IEGA

In 2003, Andalusia introduced its Tax on the Emission of Gases into the Atmosphere (*Impuesto Sobre la Emisión de Gases a la Atmósfera*[12] *or IEGA*), which deals both with GHG and air pollutant emissions. It covers direct and indirect[13] emissions of CO_2 and of two important air pollutants, NOx and SOx. The tax applies to installations in the industry, electricity and agriculture sectors.[14] Inclusion thresholds for covered installations exist. They depend on physical characteristics of installations, such as levels of thermal power, impact energy of material used, production capacity, volume, treatment capacity, storage capacity, quantity dealt with or surface. Emissions from landfills and facilities for the intensive rearing of animals as well as those from the combustion of biomass and biofuel are exempt. Since 2005, the exemption has been extended to CO_2 emissions beyond free allocation of installations subject to the EU ETS, "except for the excess that entails non-compliance with the obligation to surrender allowances under that legislation". There are tax deductions for firms investing in emissions reduction, called investment deductions. The formal design of the IEGA is presented in Box 3.3.

The revenue from Andalusian ecological taxes, such as the IEGA is meant to be used to finance the actions of the Administration of the Junta de Andalucía in matters of environmental protection and conservation of natural resources.[15] Moreover, 5% of the revenue collected annually is to constitute a reserve fund to attend to emergency situations caused by environmental catastrophes. In 2020, this tax generated about EUR 1.96 million for Andalusia.

In 2019, the tax covered about 70 installations, all in the industry and electricity sectors.[16] In total, these represented about three quarters of CO_2 emissions in these sectors.[17] The firms they belonged to had an average of 543 employees, and average sales of about 559 million euros. About 40% of installations belonged to the electricity sector or to the autoproduction of electricity (subsector of industry) and the other 60% were all in the manufacturing industry.

Box 3.3. The IEGA design, polluting units and reference values

The IEGA design and polluting units

The tax schedule is a function of "polluting units", which bundle together CO_2, NOx and SOx emissions, according to "reference values". More precisely, the polluting units are calculated as follows. First, each substance has been assigned a yearly reference value. For CO_2, this is 200 000 tonnes[1], for NOx, 100 tonnes and for SOx, 150 tonnes. Second, each tonne of substance emitted is divided by its respective reference value. Third, these resulting polluting units are added up to form one taxable base. The tax schedule is then progressive. An exemption bracket has been added to the statutory schedule, such that below 3 polluting units, the effective marginal rate is of 0. The effective base and marginal tax rates are referenced in Table 3.1.

Table 3.1. Effective tax rates and brackets for the Andalusian Tax on the Emission of Gases into the Atmosphere

Base (in polluting units)	Effective marginal rates (in EUR per polluting unit)
0-3	0
3.001-13	5 000
13.001-23	8 000
23.001-33	10 000
33.001-53	12 000
More than 53	14 000

Source: Article 32 of BOE (2004[20]).

The IEGA reference values and European Pollutant Emission Register (EPER) threshold levels

The yearly reference values are based on the threshold levels set in the European Pollutant Emission Register (EPER) Decision,[2)] which specify the lower bound beyond which firms have to declare their emissions. Those threshold levels are either the same or of the same order of magnitude as the Andalusian tax's reference values; they are of 100,000 tonnes for CO_2, 100 tonnes for NOx and 150 tonnes for SOx. These thresholds do not constitute emission limit values (Cañón-de-Francia, Garcés-Ayerbe and Ramírez-Alesón, 2008[21]): they have been set to capture the majority of emission sources and limit administrative burden (European Commission, 2017[22]).

Notes:
1. This was 100 000 from 2003 to 2005.
2. 2000/479/EC.
Source: Exchanges with Junta de Andalucia and https://www.boe.es/buscar/act.php?id=BOE-A-2004-1739&p=20100809&tn=1&se-9.

Installations subject to the IEGA are rather heterogeneous in the emissions they declare. In 2019, they declared CO_2 emissions of 179 301 tonnes on average, ranging between 0 and 1.7 million tonnes.[18] For NOx, the average was of 399 tonnes, ranging between 0 and 2.3 thousand tonnes. And for SOx, the average was of 234 tonnes, ranging between 0 and 3 thousand tonnes.

Only two installations received investment deductions in 2019. This stands in contrast to the first years of the IEGA where many more installations invested in relevant emission reductions.[19]

Despite the heterogeneity in declared emissions, in 2019, half of installations (50%) end up falling into the first tax bracket of the tax schedule presented in Table 3.1 (i.e., polluting units lower than 3). 37% fall into the second bracket. This implies that their tax burden is at most EUR 65 000, which represents less than 0.013% of average annual sales. About 9% of installations then fall into the third tax bracket and 4% into the fourth and fifth. Figure 3.4 presents the distribution of installations according to the IEGA tax brackets.

Figure 3.4. Number of installations per bracket of polluting units

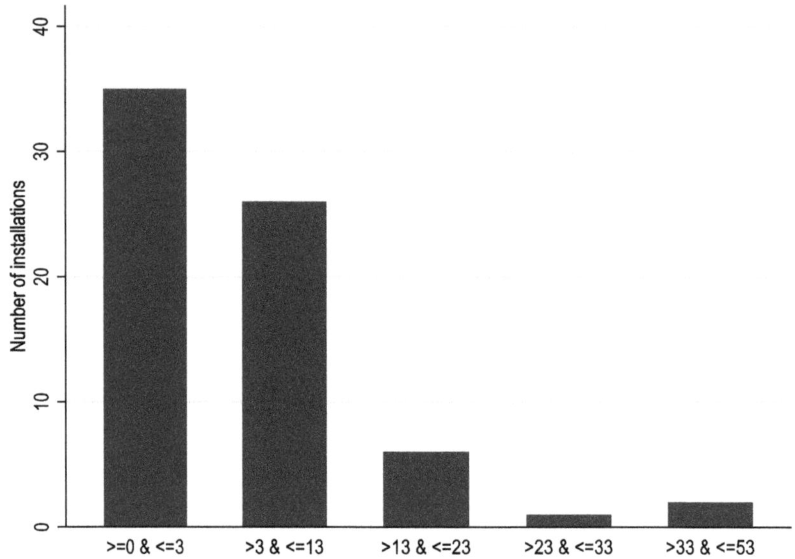

Source: Statistics provided by Junta de Andalucía.

StatLink https://stat.link/lmoskq

The administrative organisation of the tax reveals good practice in the **coverage of installations**. Indeed, the inclusion of installations into the base of the tax is based on physical characteristics, which are more straightforward to verify than emissions, for instance. Moreover, the activities covered are very clearly specified, and avoid confusion.

Andalusia plays a **pioneering role** in setting up a tax tackling air pollution, which is an important issue for environmental and health reasons, as well as economically. Its tax was set up in 2003, following Galicia, which introduced such a tax in 1995. Other Autonomous Communities have since set up similar taxes, including Murcia and Castilla-La-Mancha in 2005, Aragon in 2007, Valencia in 2012 and Catalonia in 2014. The implementation of the tax has helped Andalusia gain the administrative capacity to manage and collect such an environmental tax. This can be important even in the context of a generalisation of such taxes at a national level. Indeed, in 2014, an expert committee (CERSTE-*Comisión de Expertos para la Reforma del Sistema Tributario Español*)[20] had suggested a state-level tax on air pollutant emissions, which could be ceded to the Autonomous Regions for management and collection.

Moving on to points for improving the design of the IEGA, the remarks start with the use of **reference values**, which don't appear to be used correctly. The reference values are based on the threshold levels set in the EPER Decision, but it is not clear why such reference numbers should be used to divide the emissions amount to be taxed. Given the way the taxable base is calculated, the division of emissions by these reference values ensures that this tax applies lower rates to the release of one tonne of CO_2 into the air than to that of SOx or NOx. However, there does not seem to be any scientific reason to implement such relative rates, given that the reference values are based on numbers that are not related to relative harmfulness of different gases. Moreover, the exemption threshold then ensures that if CO_2, NOx and SOx are emitted below these reference values then the installation faces no tax liability. As a reminder, the EPER threshold levels were set for countries to report the majority of their emissions and limit administrative burden. They do not constitute a limit of acceptable amounts of emissions (see Box 3.3.).

Second, there is no clear rationale in the documentation surrounding the design of the IEGA as to why the **calculation of the taxable base bundles all three gases**. A potential reason – but not explicitly mentioned – might be to ensure, for example, using the progressive rate structure, that the release of one tonne of SOx is taxed more highly when occurring on top of the threshold of 2 million tonnes of CO_2 emissions than on top of no CO_2 emissions. Indeed, in the first case the additional tonne would then be taxed at EUR 53.3 and in the second case at EUR 33.3. If this was indeed the explanation, it would be better justified if complemented by scientific evidence. In fact, as highlighted when discussing the interactions between GHG and air pollutant emissions, it does not appear that the negative impact of one of the gases was worse when released in the presence of the other. Moreover, such a tax base provides the possibility for offsetting the tax on the emissions of one gas with the decrease in emissions of another gas. Instead, covering all three gases (with separate taxes or bases) can make sense from an environmental point of view, to avoid effects whereby, in abating one type of emissions, a firm does not pay attention to the potential increases in another type of emissions.

Third, the **progressivity of the rates** is not grounded in classical environmental economics principles, especially for GHG emissions (i.e. here, CO_2). For GHGs, the harmfulness of one extra tonne of emissions does not depend on how many tonnes have been previously emitted. Moreover, the progressivity of rates does not ensure the cost-effectiveness of environmental taxes, which calls for an alignment of tax rates in order to encourage abatement cost minimisation. Unless the progressive structure can be justified on an efficiency[21] or equity[22] ground, it is not clear that the principle of progressivity should be applied in this context.

Fourth, the whole design of the IEGA is **complex** and reduces its **salience**. Recent research shows that complexity in tax systems can make incentives harder to understand and undermine their efficiency (Boccanfuso and Ferey, 2021[23]). Regarding salience, the tax structure does not highlight the rate paid for each tonne of CO_2, NOx or SOx released into the air. For example, a firm emitting 10 000 tonnes of CO_2, 3 000 tonnes of NOx and 5 000 tonnes of SOx cannot know how much it is paying for each gas separately. The division of the emissions by the reference values and the bundling of the three gases into a same base hence affects the salience of the tax, potentially limiting firms' responses to it. Indeed, evidence finds that salience is key to ensuring responsiveness to taxes (Chetty, Looney and Kroft, 2009[24]).

Fifth, the **reduction of the taxable base** is not necessarily grounded in environmental economics or scientific evidence. The reduction is meant to align with the regulatory field of the EU EPER and is calculated based on the three reference values used to compute the polluting units. Indeed, it sets an effective tax-free threshold at 3 polluting units, which stems from a rationale of providing one polluting unit for each gas free of tax. This, however, even if aligned with the EPER threshold levels, is not aligned with their rationale (see Box 3.3.). They are set to capture the majority of countries' emissions and ease administrative burden, not to provide an order of magnitude of minimum levels of acceptable emissions. Moreover, the bundling of the three gases into one base results in this reduction calculation having cross-effects on exemptions for different gases. Taking the example of an installation emitting 50 000 tonnes of CO_2, 700 tonnes of NOx and 150 of SOx, the way the reduction of the base is calculated along with the bundling of the bases provides additional polluting rights in terms of NOx due to the fact that the CO_2 emissions of the installation are much below 200 000 tonnes. Indeed, in this example, the polluting units due to CO_2 are 0.25, to NOx, 7 and to SOx, 1. Hence, the reduction of 3 polluting units in the base "cross-subsidises" NOx emissions because of the fact that CO_2 emissions are lower than 200 000 tonnes. Because of this, emissions from NOx have a tax-free threshold of 1.75 instead of 1.

3.2.2. Carbon pricing in Andalusia

The focus of the GHG pricing analysis for stationary sources in Andalusia is on CO_2 emissions from energy use. This is for at least two reasons. First, carbon emissions are the main source of GHG emissions in Andalusia. Second, the stationary sources of emissions covered by the Andalusian Tax on the emission of gases into the atmosphere are from the industry and electricity sectors, which are mostly responsible for CO_2 emissions (see Figure 3.2), with a majority of those being from energy use. A discussion on ways forward for coverage of CO_2 and other GHG emissions from the agriculture sector, which is responsible for the emission of CO_2, CH_4 and N_2O in similar proportions, is present at the end of the subsection on stationary sources.

In Andalusia, four instruments price (directly or indirectly) CO_2 emissions from stationary sources in the industry and electricity sectors occurring at three different levels of governance. At the Andalusian level, as just described, the Andalusia Tax on the Emission of Gases into the Atmosphere acts as a carbon tax. At the Spanish level, the Tax on Hydrocarbons and the Special Tax on Coal are both fuel excise taxes. At the European level, the EU ETS applies to certain GHG emissions from the two sectors.

At the Spanish level, fuel use in stationary sources is subject to two fuel excise taxes: a Tax on Hydrocarbons (*Impuesto sobre Hidrocarburos*) and a Special Tax on Coal (*Impuesto Especial sobre el Carbón*). The Tax on Hydrocarbons applies to specified uses of liquid and gaseous fuels, including biofuels, coal tar, crude oil, waste oils, coal and coke-related gases. Hydrocarbons are untaxed when used to produce electricity in power plants or to cogenerate electricity and heat in combined power plants. The Special Tax on Coal applies to specified uses of coal and coke products, excluding peat. Together these results in the selected rates presented in Table 3.2, which leaves out exemptions. As in many countries, however, fuels used for electricity generation and some industry sectors are exempted from taxation. Exemptions are covered (but not enumerated) in the subsequent analysis.

Table 3.2. Selected excise tax rates for stationary sources in Spain, 2021

Fuel	Rate in EUR per unit	Unit
Biodiesel and diesel (agriculture, heating and stationary motors)	96.71	1000L
Biogases and natural gas (non-industry heating, stationary motors)	0.65	GJ
Biogases and natural gas (agriculture)	1.15	GJ
Biogasoline (heating, stationary motors)	472.69	1000L
Coal and coke (CHP heat, residential)	0.65	GJ
Coal and coke (agriculture, business and stationary motors)	0.15	GJ
Fuel oil (heating, stationary motors) and waste oils	17	1000Kg
Gasoline (heating, stationary motors)	503.92	1000L
LPG (heating)	15	1000Kg
LPG (stationary motors)	57.47	1000Kg

Note: Taxes as of 1 April 2021.
Source: Taxes in Europe database, https://boe.es/buscar and OECD (2022[25]).

At the European Union level, the EU ETS was introduced in 2005 (Section 2.1.1). It covers CO_2, N_2O, PFCs emissions from the industry and electricity sectors[23] in all EU countries as well as Iceland, Liechtenstein and Norway. Inclusion thresholds vary with the type of installation. In 2020, the EU ETS covered 828 stationary installations in Spain and 81 in Andalusia – respectively 831 and 82 in 2019. This represented 81% of emissions from energy use in the Andalusian industry and electricity sectors.

The different carbon-pricing instruments are summarised in the OECD effective carbon rate (ECR) (see Box 3.1 focusing on three components: carbon taxes, fuel excise taxes and permit prices from emissions trading systems. Figure 3.5 presents the price that applies to CO_2 emissions from energy use in the industry and electricity sectors in Andalusia.[24] The emissions base is divided into the fuel types used in the sectors. The width and height of the different blocks depict coverage and rates, and their colour, the type of instrument used. In blue are fuel excise taxes and in green the EU ETS. Rates of the Andalusian Tax on the emission of gases into the atmosphere are too low to be visible on the figure, but the tax, its rates and overlap with the EU ETS are described and analysed further in the following.

The ECR profile enables to analyse the carbon pricing instruments used, the effective coverage and rates of carbon pricing in Andalusia, and to compare these with benchmark costs (see Box 3.5). After an analysis of federal and EU-level carbon pricing instruments, the analysis turns to the Andalusian tax on the emissions of gases into the atmosphere. It highlights what it adds to the existing national and EU-level carbon pricing instruments and develops recommendations on whether or how to improve this tax – as reforming this tax is in the legal competence of Andalusia.

Figure 3.5. Effective Carbon Rates in the industry and electricity sectors, Andalusia

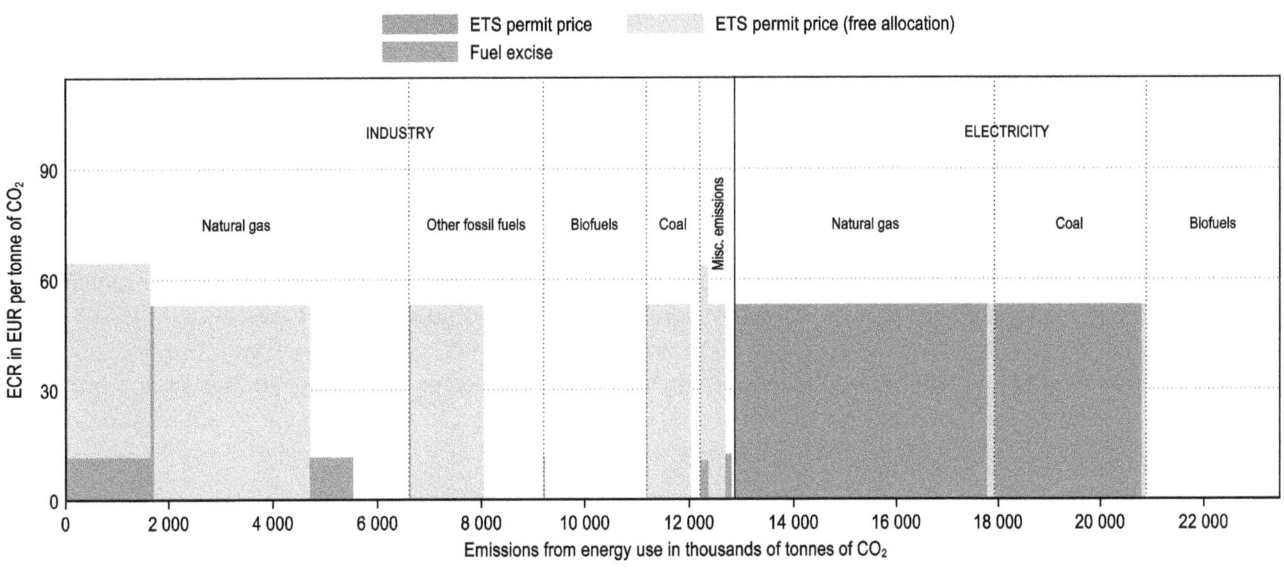

Note: This figure shows CO_2 emissions from energy use in Andalusia taken at the point of combustion and the effective carbon rate they are subject to in the industry and electricity sectors. "Misc." groups together fuels that each represent less than 5% of total energy use from combustible fuels in the sector. In the industry sector, "Misc." is composed of emissions from diesel, fuel oil and LPG. In the electricity sector, "Misc." is too small to be represented on the graph. It is composed of diesel and fuel oil, the emissions of which account for less than 1% of sectoral emissions when taken together. CO_2 emissions are calculated based on energy use data for 2019 from IEA (2020[26]), World Energy Statistics and Balances as well as the Andalusia energy balances. Fuel excise taxes are for 1 April 2021 and permit prices are the average over 2021. Coverage is for 2021. The methodology to estimate the overlap of taxes and ETS permit prices is explained in detail in OECD (2016[6]).
Source: OECD.

StatLink https://stat.link/vr6mcl

Industry is the largest emitting sector from stationary sources in Andalusia in terms of CO_2 emissions from energy use, representing 29% of CO_2 emissions from energy use. Firms in the industry sector mainly face a price signal from the EU ETS, with a permit price of about EUR 53 per tonne of CO_2 on average over 2021. The EU ETS cover roughly 60% of emissions in the industry sector (70% when leaving out emissions from biofuel combustion), i.e. 40% (resp. 30%) of emissions are not covered by the EU ETS. However, 97% of EU ETS emission permits were allocated for free in 2021 (light green). In addition, fuel excise taxes apply to very few fuel categories, with several exemptions from the tax, and at comparatively lower levels per tonne of CO_2 than permit prices. Emissions from the use of natural gas, which constitutes the main fuel category in this sector (51%), face fuel excise rates of about EUR 11.6 per tonne of CO_2, when they are not exempted; more than half of natural gas emissions are exempt. Diesel and LPG used in industry are subject to fuel excise rates of about EUR 19 per tonne of CO_2 on average but taken together represent only a minor share of CO_2 emissions from energy use in industry (less than 1%). The rest of industrial emissions (28%, mainly fuels belonging to the category "coal and other solid fossil fuels" or "other fossil fuels") face excise rates lower than EUR 2 per tonne of CO_2 (1.41 on 83% of emissions coal and other solid fossil fuels – not visible on Figure 3.5) or are not covered by fuel excise at all. The vast majority of emissions from biofuels (99%) in the industry face no excise tax, while 1% face an excise tax rate of about EUR 12 per tonne of CO_2.

The electricity sector (which consists here in plants for which the main activity is to produce electricity[25]) is responsible for 24% of CO_2 emissions from energy use in Andalusia and its emissions from fossil fuels are exclusively covered by the EU ETS. In 2019, free allocation represents about 2% of verified emissions. Emissions from biofuel combustion are not subject to the EU ETS. No fuel excise taxes apply. In Andalusia, CO_2 emissions from energy use in electricity plants stem mainly from natural gas use (48%) and then in almost equal shares from coal and biofuel combustion (respectively 28% and 24%). Compared to the industry sector, the EU ETS provides stronger long-term investment incentives in the electricity sector, where only 4% of emission permits were allocated for free in 2021. In addition, a specific electricity tax applies in Spain. Because it does not send a specific carbon pricing signal, it is not discussed here (Box 3.4.).

Box 3.4. The Spanish electricity tax and the electrification using clean power sources

The current design of the Spanish electricity tax (*Impuesto sobre el valor de la producción eléctrica*) applicable in Andalusia presents several misalignments with the objective to achieve a reduction of 39% of GHG emissions in Andalusia by 2030 compared to 2005 levels. On the one hand, the tax reduces incentives to electrify the economy by increasing the relative prices of electricity. On the other hand, the current design of the tax does not directly encourage producers to switch towards clean sources of electricity production and to decarbonise the power sector.

The tax design does not provide specific incentives for decarbonisation because the tax rate is not differentiated by the type of energy used in electricity production. Thereby, it increases the Terajoule (TJ) price of electricity also when it is produced from clean sources like solar, wind and ocean energy. Figure 3.6 represents the effective electricity price deriving from energy taxation and EU ETS permit prices in the electricity sector in Andalusia, by mapping policy instruments to the amount of electricity consumed in production and to the final user.

Figure 3.6. Effective energy rates in the electricity sector, Andalusia 2021

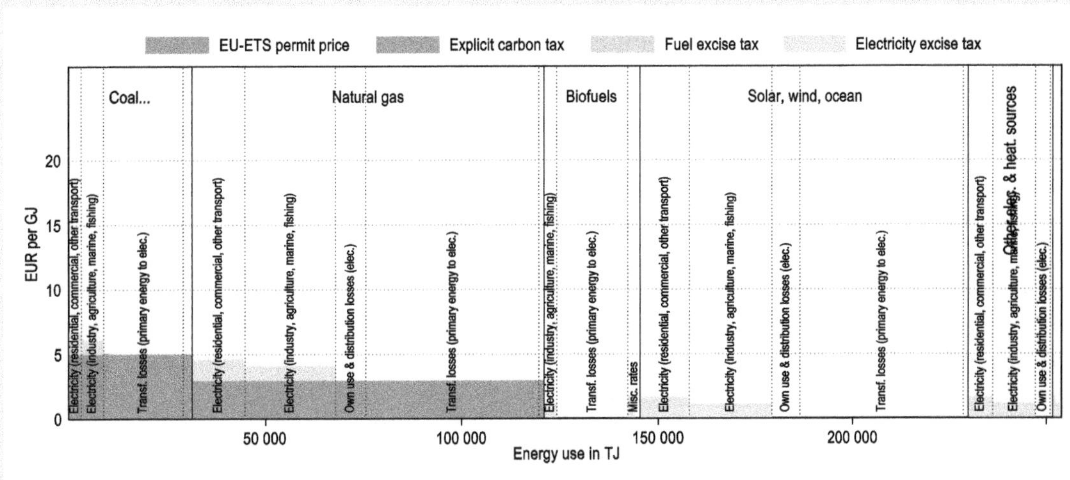

Note: Electricity taxes are for 1 April 2021 and EU-ETS permit prices are the average auction price over 2021. The ad-valorem rate of the electricity tax is translated into effective rates based on information from the European Commission's TEDB database. Energy use data is for 2019 and from IEA (2020[26]), World Energy Statistics and Balances as well as the Andalusia energy balances.

Since 1 January 2019, all hydrocarbons that are used to produce electricity in power plants or to cogenerate electricity and heat in combined power plants are exempted from the fuel tax (*Impuesto sobre Hidrocarburos*). The Andalusia carbon tax (*IEGA*) is not visible in the profile due to the low rates that currently apply.

Source: OECD Taxing Energy Use model, data provided by Andalusia Energy Agency (*Agencia Andaluza de la Energía*).

StatLink https://stat.link/0qsceo

On the horizontal axis, the figure displays electricity use in Andalusia in TJ split into the main primary energy carriers that are used to produce electricity (coal, natural gas, biofuels, renewable energies, and other sources), as well as the main electricity users (residential, commercial, industry, agriculture, etc.). Note that a large part of electricity is lost through processes that transform primary energy into electricity ("Transf. Losses") plus electricity used at plants and distribution losses ("Own use & distribution losses"). On the vertical axis, the figure depicts the price level of policy instruments that electricity users in Andalusia pay in EUR per TJ: the electricity tax (grey bars) and the price signal deriving from the EU ETS (green bars). No fuel excise tax applies in the electricity sector. Combining information on rate and base, the profile gives an indication of the effective price that applies in the sector.

The figure differs from the figure used in the main text in that it is based on TJ, instead of CO_2 emissions which helps observing two additional features. The focus on TJ allows including zero-carbon sources in the energy base that were not part of the profile based on CO_2 emissions, but that become visible in a TJ profile. It also allows mapping the electricity tax to the energy base, which was not depicted in the profile based on CO_2, because it is not considered a carbon pricing instrument.

Figure 3.6 shows that the Spanish electricity tax risks discouraging the electrification of the economy, as it applies to electricity use in the commercial, residential, industry, agriculture and transport sectors, independently of the energy source. The tax rate and coverage (grey part) do not change with the energy carrier used to produce electricity (coal, natural gas and renewables). The opposite is the case for the carbon pricing signal deriving from the EU ETS (green part) which prices electricity depending on the CO_2 intensity of the TJ produced, thereby encouraging the use of clean sources in electricity production.

The electricity tax is set at the national level, and the ETS at the EU level, which leaves only limited leeway for adjustments to the region. The IEGA could be used to further strengthen the carbon price in the electricity sector and encourage the use of clean energy sources. However, the IEGA strongly overlaps with the EU ETS in terms of coverage, and because the IEGA rates are currently low it does not send a significant additional price signal to encourage decarbonisation (see discussion in main section). Other countries use national carbon pricing systems to put a floor price on emissions from electricity generation covered by the EU ETS. Yet, the current rates of the IEGA fall well below the EU ETS permit price and therefore do not serve this goal (see Box 3.6.).

Removing the Spanish electricity tax could help strengthen signals for clean electrification of the economy, as also suggested in the White Book for Tax Reform in Spain. To avoid conflicts between environmental and fiscal objectives (i.e. revenue raising), the phasing-down of the electricity tax could be co-ordinated with the phasing-in of an effective carbon floor price in electricity and the removal of energy tax exemptions on fossil fuel use to generate additional revenue. Eventually, as the energy system is approaching full decarbonisation, electricity taxes could be reintroduced if so desired, e.g. for revenue raising reasons or to incentivise savings.

Electricity taxation still incentivises electricity savings in general. In liberalised power markets, fossil fuel powered generators are frequently the marginal electricity producer. Energy savings induced by electricity taxes could thus indirectly decrease emissions. Electricity taxes also have the advantage that they can be levied on electricity imported from abroad.

Source: Author's own elaboration, based on OECD (2019[5]).

The lack of complementary policies to the EU ETS pricing of Spanish (and hence Andalusian) carbon emissions may be an issue given free permit allocation for the industry sector and price volatility. These issues are further developed in the following.

The EU ETS covers a large part of fossil fuels in the industry and electricity sector, but extensive free allocation in the industry sector erodes the average price signal. Effective carbon rates are typically expressed in *marginal rates*, which means that these are rates faced by fuel users for an extra tonne of CO_2 emissions. Marginal rates assign permit prices to the respective emissions base independently on whether allowances are auctioned or freely allocated. As such, the ETS price component (green area in the Figure 3.5) should be understood as the opportunity cost of emitting an extra unit of CO_2 for firms (see Box 3.1) which provides an incentive to contain emissions at the margin. Figure 3.5 thus partitions the price signal deriving from the EU ETS (green area) and provides an estimate of how much of the EU ETS emissions are covered by an auctioned (dark green) or freely allocated emissions allowance (light green).

By driving a wedge between the marginal and the average carbon price faced by firms, freely allocated emissions permits can affect long term decision making in imperfectly competitive markets. Indeed, they can affect investment decisions since they can discourage investment of firms in low-carbon technologies (Flues and Van Dender (2017[26])). Other evidence also highlights lower green innovation in firms where a larger share of allocations are distributed for free (Martin, 2013[27]).

Free allocation shares are gradually being decreased in the EU ETS. Free permits do help alleviate carbon leakage and competitiveness concerns of energy-intensive and trade-exposed firms. Under current discussions at the EU level, in particular in the context of a potential carbon border adjustment mechanism (CBAM), there are increasing discussions to phase-out free permits going forward.

Permit prices alone may not provide a stable price signal for investment decisions. Despite the dramatic increase of EU ETS permit prices over 2021 and 2022 (having reached about EUR 78/tCO_2 in May 2022 from EUR 34/tCO_2 in January 2021[26]) which has strengthened the carbon price signal faced by firms under EU ETS, their volatility might weaken this signal as it results in uncertainty for investors. This uncertainty lowers incentives for firms to invest in low-carbon technology and projects (Flues and van Dender, 2020[28]).

The difficulty to predict prices for the following years, in turn, also reduces the possibility for firms to plan, adapt and avoid investing in projects that a few years later may cause them to have stranded assets. Despite the introduction of the EU ETS Market Stability Reserve (MSR), carbon price support mechanisms such as those in the United Kingdom (UK) or the Netherlands (see Box 3.6.) may be useful to further address permit price volatility.

A strength of emissions trading systems is that they impose a uniform carbon price on emissions from different fuels and sectors. Contrary to existing fuel excise taxes, which are generally fuel-specific and are set per physical unit or per unit of energy and include generous exemptions, emissions trading systems permit prices are expressed per tonne of CO_2, so result in all fuels within the covered share of the sector facing the same carbon price. This can help avoid switching to fuels that may be less polluting, but remain carbon-intensive all the same, and increases efficiency, by leaving it up to the polluters themselves to decide on which fuel to cut emissions in the least costly manner. Note however, that this is not to say that fuel excise taxes cannot result in the same rate per tonne of CO_2. If first expressed per tonne per CO_2 and transformed per litre or GJ for example, this could be the case. However, this is generally not how fuel excise tax rates are set.

It is also worth stressing that the many exemptions from the Spanish fuel excise which are depicted in Figure 3.5 (no blue bar) can lead to inefficiencies and distributional concerns across firms. For example, fuels used for chemical reduction are all exempt from the national fuel excise tax. This results in a lack of incentives for mitigation emissions for that activity even though it might be highly emitting. Often such exemptions are included to address competitiveness concerns that domestic users may face compared with firms in countries where energy taxes are lower. However, the current structure of the fuel excise does not provide relief based on the actual exposure of a sector to international competition. Alternatively, in the EU ETS, measures to address competitiveness concerns relate to the trade-exposure and energy-intensity of production. Additionally, this may generate distributional concerns between firms if firms conducting chemical reduction are larger than others. Moreover, the low rates observed for coal and hence coal emissions are not aligned with the high emission intensity of this fuel. The much lower rates observed for coal (EUR 1.6/tCO_2) than for natural gas (EUR 11.6/tCO_2) do not incentivise switching to cleaner fuels. Finally, the exemption of most biofuels from fuel excise taxes is generally justified through a life-cycle perspective on biofuels. Indeed, if sustainably sourced, biofuels may be carbon-neutral over the life cycle.[27] However, biofuel combustion raises other issues such as air pollution, which are further discussed in section 3.2.3 of the analysis.

ECRs in Andalusia in 2021 deriving from national fuel excise taxes and the EU ETS were already more or less aligned with price levels that are either consistent with attaining 2030 emissions-mitigation goals or that reflect the externalities caused by CO_2 emissions. This is even more so with recent EU ETS permit prices going beyond EUR 70/tCO_2.[28] Such benchmark prices are further discussed in Box 3.5 showing that several studies find that carbon prices of EUR 30/tCO_2 in 2021, of at least EUR 60 in 2025 and around EUR 125 in 2030 would be consistent with carbon neutrality goals – under complementary policies and technological development and deployment assumptions. Regarding external cost pricing, a recent study by the European Commission (Mottershead et al., 2021[29]) highlights a central estimate for the social cost of carbon (SSC) of EUR 100/tCO_2.

Focusing on the low-end EUR 30/tCO_2 benchmark in 2021, the analysis above shows that priced emissions in both sectors of interest go beyond this benchmark, but this stems from the EU ETS. In the industry sector, about 60% of emissions are covered by the EU ETS and 76% in the electricity sector – respectively 70% and 100% when leaving out emissions from biofuel combustion. However, the price signal stems almost exclusively from the EU ETS, raising potential issues discussed above. First this implies that the benchmark is only reached for marginal rates and not average rates that take into account free allocation – especially in the industry sector, where also generous tax exemptions are prevalent. Second, price volatility may result in lower prices – and therefore low incentives for decarbonisation – in the future. This is difficult to control without a carbon price floor. Moreover, even when leaving out emissions from biofuel combustion, 23% of CO_2 emissions in the industry sector remain unpriced, and 8% priced at an average rate of about EUR 12/tCO_2. Hence, about a quarter of emissions in the industry sector face no price induced signal to mitigate emissions, and the remaining emissions face a price signal that is too low to trigger the required level of emissions mitigation.

In the coming years, the EUR 60 benchmark would be reached on emissions subject to the EU ETS if permit prices stabilise or continue increasing at the same rate – at least for marginal prices. If they increase, they could enable attaining the EUR 100 social cost of carbon estimate. However, fuel excise rates on emissions not covered by the EU ETS remain too low to induce the transformational changes that would need to take place in the industry sector. Moreover, for emissions subject to the EU ETS they provide no underlying price stability or average price signal. While this could be reformed at the national level – and more effectively and efficiently so, Andalusia could use its regional tax to help achieve these goals.

Box 3.5. Benchmark costs for carbon pricing

Externalities and net-zero targets

As a result of the impact of GHG emissions on climate, any activity involving GHG emissions results in a climate externality imposed on others. However, emitters do not necessarily internalise the full costs that their behaviour imposes on others in their decision-making and might hence produce more emissions than socially optimal.

Moreover, steadily increasing global warming caused by these GHG emissions could ultimately result in crossing tipping points beyond which sever and disruptive changes to human society would become irreversible. In line with this, the objective of the Paris Agreement is to face the threat of climate change by keeping the increase in the global average temperature to well below 2°C above pre-industrial levels and to preferably limit the increase to 1.5°C above pre-industrial levels.[1] In order to implement this objective, countries are seeking to attain carbon neutrality by 2050 with, possibly, mid-term objectives to 2030 such as the European Union's Fit for 55 proposal.

Carbon pricing benchmarks

Related costs for GHG emissions can be established in two ways. The first relies on the calculation of the social cost of carbon (SSC) and the second on the calculation of the price of carbon that is compatible with a specific target of emission reductions (e.g. keeping the rise in global temperature from pre-industrial levels below 1.5 degrees Celsius).

A recent study by the European Commission (Mottershead et al., 2021[29]) focuses on calculations of the SSC[2] and, based on a wide range of studies, highlights a central value of EUR 100/tCO_2 through 2030.

Several studies use models to establish carbon prices consistent with mid-term or longer-term emission reduction objectives. These models depend on assumptions about energy price pathways, current and future technologies, complementary policies, and carbon capture and storage development and deployment. Kaufman et al. (2020[30]) find that for the United States, carbon prices to reach 2030 goals should be between USD 34 and 64/tCO_2 in 2025 and at USD 77 and 124/tCO_2 in 2030. These figures are slightly lower than the IEA's latest carbon price trajectory for the electricity, industry and heat sectors in advanced economies (IEA, 2021[31]), which finds prices at EUR 75/tCO_2 in 2025 and EUR 130/tCO_2 in 2030.

Notes:

1. 2°C has been established as a critical global temperature after which changes may become dramatic and irreversible; 1.5°C would further reduce the risks and impacts of climate change.
2. The SCC is defined by Nordhaus (2014[32]) as the economic cost caused by an additional tonne of CO_2 emissions or its equivalent; it rests on the concept of internalising externalities and includes considerations on inter- and intra-generation equity.

The Andalusian Tax on the Emission of Gases into the Atmosphere hardly adds to EU-wide and national-level price signals both because of its coverage and because of its rate levels. Indeed, its marginal rate can never be higher than EUR 0.07/tCO_2.[29] In practice, no installation goes beyond the fourth bracket of the effective schedule, implying a maximum marginal rate of EUR 0.06/tCO_2. Out of about 90 installations covered by the EU ETS or the IEGA, 59% are covered by both, 13% only by the IEGA and 28% only by the EU ETS. Only one installation covered by the IEGA faces a positive tax liability. The average (weighted by CO_2 emissions) marginal rate faced by installations covered by both is EUR 0.036/tCO_2. Out of the 13% of installations covered by IEGA and not the EU ETS only one installation faces a positive tax liability (with a marginal rate of EUR 0.025/tCO_2) – the others face a null tax.

Many reasons could underlie the introduction of the IEGA: (i) base broadening; (ii) increasing carbon price levels to benchmark costs; (iii) providing a backstop to volatile EU ETS permit prices or (iv) raising revenue. The first three would be aligned with environmental considerations and are discussed below.

Base broadening would increase carbon pricing coverage of emissions to smaller firms or other sectors but given the large overlap between EU ETS covered firms and firms subject to the IEGA, the tax has not achieved such an objective. Moreover, the rates faced by firms covered only by the IEGA are almost all null, and the IEGA thus does not strongly extend carbon-price coverage to emissions in the industry sector that currently do not face a carbon price. As highlighted in the above analysis, in effect, it does not extend coverage to other sectors (e.g., agriculture) either.

Given the very low marginal rates, the IEGA hardly increases carbon price signals either. These rates do not bring marginal price levels close to benchmark costs, nor do they provide enough incentives to decrease emissions by a significant amount. Recent evidence (D'Arcangelo et al., 2022[33]) shows that a EUR 10/tCO_2 increase in effective carbon rates would lead to a decrease of about 4% of emissions in the industry and electricity sectors in the long run. As a reminder, the average (weighted by CO_2 emissions) marginal IEGA tax rate faced by installations covered by both the IEGA and the EU ETS is of about EUR 0.04/tCO_2. However, the responsiveness estimates just mentioned imply that an increase in rates of EUR 0.04/tCO_2, imply an decrease in emissions for these installations of 0.016% in the long run – much below the efforts currently required to reach net zero emissions.

At such low rates, the IEGA cannot provide a backstop to volatile permit prices. Indeed, as highlighted above, such rates cannot provide a strong, stable and complementary price signal to the EU ETS. Moreover, its design does not lend itself to such an opportunity. This could occur if it were designed with similar features to the UK carbon price floor or the Dutch carbon levy described in Box 3.6. , with credible price signals, aligned to a certain extent with marginal abatement costs in these sectors. Moreover, the price signal could gradually increase over time to enable firms and investors to adapt and plan.

The opportunities for an Andalusian-level carbon tax of increasing base coverage (to smaller firms for instance), of increasing price levels or of providing a strong, stable and complimentary price signal to the EU ETS could come with political, competitiveness, leakage and administrative costs. The extension to smaller firms could engender high administrative costs if the tax were to be applied downstream, as the emissions measurement costs could be very high, given that their emissions are currently not measured – neither for the EU ETS nor for EPER. Moreover, the difficulty of introducing a significant unilateral carbon tax in smaller jurisdictions must be acknowledged. First, given that climate change is a global issue, the impacts of which are not necessarily felt locally, political support for an increase in rates may be limited. Second, competitiveness concerns for industries that are highly emitting already exist at a national level, and may be exacerbated at a regional level, where firms could relocate easily to neighbouring regions (which would result in carbon leakage). Combining carbon pricing with complementary policy measures can help alleviate competitiveness concerns, while keeping the incentive to mitigate emissions in place, as discussed below. Finally, in terms of effectiveness, climate change and GHG mitigation are best dealt with at a national or even supranational level. Indeed, this enables emissions cuts where they are the cheapest at a much larger scale and can help avoid carbon leakage. Hence, taxing greenhouse gas emissions is not necessarily recommended at a regional level.

The potential administrative issues highlighted above may be tackled through various means. The administrative burden of monitoring, reporting and verifying that would be faced if a downstream tax were to be applied to smaller firms could be tackled through an upstream implementation of the Andalusia tax. This could also be done through the introduction of a carbon tax component on fuel taxes, aligned with their CO_2 emissions. However, this may only be possible at a national level.

Potential political, competitiveness and relocation concerns from unilateral carbon pricing were recently tackled by the Netherlands who introduced a gradually increasing carbon price floor in industry, through a careful phase in of base and rates, and the provision of (costly) subsidies. The Netherlands introduced a national carbon levy in 2021 (Box 3.6.). Political hurdles were addressed by engaging in dialogue with key stakeholders in the industry. Competitiveness issues were addressed by a careful and pre-announced phase-in of base and rates (see Table 3.3). This decreases uncertainty for investors and enables firms to adapt and plan, in order to switch to cleaner modes of production. Going forward, this also avoids the risk of stranded assets, and ensures firms remain competitive in a cleaner production environment. Finally, the careful use of technology subsidies to ease the transition for firms was another way competitiveness concerns were tackled by the Netherlands. Such subsidies can be at the research and development (R&D) or at the adoption and deployment level. These could also help deal with affordability concerns facing firms, especially if the carbon price were increased.

Finally, the tax could cover other GHGs not covered by a national tax, such as methane or nitrous oxide, but the issue of the lack of effectiveness of tackling climate change at a regional level would remain. It may make more sense to try and tackle more local issues, such as air pollution, which is what this Section now turns to. The possible extension of the tax to farming is addressed last.

Box 3.6. Carbon pricing floors in practice

The Carbon Price Floor in the United Kingdom

In 2013, the United Kingdom (UK) introduced a carbon price floor (CPF) for fossil fuel emissions in the electricity sector covered by the EU ETS (and now covered by the UK ETS). The CPF consists of two elements: the ETS allowance price and a carbon price support (CPS) mechanism, which is a fixed element charged on top of permit prices. In 2013, the CPS was at GBP 9/tCO_2 emissions and rose to GBP 18 in 2015 (Hirst, 2018[35]). In 2018, this allowed the average effective carbon rate in the electricity sector to reach about EUR 26/tCO_2 while the average EU ETS permit price over 2018 was at about EUR 16/tCO_2.

Leroutier (2022[34]) finds that the UK CPS induced emissions from the UK power sector to drop by 20% to 26% per year on average between 2013 and 2017.

The Dutch carbon levy

The Netherlands, as part of its 2020 Climate Agreement, implemented a new carbon levy for industry on 1 January 2021. The new carbon levy complements the permit prices from the EU ETS and effectively puts a domestic price floor for Dutch industrial emissions. It consists of a floating contribution added on top of the EU ETS price – so that if the price of emissions allowances exceeds the floor price, the floating contribution becomes zero. The total price (EU ETS price plus carbon levy) is intended to increase gradually over time from EUR 30/tCO_2 in 2020 to EUR 125/tCO_2 in 2030, as shown in Table 3.3. The carbon price path was designed based on current and planned abatement cost curves in the Dutch industry sector.

This carbon levy was implemented in the industry sector, where the risk of EU ETS price drops threatens investment in low-carbon assets. The price path was announced from the start of its implementation (with a foreseen review after five years) to allow firms to plan and invest accordingly. To give firms additional lead time, the levy base phases in over time.

Table 3.3. The Dutch carbon price path for industrial emissions

Year	2021	2022	2023	2024	2025	2026	2027	2028	2029	2030
Floor price (in EUR per tonne of CO_2)	30	40.56	51.12	61.68	72.24	82.80	93.36	103.92	114.48	125.04

Source: Adapted from Anderson et al. (2021[35]).
Note: Additional details on the approach taken in the United Kingdom and the Netherlands are provided in Annex Table 3.B.3 and Annex Table 3.B.6.

3.2.3. Air pollution pricing in Andalusia

In Andalusia, the only pricing instrument that applies directly to air pollutant emissions is the IEGA. By pricing GHG emissions, the EU ETS and national fuel excise taxes affect fuel consumption and hence air pollution (OECD, 2019[5]), but the effect is indirect. Hence these latter instruments are not discussed in this section and the focus is on the IEGA.

The calculation of the taxable base implies that marginal rates faced by an additional tonne of NOx (resp. SOx) range between EUR 0 and EUR 140 (resp. EUR 0 and EUR 93.3). Table 3.4 presents these rates according to the bracket which they belong to. The analysis of the 70 installations facing the IEGA in 2019 shows that in practice, 50% of installations face zero marginal rates for their NOx and SOx emissions and that 37% of installations face marginal rates of EUR 50/tNOx of EUR 33.3/tSOx. The maximal marginal rates faced are of EUR 120/tNOx and EUR 80/tSOx. This results in emissions-weighted average marginal rates of about EUR 40/tNOx and EUR 21/tSOx.

Table 3.4. Effective marginal rates faced by the emission of one extra tonne of NOx or SOx, and respective share of firms subject to these rates

Base (in polluting units)	Marginal rate faced by the emission of one extra tonne of NOx (in EUR/tonne of NOx)	Marginal rate faced by the emission of one extra tonne of SOx (in EUR/tonne of SOx)	Estimated share of installations subject to the specific marginal rate in 2019
0-3	0	0	50%
3.0001-13	50	33.3	37%
13.0001-23	80	53.3	9%
23.0001-33	100	66.7	4%
33.0001-53	120	80	
More than 53	140	93.3	

Source: Author's own calculations based on data provided by the Junta de Andalucía.

Usual estimates of NOx and SOx-associated costs generally show higher costs for NOx than for SOx. However, recent estimates provided by the European Commission (Mottershead et al., 2021[29]), find NOx costs of EUR 6/kg/year on average (i.e. EUR 6 000/t/year) and SOx costs of EUR 7.9/kg/year on average (i.e. EUR 7 900/t/year). However, these costs alone should not impact the level of tax set for these pollutants. The price elasticity of these emissions should also be accounted for.

Few elasticity estimates of air pollutant emissions to taxes exist. Descriptive evidence and models find that such taxes do coincide with decreases in emissions (Juřík and Braathen, 2021[36]; Mardones and Cabello, 2019[37]). Moreover, with increasingly emerging abatement technologies and the decrease in their price, elasticities are bound to increase in the coming years. Decreasing air pollutant emissions from stationary sources can be done through fuel switching, through the adoption of technologies, in particular abatement technologies, through efficient production processes or through decrease in production. The first three options are increasingly within reach for firms and enable them to maintain their output while decreasing local air pollution. Shapiro and Walker (2018[38]) find that the decrease by 60% of manufacturing firms' air pollution in the United States was accompanied by a substantial increase in manufacturing output. They show that these emissions reductions were primarily driven by changes in emissions intensity.

The Andalusian **rates** (EUR 63/tNOx and EUR 56/tSOx on average – i.e. weighted by emissions) are **in the lower range of air pollution tax rates** in other countries but similar to other rates observed in Spain. In Catalonia, they are of EUR 45/tSOx and EUR 75/tNOx and in Aragon they are equal to EUR 50/t for both SOx and NOx. In the Czech Republic, they are of EUR 152/tNOx and EUR 191/tSO_2. They are of the same order of magnitude than minimum rates in Chile (see Box 3.7). Increasing these rates might be considered an option, but this would also depend on technologies available and their costs.

Air pollution health externalities are local and depend in particular on local population density. Population density is very heterogeneous in Andalusia, mostly concentrating around the largest cities and on the Southern coast (Figure 3.7). Hence, at the same emissions level, the air pollution impact on human health of a firm located in the Northern part of Andalusia should be lower than that of a firm located, for example between Malaga and Marbella.

Figure 3.7. Population density in Andalusia, 2020

Source: https://www.juntadeandalucia.es/institutodeestadisticaycartografia/VisorGrid/visor.htm#, generated on 15 November 2022.

The differential impacts of air pollution depending on population density are reflected in the Chilean green taxes on PM, SO_2 and NOx from stationary sources that were introduced in 2015. Box 3.7 provides additional detail on the design of this tax, and the different local conditions taken into account. Of course, many effects may be accounted for when considering the impacts of air pollution (wind for example), but at the first order, population density matters for health-related issues. The Chilean green taxes also account for levels of pollutant concentration: the higher the initial level of air pollutant concentration, the worse is the impact of the emission of an extra tonne of air pollutant.

Box 3.7 The Chilean Green Taxes on PM, SO_2 and NOx

Tax design

The Chilean taxes on PM, SO_2 and NOx were introduced as part of Chile's General Tax Reform Bill (Ley 20.780) passed in September 2014.

Each tax base consists in annual emissions of liable facilities. Rates were determined in terms of the respective marginal costs of each pollutant. They also depend on how "saturated" a zone is and on population density. These are two main elements in determining the health damages imposed by these air pollutants.

For each pollutant "i", tax rates depend on both characteristics of the pollutant and of the municipality "j" where it is emitted:

$$T_{ij} = 0.1 \times AQ_j \times SC_i \times Pop_j$$

Where:

- $\mathbf{T_{ij}}$ –tax rate on pollutant "i" in municipality "j" in USD/tonne.
- Characteristics of pollutant "i" are
    i. $\mathbf{SC_i}$ –social cost of pollutant "i", presented in Table 3.5.
- Characteristics of municipality "j" are
    i. $\mathbf{AQ_j}$ –air quality coefficient in municipality "j". AQ is equal to 1.1 if the municipality is in a latent zone, and 1.2 if in a saturated zone.
    ii. $\mathbf{Pop_j}$ –population in municipality "j".

Table 3.5. Social costs of pollutants

Pollutant	PM	SO2	NOx
Cost (USD/tonne)	0.9	0.01	0.025

Source: Pizarro (2019[39])

Chile also has a carbon tax, which does not depend on local characteristics as climate change impacts are not local.

Saturated and latent zones

In Chile, areas that exceed the air pollution standards as defined by the Chilean Air Quality Standards of CONAMA (*Comisión Nacional del Medio Ambiente*) are classified as non-attainment areas (similar to the United States). An area is then designated as a "latent" non-attainment area when pollutant concentrations are between 80 and 100% of the standard, and as a "saturated" non-attainment area, when pollutant concentration exceeds the set standard (Díaz-Robles et al., 2011[40]).

Comparison of tax liabilities for two firms located in different density areas

The tax structure implies that two firms located in municipalities respectively of 20 000 inhabitants and 500 000 inhabitants would have very different tax liabilities, even if they emitted the same amount of air pollutants. The latter firm's total tax liability would be 25 larger than the former's. For NOx, the rate would go from USD 55/tNOx to USD 1 375/tNOx and for SO_2, it would go from USD 22/tSO_2 to USD 550/tSO_2.

Note: Additional details are provided in Annex Table 3.B.1.
Source: Pizarro (2019[39]), Diaz-Robles et al. (2011[40]).

Taking into account **population density or levels** helps to better price external costs associated to local air pollutants and can discourage firms to settle in densely populated areas. Better pricing of environmental externalities implied by air pollution by adapting the tax to local characteristics is important. A price signal aligned with local population levels can help bring pollution to levels in line with how harmful they are. Moreover, this may be more easily sustained politically as well because air pollution impacts are generally very localised and occur on a shorter time horizon than climate change, so are felt more strongly by the population. Another effect can arise, which is to go beyond a reduction in existing firms' emissions and discourage new firms from settling in populated areas, where their activity would be much more harmful than in low density areas. As can be seen in the comparison presented in Box 3.7 , such a design of air pollutant taxes can make it prohibitively expensive for firms to settle in such areas. Hence such an adjustment to the tax could allow both intensive (reduction of emissions in a location) and extensive margins (less new polluting firms in a location) adjustment to be at play.[30] Accounting for **pre-existing air pollution density levels** (based on indicators such as the US AQI) would also help the design of the tax to better aligned with external costs and have similar effects as those just described.

Regional environmental authorities could provide high-resolution baseline air pollution maps, which would allow these additional factors to be incorporated in the design of air pollutant taxes in Andalusia. Dispersion studies, which would identify the exact areas affected by pollution, could be combined with the sophisticated population georeferencing that is maintained by the Institute of Statistics and Cartography of Andalusia (IECA). In addition to population values, the IECA provides details on other demographic, health, economic and social variables that would enable air pollutant taxes to account for other parameters related to population vulnerability that influence the estimation of the health impact of pollution. However, careful attention should be given to balancing design and administrative complexity with the precise alignment of rates with environmental and health externalities. Indeed, many factors influence the health and environmental impact of air pollution and accounting for all of them would make these taxes unmanageable – so a focus on the main factors is recommended.

Extending the base of the tax to cover PM emissions and emissions from the combustion of biofuels and biomass could be considered as options. Indeed, given the tax already covers NOx and SOx, which are closely linked to PM (see section 3.1.2), the extension to PM would be straightforward to implement. This could have a sizeable impact on one of the most harmful air pollutants on human health. Moreover, while the exemption of biofuel combustion from a tax on CO_2 emissions can be justified from a life-cycle perspective (see section 3.2.2), this is not the case concerning air pollutants. Indeed, biomass combustion may worsen local air pollution, especially from particulate matter (PM) and nitrogen oxides (NOx) emissions, which is not compensated for from a life-cycle point of view. The 2021 proposed revision to the EU ETD (European Commission, 2021[41]) goes in this direction, by considering minimum taxation rates for biofuels.

The rates might have been set to ensure progressivity between more or less polluting firms if this is linked to their size, but **distributional impacts or equity considerations can, and generally should, be addressed through other policy instruments**. Revenue recycling options could be considered, such as support to firms for adoption of abatement technologies. For example, the revenues of the French tax on air pollution were earmarked for abatement subsidies and the financing of air quality surveillance systems (Millock and Nauges, 2003[42]).

Regarding potential competitiveness issues, the Spanish context is such that these might be limited. Indeed, the examples of Aragon and Catalonia show that higher rates can be applied in the long term. Moreover, the higher rates or similar rates observed in other Autonomous Communities also alleviate competitiveness concerns for Andalusia with respect to other Spanish firms. In this respect, the White Book for Tax Reform in Spain recommends maintaining existing regional taxes and introducing a national tax that sets a minimum tax base and tax rates.

Finally, political hurdles might be easier to address in the context of local air pollution, as, contrary to GHG emissions, effects are very local and can be felt in the short term. Moreover, as highlighted in section 3.1.2, reducing air pollution may also be helpful for firms' economic output.

3.2.4. Pricing emissions from the agricultural sector

In Andalusia, the agricultural sector is responsible for a major share of air pollutant emissions, especially for NOx, NH_3 and PM emissions, as well as of GHG emissions other than CO_2 – it is the main source of N_2O and CH_4 emissions. Managing emissions from the farming sector hence requires the coverage of different emissions. Pricing these air pollutant emissions as well as N_2O and CH_4 emissions in this sector would be important to align with the Polluter Pays principle. Moreover, the growth of farming areas and the expansion of urban centres increases the exposure of the Andalusian population to these local air pollutants.

Given the difficulty of measuring emissions in this sector, which are generally diffuse (as opposed to point source), the administrative organisation of managing emissions in agriculture may need to be different from other stationary source sectors.

Given the economic importance of the agricultural sector especially in Andalusia, political hurdles may be important. This stresses the importance of building a strong dialogue and cooperation with this sector. Agriculture holds an important part in the Andalusian economy. It made up 6.7% of Andalusian Gross Value Added (GVA) in 2021[31] and represented 30.8% of Spanish agricultural GVA (INE, 2023[43]) – and agricultural areas in Andalusia have been increasing in the past years (Junta de Andalucia, 2019[44]). The NOx and direct PM emissions from the agricultural sector in Andalusia mostly stem from fuel combustion. This is best managed by fuel excise duties that reflect the fuels' environmental damage, as recently put forward by the proposed EU ETD reform. In the face of such high fuel emissions, reduced rates for the agricultural sector should be avoided.

NH_3 emissions in Andalusia almost entirely stem from agricultural activities. These include livestock waste and the heavy use of nitrogen fertilisers. For poultry farms, manure is the main NH_3 emitter. NH_3 then combines with other air pollutants from combustion (NOx and SOx) to create $PM_{2.5}$ (Bauer, Tsigaridis and Miller, 2016[45]; Lelieveld et al., 2015[46]). Some researchers point to the need of reducing nitrogen fertiliser use, while other researchers argue that the decrease in NOx and SOx emissions would be enough to limit the creation of PM and hence limit health damage of NH_3 emissions.[32] While in many regions and countries, NOx mostly stems from the road transport, electricity and industry sectors, in Andalusia, this gas is also a result of agricultural activities. Hence, limiting PM emissions in Andalusia could not only rely on other sectors, and the agricultural sector would have to be involved. Moreover, NH_3 also impacts soil and water acidification (see Part III, Section 6) and may harm animals themselves, resulting in short and long-term losses for farmers themselves.

NH_3 emissions might be better managed through the taxation of intrants and livestock numbers or through regulation and promotion of different agricultural practices. Indeed, NH_3 emissions are complex to measure directly (Herrero et al., 2021[47]). Hence, these emissions could for instance, be better managed through taxation of nitrogen fertilisers. A tax on livestock numbers, however, might not give the right incentives to decrease emissions and a tax on nitrogen fertilisers could be avoided by purchasing this intrant outside of regional borders. Regulation, through the promotion of certain agricultural practices could also contribute to decreasing emissions, through for example livestock waste management methods which are less polluting. Moreover, the type of manure management system that is used in livestock and poultry production can also affect emission levels (Dunkley and Dunkley, 2013[48]). Promoting the use of less polluting manure compositions and management can constitute a key element in decreasing NH_3 emissions in the sector. Finally, sustainable management practices to enhance nitrogen use efficiency are also key to mitigating NH_3 as well as N_2O emissions. Pan et al. (2022[49]) propose options.

Regarding other GHG emissions, CH_4 mostly stems from livestock, while N_2O emissions result from both livestock and soil management. NH_3 is also a precursor to N_2O.

Current proposals for the taxation of farm-level emissions include considerations on nitrogen fertiliser application as well as livestock rearing. Based on the GHG footprint of mineral fertilisers, Anderson and Bonnis (2021[50]) propose an average rate of EUR 1 to 2 for a tax on the surplus application of nitrogen. New Zealand is, at the time of writing, one of the first countries to consider taxing GHG emissions at a farm level. This is taking place within a long-term process of cooperation and dialogue with farmer associations,[33] and in a context where agriculture is responsible for about half of nation-wide emissions. The current consultation document (Ministry for the Environment and Ministry for Primary Industries, 2022[51]) proposes a model which accounts for farm area, stock reconciliation, livestock production data and total synthetic nitrogen fertiliser use. Such an approach could also be interesting for the taxation of NH_3 emissions.The risk of relocation to other Autonomous Communities or Portugal is limited. Political hurdles, however, may be important, as can be seen with the protests taking place in New Zealand following the government's confirmation of plans to price farm-level GHG emissions.[34] This is especially so when the sector is an important backbone of the local economy.

This stresses the importance of accompanying farmers through the transition, of enabling them to measure their emissions and to propose viable solutions for them to decrease emissions. A slow phase-in of tax rates can enable farmers to plan and adapt. Programs such as OverseerFM[35] can help farmers better manage their intrants and get a better grip of their environmental impacts. The promotion of new technologies and of better farming practices can also provide options for farmers to switch to less emitting practices. The New Zealand proposal also includes payments to farmers using approved mitigation technologies or approved on-farm vegetation. In the long run, it also includes revenue recycling in part to funding for R&D to lower on-farm emissions.[36] Improvements would be built into the system, as can already be seen with the consultation process, which leaves many questions open for farmer organisations to get a say (Ministry for the Environment and Ministry for Primary Industries, 2022[51]). Payments to farmers could also be made on the basis of adoption of recommended farming practices and could be based on proceeds of the tax. The recently published White Book for Tax Reform in Spain,[37] suggests a gradual introduction of such taxes along with a share of the revenues dedicated to technological improvements in the sector to facilitate their introduction.

Regarding GHG emissions, the recommendation for dealing with such emissions at least at the national level remain, though an engagement with farmers at this stage would be an important step for future pricing or regulation measures to be introduced in this sector. Moreover, given that the Andalusian agricultural sector represents an important share of the Spanish agricultural sector, dealing with GHG in this sector at the regional level could be justified.

3.3. Key findings and strategic recommendations

The Andalusian tax on the emission of gases into the atmosphere (IEGA) follows good administrative practice in designating covered entities through physical characteristics and plays a pioneering role in air pollutant emissions pricing in Spanish regions. It also presents an interesting feature through its effort to cover CO_2 emissions as well as NOx and SOx emissions.

However, the IEGA presents a design that is complex, which might mute its price signal and provide unintended incentives. This is mainly due to the bundling of all three gases into one single base, and through the calculation of polluting units. An application of a tax for each gas, applied per tonne of emission would be more straightforward, would make the price more salient, and would enable a better alignment of price levels with environmental costs and mitigation targets.

The current progressivity of rates as a function of a firm's emissions is not aligned with environmental economic principles (in particular cost-efficiency). According to such principles, the tax schedules should be flat – i.e. have a single rate with no exemption threshold (but could depend on location for air pollution). The progressivity of rates might be to deal with affordability or equity considerations, by giving a minimum emission right to all installations and making each tonne of emission more costly above certain thresholds. However, equity and affordability concerns are best dealt with by complementary instruments providing support to firms, which can be direct or indirect. Indirect support could include a time-progressive phase in of base and rates. Direct support could include subsidies for green technology adoption. To ensure equity, subsidies could be tailored to firm size. Such measures are costly and could be implemented using the general budget or the revenue from green taxes (revenue recycling). Such measures could also help deal with competitiveness issues.

Given that GHG emissions are a global issue, the regional level may not be the most suitable governance level for regulation in this area. CO_2 emissions in the industry and electricity sectors are already covered at the European level by the EU ETS and at the national level by fuel excise taxes. While the level of fuel excise taxes could be reformed to better align with benchmark carbon costs, this should be done at the national level.

Air pollutant emissions are principally a local issue, which makes them a suitable target for mitigation for regional level action. Current tax rates levied in Andalusia are on average similar to other rates observed in Spain and in the lower range when compared to other countries with similar taxes. This is useful for coordination with other Spanish Autonomous Communities but may be too low all the same to encourage enough abatement efforts. Having a better idea of target levels for SOx and NOx emission reductions as well as available mitigation technologies and costs could help adjust the price levels to reach such targets. If the objective is to reflect external costs for health in tax rates, Andalusia could consider including population density and pollution levels in the calculation of tax rates, similar to Chile. This would better align price levels with health and environmental costs (which are higher in more populated areas) and possibly discourage firms from settling in densely populated areas – where the negative impact of air pollution is higher – going forward.

An extension of the tax to PM emissions from industrial and electricity sector stationary sources could be considered. This would be relatively straightforward to implement given that NOx and SOx emissions are already taxed. Moreover, this would deal with one of the most harmful air pollutants for human health.

Finally, an extension of the tax to the farming sector would entail extending the coverage to other pollutants, such as NH_3 and to other GHGs such as N_2O and CH_4 as well as adapting the emissions measurement methods to this sector. This would require dialogue and engagement with stakeholders, proposals for and existence of alternatives, possibility of measurement of farm-level emissions and support for farmers in the transition. Examples based on the New Zealand 2022 proposal for taxing farm-level emissions are exposed. Dialogue with farmers should also stress the benefits that better air quality and mitigated climate change would have on their sector and employees.

References

Ainzúa et al. (2020), *Capacity-building for the implementation of Chile's green tax*. [54]

Andalucia, J. (2021), *Inventario de emisiones a la atmósfera de la Comunidad Autónoma de Andalucía. Serie 2003-2019*. [17]

Andersen, M. and G. Bonnis (2021), "Climate mitigation co-benefits from sustainable nutrient management in agriculture: Incentives and opportunities", *OECD Environment Working Papers*, No. 186, OECD Publishing, Paris, https://doi.org/10.1787/a2960c54-en. [50]

Anderson, B. (2021), "Policies for a climate-neutral industry: Lessons from the Netherlands", *OECD Science, Technology and Industry Policy Papers* 108. [35]

Arneth, A. et al. (2009), "Clean the Air, Heat the Planet?", *Science*, Vol. 326/5953, pp. 672-673, https://doi.org/10.1126/science.1181568. [18]

Bauer, S., K. Tsigaridis and R. Miller (2016), "Significant atmospheric aerosol pollution caused by world food cultivation", *Geophysical Research Letters*, Vol. 43/10, pp. 5394-5400, https://doi.org/10.1002/2016gl068354. [45]

Boccanfuso, J. and A. Ferey (2021), *Inattention and the taxation bias (Working paper)*. [23]

BOE, B. (2004), "Ley 18/2003, de 29 de diciembre" 26, https://www.boe.es/boe/dias/2004/01/30/pdfs/A03889-03925.pdf. [20]

Cañón-de-Francia, J., C. Garcés-Ayerbe and M. Ramírez-Alesón (2008), "Analysis of the effectiveness of the first European Pollutant Emission Register (EPER)", *Ecological Economics*, Vol. 67/1, pp. 83-92, https://doi.org/10.1016/j.ecolecon.2007.11.016. [21]

Chetty, R., A. Looney and K. Kroft (2009), "Salience and Taxation: Theory and Evidence", *American Economic Review*, Vol. 99/4, pp. 1145-1177, https://doi.org/10.1257/aer.99.4.1145. [24]

D'Arcangelo, F. et al. (2022), "Estimating the CO2 emission and revenue effects of carbon pricing: New evidence from a large cross-country dataset", *OECD Economics Department Working Papers*, No. 1732, OECD Publishing, Paris, https://doi.org/10.1787/39aa16d4-en. [33]

Dechezleprêtre, A., N. Rivers and B. Stadler (2019), "The economic cost of air pollution: Evidence from Europe", *OECD Economics Department Working Papers*, No. 1584, OECD Publishing, Paris, https://doi.org/10.1787/56119490-en. [10]

Deryugina, T. et al. (2019), "The Mortality and Medical Costs of Air Pollution: Evidence from Changes in Wind Direction", *American Economic Review*, Vol. 109/12, pp. 4178-4219, https://doi.org/10.1257/aer.20180279. [12]

Díaz-Robles, L. et al. (2011), "The Air Quality in Chile: Twenty Years of Challenge", *Environmental Management (EM), Air & Waste Management Association*, Vol. 3. [40]

Dunkley, C. and K. Dunkley (2013), "Review - Greenhouse Gas Emissions from Livestock and Poultry". [48]

European Commission (2021), "Ensuring that polluters pay - The Netherlands". [56]

European Commission (2021), *Proposal for a COUNCIL DIRECTIVE restructuring the Union framework for the taxation of energy products and electricity - COM(2021)563*. [41]

European Commission (2017), "REFIT evaluation of Regulation (EC) No 166/2006 concerning the establishment of a European Pollutant Release and Transfer Register (E-PRTR)", https://eur-lex.europa.eu/legal-content/EN/TXT/PDF/?uri=CELEX:52017SC0710&from=en. [22]

Flues, F. and K. van Dender (2020), "Carbon pricing design: Effectiveness, efficiency and feasibility: An investment perspective", *OECD Taxation Working Papers*, No. 48, OECD Publishing, Paris, https://doi.org/10.1787/91ad6a1e-en. [28]

Flues, F. and K. van Dender (2017), "Permit allocation rules and investment incentives in emissions trading systems", *OECD Taxation Working Papers*, No. 33, OECD Publishing, Paris, https://doi.org/10.1787/c3acf05e-en. [26]

García Bernal (2018), "Implementación del Impuesto Verde en Chile". [52]

Herrero, E. et al. (2021), "Towards robust on-site ammonia emission measuring techniques based on inverse dispersion modeling", *Agricultural and Forest Meteorology*, Vol. 307, p. 108517, https://doi.org/10.1016/j.agrformet.2021.108517. [47]

Hirst (2018), "Carbon Price Floor (CPF) and the price support mechanism". [61]

Hirst, D. (2018), *Carbon Price Floor (CPF) and the price support mechanism*, https://researchbriefings.files.parliament.uk/documents/SN05927/SN05927.pdf. [63]

IEA (2021), *Net Zero by 2050 - A Roadmap for the Global Energy Sector*. [31]

IEA (2020), *Extended world energy balances (database)*, https://doi.org/www.iea.org/statistics/topics/energybalances. [62]

INE (2023), *Regional Accounts of Spain. Series 2016-2021..* [43]

IPCC (2014), "IPCC, 2014: Climate Change 2014: Synthesis Report. Contribution of Working Groups I, II and III to the Fifth Assessment Report of the Intergovernmental Panel on Climate Change", *[Core Writing Team, R.K. Pachauri and L.A. Meyer (eds.)]. IPCC, Geneva, Switzerland, 151 pp*, https://www.ipcc.ch/site/assets/uploads/2018/02/SYR_AR5_FINAL_full.pdf. [1]

Junta de Andalucia (2019), "Basic Environmental Data of Andalusia 2019", http://www.juntadeandalucia.es/medioambiente/ddbb19. [44]

Juřík, R. and N. Braathen (2021), "Assessment of the air pollution tax and emission concentration limits in the Czech Republic", *OECD Environment Working Papers*, No. 174, OECD Publishing, Paris, https://doi.org/10.1787/172ad5b9-en. [36]

Kaufman, N. et al. (2020), "A near-term to net zero alternative to the social cost of carbon for setting carbon prices", *Nature Climate Change*, Vol. 10/11, pp. 1010-1014, https://doi.org/10.1038/s41558-020-0880-3. [30]

Lelieveld, J. et al. (2015), "The contribution of outdoor air pollution sources to premature mortality on a global scale", *Nature*, Vol. 525/7569, pp. 367-371, https://doi.org/10.1038/nature15371. [46]

Leroutier (2022), "Carbon Pricing and Power Sector Decarbonization: Evidence from the UK". [57]

Leroutier, M. (2022), "Carbon Pricing and Power Sector Decarbonisation: Evidence from the UK", *Journal of Environmental Economics and Management*, Vol. 111, https://doi.org/10.1016/j.jeem.2021.102580. [34]

Leroutier, M. and H. Ollivier (2022), "The Cost of Air Pollution for Workers and Firms", *Working Paper*, https://marionleroutier.github.io/files/LeroutierOllivier_2022_cost_pollution_jmp.pdf. [11]

Lobell, D., S. Di Tommaso and J. Burney (2022), "Globally ubiquitous negative effects of nitrogen dioxide on crop growth", *Science Advances*, Vol. 8/22, https://doi.org/10.1126/sciadv.abm9909. [14]

Mardones, C. and M. Cabello (2019), "Effectiveness of local air pollution and GHG taxes: The case of Chilean industrial sources", *Energy Economics*, Vol. 83, pp. 491-500, https://doi.org/10.1016/j.eneco.2019.08.007. [37]

Martin, R. (2013), "Carbon markets, carbon prices and innovation: Evidence from interviews with managers.", *Paper presented at the Annual Meetings of the American Economic Association, San Diego.*. [27]

Millock, K. and C. Nauges (2003), "The French Tax on Air Pollution: Some Preliminary Results on its Effectiveness", *SSRN Electronic Journal*, https://doi.org/10.2139/ssrn.419082. [42]

Ministry for the Environment and Ministry for Primary Industries (2022), *Pricing agricultural emissions: Consultation document.*. [51]

Mink, J. (2022), "Putting a price tag on air pollution: the social healthcare costs of air pollution in France", *Working Paper*. [13]

Mottershead, D. et al. (2021), *Green taxation and other economic instruments. Internalising environmental costs to make the polluter pay*, European Commission, Directorate-General for Environment, Brussels, https://environment.ec.europa.eu/publications/green-taxation-and-other-economic-instruments-internalising-environmental-costs-make-polluter-pay_en. [29]

Nordhaus, W. (2014), "Estimates of the Social Cost of Carbon: Concepts and Results from the DICE-2013R Model and Alternative Approaches", *Journal of the Association of Environmental and Resource Economists*, Vol. 1/1/2, pp. 273-312, https://doi.org/10.1086/676035. [32]

OECD (2022), *Air pollution effects* (indicator), https://doi.org/10.1787/573e3faf-en (accessed on 29 November 2022). [7]

OECD (2022), *Pricing Greenhouse Gas Emissions: Turning Climate Targets into Climate Action*, OECD Series on Carbon Pricing and Energy Taxation, OECD Publishing, Paris, https://doi.org/10.1787/e9778969-en. [25]

OECD (2021), *Effective Carbon Rates 2021: Pricing Carbon Emissions through Taxes and Emissions Trading*, OECD Publishing, Paris, https://doi.org/10.1787/0e8e24f5-en. [4]

OECD (2021), *Policies for a Carbon-Neutral Industry in the Netherlands*, OECD Publishing, Paris, https://doi.org/10.1787/6813bf38-en. [55]

OECD (2020), *Environment at a Glance 2020*, OECD Publishing, Paris, https://doi.org/10.1787/4ea7d35f-en. [15]

OECD (2019), *Taxing Energy Use 2019: Using Taxes for Climate Action*, OECD Publishing, Paris, https://doi.org/10.1787/058ca239-en. [5]

OECD (2016), *Effective Carbon Rates: Pricing CO2 through Taxes and Emissions Trading Systems*, OECD Publishing, Paris, https://doi.org/10.1787/9789264260115-en. [6]

P.R. Shukla, J. (ed.) (2019), *Annex I: Glossary*, In Press, https://www.ipcc.ch/site/assets/uploads/2019/11/11_Annex-I-Glossary.pdf. [8]

Pan, S. et al. (2022), "Addressing nitrogenous gases from croplands toward low-emission agriculture", *npj Climate and Atmospheric Science*, Vol. 5/1, https://doi.org/10.1038/s41612-022-00265-3. [49]

Pizarro (2019), *Lessons from the Carbon Tax in Chile*. [53]

Pizarro, R. (2019), *https://greenfiscalpolicy.org/blog/environmental-fiscal-reform-in-chile-a-way-forward-for-developing-countries/*. [39]

Portugal-Pereira, J. et al. (2018), "Interactions between global climate change strategies and local air pollution: lessons learnt from the expansion of the power sector in Brazil", *Climatic Change*, Vol. 148/1-2, pp. 293-309, https://doi.org/10.1007/s10584-018-2193-3. [19]

Shapiro, J. and R. Walker (2018), "Why Is Pollution from US Manufacturing Declining? The Roles of Environmental Regulation, Productivity, and Trade", *American Economic Review*, Vol. 108/12, pp. 3814-3854, https://doi.org/10.1257/aer.20151272. [38]

Spanish Ministry for Ecological Transition (2020), *Nota informativa sobre el Avance de Emisiones de Gases de Efecto Invernadero correspondientes al año 2020*. [2]

U.S. Environmental Protection Agency (2018), *Technical Assistance Document for the Reporting of Daily Air Quality – the Air Quality Index (AQI)*. [16]

U.S. EPA Center for Corporate Climate Leadership (2016), *Greenhouse Gas Inventory Guidance - Direct Emissions from Stationary Combustion Sources*. [3]

United Kingdom government (2022), *Participating in the UK ETS*. [60]

United Kingdom government (2016), *Carbon price floor*. [59]

United Kingdom government (2016), "Climate Change Levy rates". [58]

Zivin, J. and M. Neidell (2018), "Air pollution's hidden impacts", *Science*, Vol. 359/6371, pp. 39-40, https://doi.org/10.1126/science.aap7711. [9]

Annex 3.A. OECD Effective Carbon Rates: additional information

Annex Table 3.A.1. Sectoral decomposition in the OECD Effective Carbon Rates database

Sector	Definition
Road	All energy used in road transport.
Electricity	All fuels used to generate electricity for domestic use (rather than the amount of energy generated from each fuel). Note that fuels used in the auto-generation of electricity are classified under industrial production.
Industry	All energy used in industrial processes, in heating (incl. inside industrial installations) and in the transformation of energy, including fuels used for auto-generation of electricity in industrial installations.
Buildings	All energy used for commercial and residential heating.
Off-road	All energy used in off-road transport (incl. pipelines, rail transport, domestic aviation and maritime transport).
Agriculture & fisheries	Energy used in agriculture, fisheries and forestry. Energy used in on-road transport in this sector is included in the road transport sector.

Source: OECD (2016[6]).

Annex Table 3.A.2. Fuel category breakdown in the OECD Effective Carbon Rates database

Energy type	Fuel category	Energy Products
Fossil fuels	Coal and other solid fossil fuels	Anthracite; Bitumen; Bituminous coal; Brown coal briquettes; Oven coke; Coking coal; Gas coke; Lignite; Oil shale; Patent fuel; Peat; Peat products; Petroleum coke; Sub-bituminous coal
	Fuel oil	Fuel oil
	Diesel	Gas/diesel oil excluding biofuels
	Kerosene	Jet kerosene; Other kerosene
	Gasoline	Aviation gasoline; Jet gasoline; Motor gasoline
	LPG	Liquefied Petroleum Gas
	Natural gas	Natural gas
	Other fossil fuels	Additives; Blast furnace gas; Coal tar; Coke oven gas; Converter gas; Crude oil; Ethane; Gas works gas; Lubricants; Naphtha; Natural gas liquids; Other hydrocarbons; Other oil products; Paraffin waxes; Refinery feedstocks; Refinery gas; White and industrial spirit
Other combustible fuels	Non-renewable waste	Industrial waste; Non-renewable municipal waste
	Biofuels	Bio jet kerosene; Biodiesels; Biogases; Biogasoline; Charcoal; Municipal waste (renewable); Non-specified primary biofuels and waste; Other liquid biofuels; Primary solid biofuels

Note: Energy products are defined as in IEA (2020[52]), World Energy Statistics and Balances.
Source: OECD (2019[5]).

Annex 3.B. Detailed case studies: stationary sources

This section presents selected case studies in the domain of greenhouse gas emissions and air pollution across the world with a focus on the industry and electricity sector

Chile: Green Tax

Annex Table 3.B.1. Green tax (Chile)

Legal bases	Law 20.780 (2014)
Objective	To tax local air pollutant and GHG emissions from stationary sources generating thermal energy.
Level of responsibility	Central government (Chile)
Tax setter(s)	Central government (Chile)
Revenue beneficiary(ies)	Central government (Chile)
Tax payer(s)	Polluting industries generating thermal energy with power capacity greater than or equal to 50 MWt.
Tax base (including main exemption(s), credits or deductions)	Annual mass emissions in tonnes for CO_2, SO_2, PM and NOx classified according to the scale of their impact. The tax levied on the CO_2 component does not apply to emitting sources using biomass.
Tax rate(s) (including their calculation)	The tax calculation is different for SO_2, PM and NO_X as compared to CO_2, as the former are have a local negative impact, whereas the latter has a global impact. For PM, NOX, and SO_2, the tax is 0.1 per tonne emitted multiplied by the social cost of pollution, the local population, and an air quality coefficient using the formula: Tij= CCAji×CSCpci×Pobj. Where Tij: tax rate per tonne of pollutant "i" emitted in municipality "j" measured in USD/ton, CCAji: air quality coefficient in municipality "j" for pollutant "i", CSCpci: social cost of pollution per capita of pollutant "i", and Pobj: population of municipality "j". The air quality coefficient applies to zones declared saturated or latent for a particular pollutant. In the former case, the coefficient is 1.2; in the latter, it is 1.1. The social costs per capita of PM, SO_2, NO_X are presented in Annex Table 3.B.2. For CO_2, the tax rate is USD 5 (EUR 5.02) per tonne.

Governance and implementation	Multiple government bodies work together in the implementation of the tax: • The Ministry of Environment establishes the methodologies and systems to monitor, report, and verify emission, • The Revenue Service receives declarations from establishments subject to the tax, • The General Treasury receives the payments. The implementation of the tax system required creating a registry system, developing and designing the Monitoring, Reporting, and Verification System (MRV) by the Ministry of the Environment. In addition, it was also necessary to promote social acceptance of the tax with taxable entities and run capacity-building workshops to instruct and support them in using the emission reporting systems.
Environmental, social & health impacts	An assessment prepared for the Ministry of Environment found a reduction of 1.1% in CO_2 emissions, of 7% in particulate matter present in the air, of 2% in NOx emissions, and of 0.01% in SO2 emissions between 2017 and 2018.

Source: (García Bernal, 2018[52]; Pizarro, 2019[53]; Ainzúa et al., 2020[54])

Annex Table 3.B.2. Social costs of pollutants per capita

Pollutant	PM	SOx	NOx
Cost (USD/tonne)	0.9	0.01	0.025

Source: (Pizarro, 2019[53])

The Netherlands: National Carbon Levy for Industry

Annex Table 3.B.3. National carbon levy for industry (the Netherlands)

Legal bases	National Climate Agreement of 2020
Objective	To supplement existing climate policy instruments in order to achieve the carbon emission reduction target of 14.3 million tonnes in industry by 2030.
Level of responsibility	Central government (the Netherlands)
Tax setter(s)	Central government (the Netherlands)
Revenue beneficiary(ies)	Central government (the Netherlands)
Tax payer(s)	Installations that are part of the EU ETS, waste incineration plants and nitrous oxide installations
Tax base (including main exemption(s), credits or deductions)	The tax base is the emission of CO_2 measured in tonnes. This mechanism follows the logic of the EU ETS system, meaning that emissions above the baseline are taxed, while emissions below the baseline can be traded. The baseline is defined by "dispensation rights", which analogues the levy to free allocation. These rights are the product of the installation's output, the EU ETS benchmark emissions and an annual reduction factor (Annex Table 3.B.4) that decreases yearly. They can be traded via bilateral contracts between entities.
Tax rate(s) (including their calculation)	The carbon levy adds a floating contribution on top of the EU ETS allowance price to yield a fixed price floor per tonne of CO_2. The total levy represents the sum of the floating national part and of the EU ETS price. It started at EUR 30 per tonne in 2021 to rise gradually to EUR 125 per tonne in 2030 with an annual increase of EUR 10.56 per tonne of CO_2 (Annex Table 3.B.5).
Governance and implementation	The national carbon levy has been developed as part of the National Climate Agreement in order to achieve the objective of greenhouse gas emission reduction of 49% by 2030 compared to 1990 levels in the Netherlands. Several stakeholders have been involved to draft the Climate Agreement, chaired by the central government, among which industry, labour unions, subnational governments, non-for-profit organisations (NGOs).
Environmental, social & health impacts	The expected environmental impact is to achieve its target of 14.3 million tonnes in industry CO2 emissions by 2030.

Source: (OECD, 2021[55]; European Commission, 2021[56])

Annex Table 3.B.4. Reduction factor to define levy-free base

	2021	2022	2023	2024	2025	2026	2027	2028	2029	2030
Reduction factor	1.2	1.14	1.09	1.03	0.97	0.92	0.86	0.8	0.74	0.69

Source: (OECD, 2021[55]).

Annex Table 3.B.5. Statutory price trajectory of carbon levy in 2021 (EUR/t CO_2)

	2021	2022	2023	2024	2025	2026	2027	2028	2029	2030
Levy rate	30	40.56	51.12	61.68	72.24	82.8	93.36	103.92	114.48	125.04

Source: (OECD, 2021[55]).

The United Kingdom: Carbon Price Floor

Annex Table 3.B.6. Carbon price floor (the United Kingdom)

Legal bases	Finance Act 2011
Objective	The Carbon Price Floor (CPF) is a United Kingdom (UK) government's tool established to supplement the EU ETS (initially) and now the UK ETS and encourage low carbon investment.
Level of responsibility	Central government (the United Kingdom)
Tax setter(s)	Central government (the United Kingdom)
Revenue beneficiary(ies)	Central government (the United Kingdom)
Tax payer(s)	Owners of electricity generating stations or operators of combined heat and power stations
Tax base (including main exemption(s), credits or deductions)	The tax base of the CPF is tonnes of CO_2 and the tax base of the Carbon Price Support (CPS) depends on the fuel (natural gas in kWh, liquified petroleum gas or other gaseous hydrocarbons in a liquid state in kg, and coal and other solid fossil fuels in GJ on Gross Calorific Value). The only exemptions apply to generators that provide electricity supplies in emergency cases (i.e. when a building's usual power supply is cut) and generators with a rated thermal input smaller than 2 MWth.
Tax rate(s) (including their calculation)	The CPS, which is specific to the UK, tops up UK ETS (initially EU ETS) allowance prices to the CPF target. It applies to fuels used for electricity generation, as shown in Annex Table 3.B.6. The UK Treasury is responsible for setting CPS rates for the three following years and indicative rates for the next two years. The rates are calculated as follows: CPS rate = (CPF – market carbon price) * (emission factor of the fuel) The difference between the CPF target and market carbon price represents the 'carbon price support rates' per tonne of CO_2. In 2021, the CPF target was GBP 18 (EUR 29.9) per tonne of CO_2e.
Governance and implementation	The initial rate of the CPF was set at around GBP 5 per tCO2e (EUR 5.8). However, in 2014, the UK government decided to freeze the CPF rate to GBP 18 per tCO2e (EUR 29.9) until 2019-2020 after business representatives expressed concerns over the competitiveness of energy-intensive industries due to electricity generators passing on the tax cost.
Environmental, social & health impacts	The tax operated via three mechanisms: (i) a decrease in emissions at the intensive margin; (ii) the closure of some high-emission plants; and a (iii) higher probability of closure for plants already at risk due to European air quality regulations. Hirst (2018[27]) reported that coal electricity generation significantly decreased between 2013 and 2016, together with the closure of several coal stations. He stressed that the doubling of the CPF in 2015 from GBP9 to GBP18 is one of the main factors that accelerated the decline in 2016. Leroutier (2022[57]) also found that emissions from the UK power sector declined by 20 to 26% per year on average between 2013 and 2017.

Source: (United Kingdom government, 2016[58]; United Kingdom government, 2016[59]; United Kingdom government, 2022[60]; Hirst, 2018[61]; Leroutier, 2022[57])

Annex Table 3.B.7. Tax rates

CPS rate commodity	Gas	Petroleum gas or other gaseous hydrocarbon in a liquid state	Coal and other solid fossil fuels
Unit	GBP (EUR) per kilowatt hour (kWh)	GBP (EUR) per kilogram (kg)	GBP (EUR) per gigajoule (GJ) on gross calorific value (GCV)
1 April 2016 to 31 March 2025	0.00331 (0.00384)	0.05280 (0.06131)	1.54790 (1.79738)

Source: (United Kingdom government, 2016[58]; United Kingdom government, 2022[60])

Notes

[1] Cap-and-trade mechanisms should be understood as such.

[2] Non-stationary sources refer to vehicles, which are covered in Section 0.

[3] A gas's radiative forcing can be understood as "the ability of a gas to absorb energy" and its lifetime as "how long they stay in the atmosphere", see https://www.epa.gov/ghgemissions/understanding-global-warming-potentials, as accessed on 12 May 2022.

[4] This period of 100 years is the most standard, but GWPs also exist for, e.g., 20 years.

[5] Calorific factors from the IEA World Energy Statistics and Balances (IEA, 2020[62]) enable common units of fuels (e.g., kilograms for solid fuels, litres for liquid fuels, cubic metres for gaseous fuels) to be converted into energy units (e.g. GJ). In turn, these can then be converted into CO_2 emissions using the IPCC emissions conversion factors (Intergovernmental Panel on Climate Change's Guidelines for National Greenhouse Gas Inventories (2019[8]), volume 2).

[6] PMs are microscopic particles of solid or liquid matter suspended in the air. Some particles, such as dust, dirt, soot, or smoke, are sufficiently large or dark to be seen by eye. Others, such as PM_{10} or $PM_{2.5}$ are not as visible. PM_{10} (resp. $PM_{2.5}$) represent inhalable particles, with diameters that are generally 10 (2.5) micrometres and smaller.

[7] https://www.who.int/news/item/25-03-2014-7-million-premature-deaths-annually-linked-to-air-pollution, as accessed on 29 November 2022.

[8] See https://www.airnow.gov/aqi/aqi-basics/ (as accessed on 25 Januray 2023) for additional detail. In particular, "Good" stands for "Air quality is satisfactory, and air pollution poses little or no risk", "Unhealthy" for "Some members of the general public may experience health effects; members of sensitive groups may experience more serious health effects" and "Hazardous" for "Health warning of emergency conditions: everyone is more likely to be affected".

[9] These figures are similar for PM_{10}.

[10] https://ec.europa.eu/environment/integration/research/newsalert/pdf/24si_en.pdf, as accessed on 12 May 2022.

[11] As defined in the OECD effective carbon rates methodology (OECD, 2016[6]).

[12] See BOE (2004[20]) or https://www.boe.es/buscar/act.php?id=BOE-A-2004-1739&p=20100809&tn=1&se-9, section 2, for additional detail. The webpage provides the possibility to access latest modifications to the legislation.

[13] Direct GHG emissions are emissions from sources that are owned or controlled by the reporting entity. Indirect GHG emissions are emissions that are a consequence of the activities of the reporting entity, but occur at sources owned or controlled by another entity (https://ghgprotocol.org/calculationg-tools-faq).

[14] Annex 1 of Law 16/2002 of July 1 provides a list of the fourteen activities covered. These refer to certain combustion installations, production and transformation of metals, mineral industries, chemical industries, waste management, industry derived from wood, textile industry, leather industry, agri-food industry and livestock farms, organic solvents, carbon industry, wood preservation industry, water treatment and capture of CO_2.

[15] Article 15 of BOE (2004[20]).

[16] In 2018, it applied to 77 installations, and in 2020, to 66 installations.

[17] Excluding CO_2 emissions from the combustion of biofuels. This figure is of about 65% if including these emissions.

[18] Andalusian firms subject to the EU ETS had average verified emissions of about 231 thousand tonnes of CO_2, ranging between less than 10 and about 1.7 million tonnes. At the Spanish level, the average is at 139 thousand tonnes, ranging between 0 and about 5 million tonnes.

[19] Information provided by the Tax Agency of Andalusia (ATRIAN).

[20] https://www.hacienda.gob.es/es-ES/Prensa/En%20Portada/2014/Documents/Informe%20expertos.pdf.

[21] E.g., more effective in mitigating emissions.

[22] E.g., if a justification were made that larger polluters should be made responsible for proportionally more of their emissions.

[23] It also applies to emissions from aviation and to a very small share of emissions in the buildings sector, but this is not discussed here.

[24] Other greenhouse gases are excluded from this analysis, as they constitute a minor part of emissions in these two sectors.

[25] I.e., it does not include auto-generation of electricity in industrial plants.

[26] https://tradingeconomics.com/commodity/carbon, as viewed on 21/07/2022. It is also worth noting that the price signal arising from the EU ETS in 2018 was much lower, at an average of EUR 16/tCO_2.

[27] Indeed, while not all biomass is carbon neutral, it can be. Taken at the point of combustion, biomass releases CO_2. However, as discussed in OECD (2019[6]), sustainably sourced biomass may be carbon-neutral over the lifecycle because before being burnt, feedstocks have previously absorbed a similar amount of CO_2 from the atmosphere.

[28] See https://ember-climate.org/data/data-tools/carbon-price-viewer/, as accessed on 28 November 2022.

[29] According to Table 3.1, the highest marginal rate is of 14 000 per unit of pollutant. Hence, at that marginal rate, one extra tonne of CO_2 is equivalent to 1/200 000 polluting unit and is hence subject to a marginal rate of 14 000/200 000.

[30] Such effects could also help go beyond the use of best available technologies promoted in Andalusia (https://eippcb.jrc.ec.europa.eu/es/reference). Indeed, the extensive margin effect would not be at play in the context of technology requirements only.

[31] Agriculture made up 2.9% of Spanish GVA in 2021.

[32] "A Major Source of Air Pollution: Farms – Global Study Shows How Agriculture Interacts with Industry", https://www.earth.columbia.edu/articles/view/3281, as accessed on 29 November 2022.

[33] See https://environment.govt.nz/news/consultaton-on-government-proposals-to-price-agricultural-greenhouse-gas-emissions/, as accessed on 08 November 2022.

[34] See https://www.reuters.com/world/asia-pacific/new-zealand-farmers-protest-agricultural-emissions-plan-2022-10-20/, as accessed on 08 November 2022.

[35] https://www.overseer.org.nz/, as accessed on 08 November 2022.

[36] https://www.bloomberg.com/news/articles/2022-10-10/new-zealand-accepts-farm-level-pricing-of-agricultural-emissions, as accessed on 30 November 2022.

[37] See https://www.realinstitutoelcano.org/en/work-document/taxation-and-ecological-transition-during-climate-and-energy-crises/ for a summary in English of the environmentally-related recommendations of the White Book.

4 Assessment: Road transport

Driving is a major source of carbon emissions, local air pollutants, congestion and noise and can contribute to road accidents. Particular in urban areas, congestion and air pollution are of key concern. Such damage on the environment and on society – commonly referred to as *external costs* – are typically not reflected in an individual's or a firm's decision to own and drive a car. Tax policy can contribute to reflecting the full costs from environmental damage in drivers' and vehicle owners' decision making process by relating a price to it, thereby improving transport and environmental outcomes.

This section assesses how taxes that apply to road transport in Andalusia (passenger vehicles) may better align with environmental and tax policy objectives, such as the management of external costs. The focus of the project is on GHG emissions and air pollution from passenger vehicles. Other external costs (e.g. accidents, congestion and noise) will also be discussed.

The section provides strategic reform suggestions for better management of external costs. It also proposes and discusses an alternative view of environmental tax reform, namely reform that supports reaching relevant policy objectives, such as Andalusia's objective to reduce GHG emissions by 30-43% in 2030 (PAAC 2021) or the nation goals to reach a zero GHG emissions fleet of passenger cars by 2050 and air pollution reduction of 62% for of NOx and 50% for PM by 2030 (compared to 2005) or the reduction in the consumption of petroleum derivatives in transport of at 30% by 2030 (compared to 2019) expressed in the Andalusia Energy Strategy 2030.

Section 4.1 sets the scene, Section 4.2 discusses the Tax policy objectives in road transport. Section 4.3 introduces the main categories of external costs and estimated magnitudes. Section 4.4 lines out the principles of sound environmental tax policy. Section 4.5 assesses the alignment of Andalusia's tax framework with sound environmental tax principles and lays out strategic reform options.

4.1. Setting the scene

While CO_2 emissions in Andalusia's road transport sector increased by 12% over 2011-2019, air pollutant emissions have followed a downward trend. Road transport overall is responsible for 27.6% of Andalusia's total CO_2 in 2019 according to the Junta de Andalucia's Institute of Statistics and Cartography (Junta de Andalucia, 2022[1]); a share that is rather stable over time (28.2% in 2011). Andalusia's NOx emissions in road transport continuously decreased from 50 960 tonnes in 2011 to 34 671 tonnes in 2019 accounting for 12.6% of Andalusia's total NOx emissions in 2019 (34.7% in 2011). Road transport was also a major source of fine particles in Andalusia, accounting for 22.7% and 21.5% of Andalusia's total PM2.5 and PM10 emissions in 2011 but decreased towards 2.7% and 3.2% in 2019. While CO_2 is a global pollutant contributing to global warming with impacts across the world (see discussion in Section 3.1), NOx and fine particles contribute to air pollution with repercussions at a local level. Although pollution from vehicles has been decreasing, it remains an important source of pollution which affects population, especially in urban environments.

In 2021, Andalusia counted 4,277,106 registered passenger vehicles ("Tourismo") of which 61% were diesel and 38% gasoline cars. The remaining 1% used other energy types, including electricity (0.12% in total) (Junta de Andalucia, 2022[1]). The emissions profiles of diesel and gasoline cars differ. While gasoline cars emit relatively more CO_2 and CO per km driven, diesel cars emit relatively more NOx, black smoke and particulates (Crawford and Smith, 1995[2]). Often diesel cars drive more and risk to emit more CO_2 in total compared to gasoline-driven cars. Due to the difference in pollution profiles, it is not straightforward to assess the relative total damage of gasoline vs diesel cars. Other car characteristics, such as weight and engine size, are to be taken into consideration too when assessing pollution outcomes, as heavier cars and cars with larger engines typically emit more per kilometre driven.

Vehicle turnover has a significant impact on how quickly new technology penetrates the fleet. The average lifespan of a car varies, and it is not straightforward to understand how quickly polluting cars may be replaced by clean ones – in particular relatively young cars. According to Held et al. (2021[3]), the mean lifespan of a car settles at 18 years in Western European countries. In Andalusia, 36% of all personal vehicles are younger than 10 years in 2021, and 21% even younger than 5 years, while 45% are older than 15 years.

Emissions in road transport do not only relate to passenger cars that are registered in the region. There also is drive-through traffic, tourist and company cars. Beyond passenger car use, there is traffic from other vehicle types, like heavy and light duty vehicles, motorcycles, etc. Looking at the Andalusia vehicle dataset 2021, registered vehicles other than passenger cars ("tourism") include trucks, trailers and semi-trailers (537 228, i.e. 8% of all registered vehicles), vans (436 002, i.e. 6.5%), motorcycles and mopeds (1 294 529, i.e. 19.4%) and other vehicles, including buses and industrial tractors (130 770, i.e. 2%).

4.2. Tax policy objectives in road transport

Tax policy objectives in road transport may be multiple. They can relate to raising revenues, increase tax system efficiency including through a better management of external costs, to manage distributional consequences or contribute to specific policy goals such as driving a transition to net-zero emissions or achieving a certain share of alternative fuel vehicles in the vehicle fleet, for example. While some of these policy objectives may reinforce each other, others may be less well aligned with each other. For example, when tax policy is used to create incentives to expedite the transition to net zero, it comes with budgetary impacts in the short and long run, either because the instrument provides for forgone revenue (e.g. CIT and PIT deductions for electric vehicles) or because they explicitly aim to erode the tax base by reducing carbon-based fuel use.

Different tax instruments and designs may be better suited to address specific objectives. For example, a carbon tax may be a good tax instrument to internalise the external costs related to emitting carbon emissions, while raising revenue in the short to medium term. However, given recent elasticity estimates, it may not be able to drive the net-zero transition by 2050 if applied in isolation.

The present analysis focuses on the project's objective to increase tax system efficiency through a better management of external cost. External costs from air pollution and GHG emissions are of particular interest to the Andalusia authorities. The analysis discusses how external costs are managed in the current tax policy framework of Andalusia and ways to improve it. Besides this focus, the analysis encompasses a discussion of other policy considerations where necessary (e.g. distributional consequences, revenues) and sets the external cost focus in context of other pressing policy goals. Focusing purely on the discussion of external cost pricing may be too narrow at a time where the transition to net-zero calls for a full review of tax (and non-tax) policy instruments in view of supporting the transition of the energy base towards clean fuels and mobility options.

In Andalusia, policy objectives on climate change relating to the road transport sector are clear and enshrined in specific targets. The region has taken a leading role in this area being the first autonomous region in Spain to develop a regional Climate Change Strategy in 2002. The most recent Andalusia Climate Action Plan (PAAC 2021) aims for a 30-43% reduction of GHG emissions by 2030 (compared to 2008 levels) in the transport and mobility sector. The PAAC is established in the context of the Spanish Climate Change and Energy Transition law, which sets the objective of reaching a zero GHG emissions fleet of passenger cars and light commercial vehicles by 2050, with the central government, the autonomous communities and local governments asked to contribute within the scope of their responsibilities. Spain also committed to put a halt on the sales of new passenger cars and light commercial vehicles that emit GHG emissions after 2040. To support autonomous communities and cities to promote electric mobility, the central government launched the Moves III Plan, which provides funding to encourage the purchase of electric vehicles and to finance the deployment of charging facilities. Prior to the Moves III Plan, the government launched the Moves I and Moves II Plans managed by the Andalusian Energy Agency.

On air pollution, no explicit regional targets exist, but the Andalusia Strategy for Air Quality builds a framework to facilitate the preparation of air quality improvement plans by local governments in Andalusia. Reduction targets exist at the national level, including a reduction commitment for NOx at 62% and for PM 2.5 at 50% by 2030 (compared to 2005). For more details on policy objectives and specific reduction targets in the area of climate change and air pollution see Section I.2).

4.3. External costs in road transport

Optimal tax theory predicts that an efficient tax system would tax behaviours that generate external costs. In the context of road transport, fuel and vehicle use as well as driving would ideally be taxed at a rate that reflects external costs induced by these behaviours. Currently, the full range of external costs from driving are under-priced in many countries (Van Dender, 2019[4]). This includes estimates of external costs which relate to using fossil fuel technologies during driving (e.g. GHG emissions, air pollution and noise), but also those costs that are unrelated to fossil fuel use and would remain present even under a fully decarbonised vehicle fleet (in particular, accidents, congestion, road damage, noise, use of public space and reduced mobility for non-drivers). Figure 4.1 summarises the different external costs in road transport and their impact categories, such as impacts on human health, ecosystems, material and building damages, etc. Section 4.4 discusses in detail the suitability of different tax types to cover these costs.

Figure 4.1. Main external costs in road transport relate to fossil fuel use, vehicle use and driving

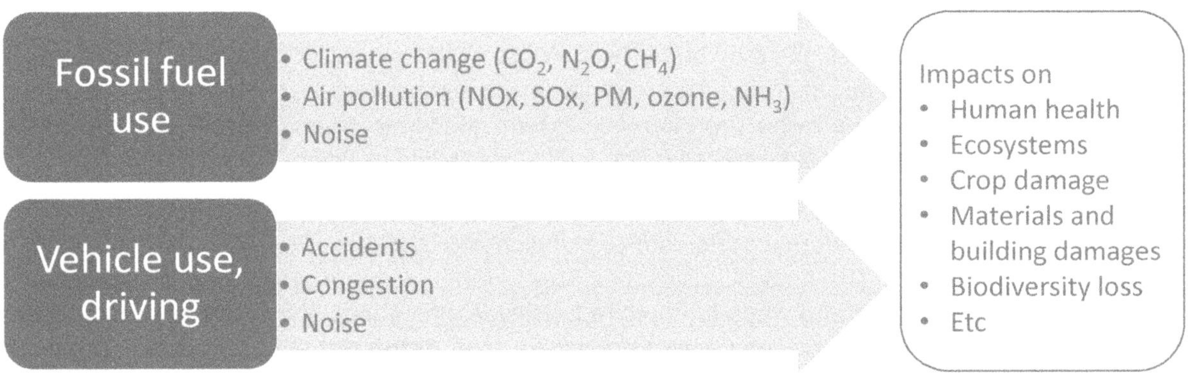

Source: Own representation based on Mottershead et al . (2021[5]) and Van Dender (2019[4]).

External costs in road transport can be substantial but are uncertain and often vary according to time and location of driving. Van Dender (2019[4]) presents and discusses the main external costs related to road transport and compares literature estimates of such costs expressed by kilometre. The study shows ranges of estimates for those costs in the European Union. Due to the large uncertainty on the estimates, it does not depict single numbers (Figure 4.2).

Figure 4.2. External cost estimates for the use of passenger cars

Summary of estimated ranges of external costs of passenger cars in the European Union (in Eurocents per vkm)

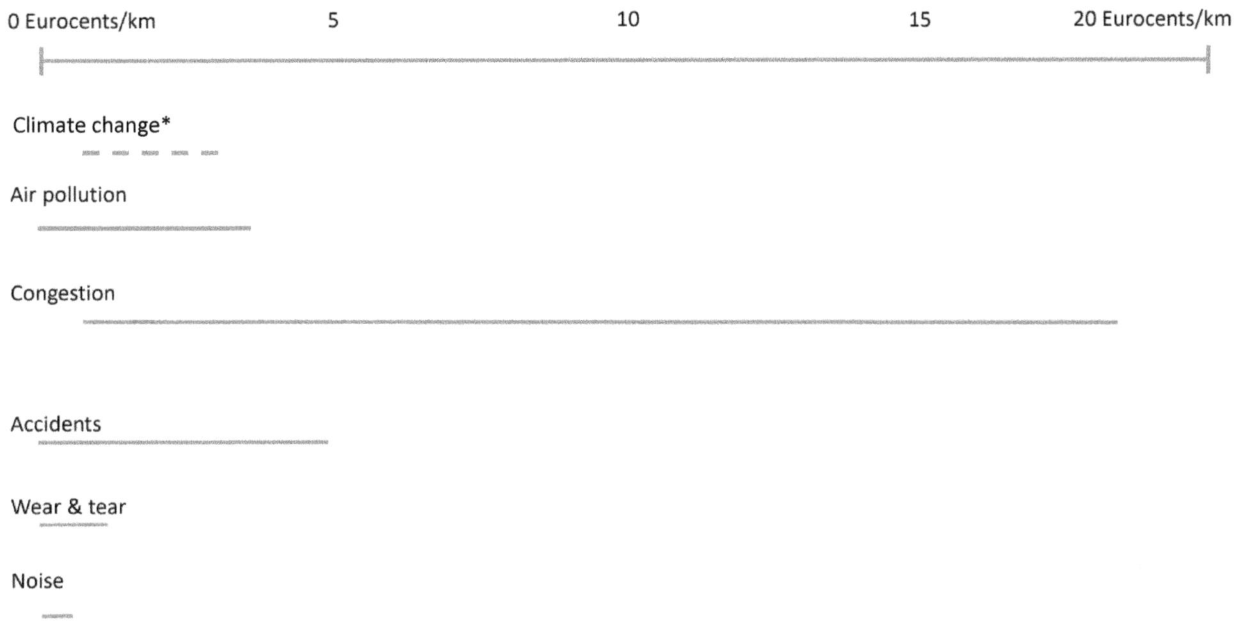

Note: Approaches to value external costs are reviewed, for example, by Mottershead et al. (2021[5]).
* For illustrative purpose only. The external cost related to climate change is proportional to fuel consumption so is more naturally expressed per litre. The range shown covers fuel consumption between 5 and 10 litre per 100 km and a social cost of carbon of EUR 60 per tonne of CO_2.
Source: Van Dender (2019[4]).

External cost estimates from driving in the European Union vary from zero to 20 Euro-cent per vehicle-kilometre (vkm) driven as indicated in the first line of Figure 4.2, which highlights the dominance of congestion costs. The high end of costs estimates is relevant to congestion cost only, while other related costs range from zero to five cents per vehicle-kilometre. For illustrative purpose, the figure includes external costs from CO_2 emissions to allow a comparison of orders of magnitudes of different cost categories only. The external cost related to climate change is proportional to fuel consumption so is more naturally expressed per litre and not in vkm.

External cost estimates are driven by assumptions on the level of congestion, on the amount of carbon and number and severity of accidents, the location of driving (urban, rural), fuel use (gasoline vs diesel) and interaction of these. For example, climate and congestion costs dominate for gasoline cars, whereas air pollution costs are more than twice as large for diesel than for gasoline cars. External costs related to air pollution and congestion vary widely with the location and time of driving (Box 4.1). Air pollution cost depend on population exposure which is different at different locations and time of day as well as its social and demographic characteristics (with children, the elderly and people with previous pathologies being particularly vulnerable). Impacts of air pollution on health accumulate in the long term and are less affected by temporary circumstances. Congestion cost (i.e. time loss) depend on levels of congestion that are different in urban and rural areas and at different times of the day. This differs from climate change which increases temperatures at a global level.

Box 4.1. Health impacts from air pollution related to driving: time and place matters

The health effects deriving from transport-related air pollution is complex and a key concern. WHO (2005[6]) provides an in-depth review of the channels and extent to which air pollution from road transport affects health. For example, road transport was estimated to cause about 70 000 premature deaths in the European Union in 2018 (European Commission, 2022[8]). Additional details are provided in Box 3.2.

Complex pattern of pollution concentration

In the Andalusia context, Adame et al. (2014[7]) evaluate air pollutant concentration (O_3, NO and NO_2, CO and PM_{10}) during 2003-2008. The study does not focus on transport-related pollutants specifically but cites traffic as one of the main sources of pollution in the region (together with industrial activities and biogenic emissions). Pollution concentration displays strong variation according to key dimensions, such as the area type where emissions are measured (urban, suburban including in industrial areas, and rural), the time of day and the day in the week.

Urban areas that are characterised by high traffic intensity and located mainly in the city centres report consistently higher CO, NO and NO_2 values (Figure 4.3). PM 10 concentration are typically high in urban centres too, in particular on weekdays. Only for O_3, concentration in urban areas appears to be lower than in all other area types.

The study also shows that pollution concentration varies according to the time of day. Focusing again on the traffic-heavy urban area category, CO, NO and NO_2 concentrations peak in the morning (9:00-11:00) and the evening hours (21:00-23:00), whereas the peak for O_3 arises between 15:00-20:00. PM10 displays a more evenly concentration over the day. Similar patterns, but of different amplitude and levels, can be observed for the other geographic areas (Figure 4.4).

Figure 4.3. Concentration of main pollutants, by area type and weekday

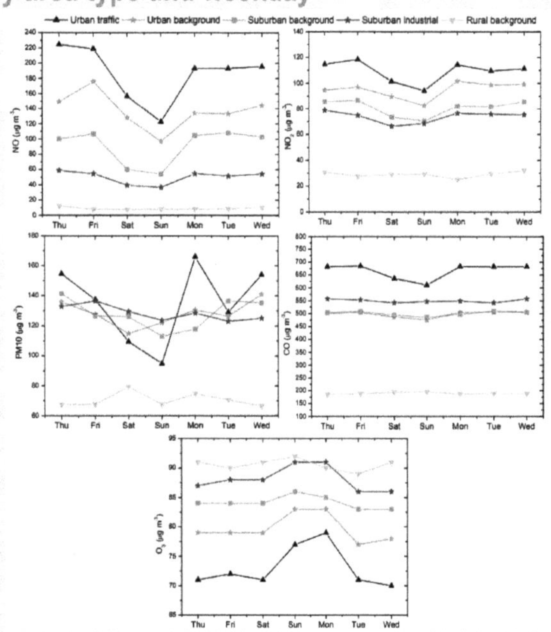

Source: Adame et al. (2014[7]).

Figure 4.4. Daily pollution concentration across area types, by air pollutant and time of day

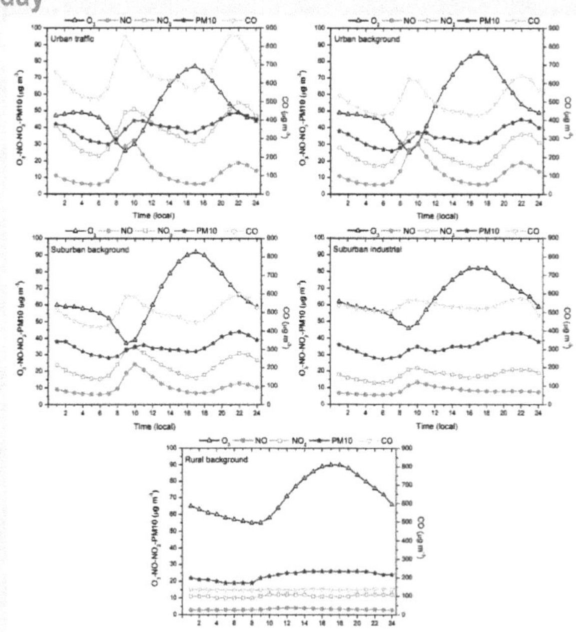

Source: Adame et al. (2014[7]).

Insights for Andalusia

When considering the external costs of air pollution concentration through health or ecosystem impacts, these numbers would need to be crossed with population and ecosystem exposure in the different geographic areas and estimates of their impact on health. Because urban areas are relatively dense, total external costs relating to the pollution profiles discussed in Adame et al. (2014[7]) are expected to be relatively higher in urban areas.

The complexities in air pollution concentration shown in the figures are not straightforward to reflect in a simple vehicle tax or an air pollution tax. A combination of instruments may reach superior results. For example, regulating cars (e.g. via air pollution standards that increase in stringency over the years, such as the Euro standards) together with well-designed congestion charges (i.e. that charge vehicles for circulating in particular areas or cities) or distance-based pricing that reflect pollution concentration, population exposure and their time effects, may be preferable.

Overall, the study shows that air pollution has very local effects which may call for policy action at the local level, where opportunities exist to follow the spatial variation of pollution and their impacts relatively better. For example, the local governance level may be well-equipped to implement congestion charges in areas where effects are likely most important, by combining information of density in different cities with traffic and air pollution data.

A recent study by the European Commission's Directorate-General Environment (Mottershead et al., 2021[5]) presents external *environmental* cost estimates in different countries per unit of emission based on literature. For driving the study focuses on costs related to air pollution and climate change. It also includes estimates of *internalisation rates* that compare the revenues collected from taxes or other economic instruments to their estimated external costs. For the household sector, which includes revenues and costs related to driving passenger vehicles, the study reports an EU-wide internalisation rate above 100%. Such levels could be interpreted to indicate that the external *environmental* costs from driving are fully covered by taxation and the polluter pays principle is met. However, certain caveats apply when interpreting this number as the actual internalisation rate may be substantially lower:

The authors of the study note that the internalisation rate as reported in Mottershead et al., (2021[5]) likely represents an over-estimate of external cost coverage in the case of road transport for several reasons. First, the study includes *external* costs related to GHG emissions and air pollution but does not cover other driving-related external costs from accidents, congestion, and noise, which can have important impacts. Second, cost estimates for air pollution are conservative because they only cover part of all toxic substances. For GHG emissions cost estimates are uncertain and can be much higher compared to those used in the comparison. Third, revenues for infrastructure drive up the revenue estimates in road transport (and thus put upward pressure on the internalisation rate) although they do not provide a signal that pollution is costly to economic actors. Finally, the EU-wide numbers may hide important differences across EU member states, regions, and cities.

The discussion in this section shows that focussing the project on "taxes for gas emissions from road transport" is narrow as it only covers part of the external costs related to road transport. External costs can be substantial but are uncertain and often vary according to time and the location of driving. Finally, action at the subnational level may be justified to cover air pollution and congestion-related external costs, because the local level may be better equipped to monitor the necessary variation in traffic and air pollution data across different cities.

4.4. Principles of sound environmental tax policy

If well-designed, taxation can signal drivers the actual costs of their behaviour and incentivise them to reduce harmful effects on society. This section summarises the different tax-policy instruments available in road transport, how they relate to the three main tax bases in the sector: energy use, vehicle stock and road use, and how they currently apply to drivers in Andalusia (Section 4.4.1). It then discusses to what extent different tax types are suitable to cover driving-related external costs (Section 4.4.2). The section closes by mentioning additional aspects of a sustainable tax policy strategy in road transport, taking behavioural effects, revenue stability and distributional consequences into account (Section 4.4.3).[1]

4.4.1. The tax framework in road transport and applications to Andalusia

A driver engaged in road transport, may be liable to pay taxes or fees on three tax bases: energy (*E*), vehicle stock (*V*) and road use (*M*), which can be summarised in the following country-wide revenue function:

$$\begin{aligned} Tax\ revenue_{\ c} &= R_E + R_V + R_M \\ &= \sum\nolimits_i (tax_i * Joule_i) + \sum\nolimits_j (tax_j * nb\ vehicles_j) + \sum\nolimits_k (tax_k * km\ driven_k) \end{aligned}$$

Revenue from energy (R_E) relates to energy purchases in country c. It includes all energy types, i, such as gasoline, diesel and electricity, used in road transport. Currently, energy use in road transport derives predominantly from fossil fuels, but may shift towards alternative fuels following technological advances, falling prices for alternative fuel vehicles and stringent climate or air pollution policies. The energy base comprises fuel that is bought (and taxed) within the country, even if combusted abroad. Fuel that is bought outside the country will not fall in the tax base of country c, even when combusted there. Countries tax energy in road transport generally via excise duties per litre of fuel or specific taxes on the carbon content of the fuel. Sometimes emissions from road transport and electricity production are instead, or additionally, covered by an emissions trading system. (OECD, 2022[8]) On top of these excise duties VAT is typically payable.

A driver in Andalusia, pays excise duty on fuels at the national level. Spain levies their hydrocarbon tax within the framework of the EU Energy Tax Directive (2003/96/EC), which sets minimum rates for the taxation of energy products in member states of the EU. For gasoline, Spain applies an excise duty of EUR 0.5078 per litre. For diesel, the excise duty amounts to EUR 0.379 per litre. Electricity use is subject to an ad-valorem tax (5.1%, which was reduced to 0.5% in 2022) that is based on the taxable amount established for the purpose of VAT. VAT on fuels is levied at the full rate of 21%.[2]

Revenue from motor vehicles, R_V, typically relate to all vehicles, j, that are registered in country c. Tax rates typically depend on a combination of specific vehicle characteristics, for example, a vehicle's type (i.e., whether the vehicle is a bus, passenger car, truck, motorcycle), engine power, weight, type of fuel used, whether the vehicle is used for commercial or personal purpose, or according to the environmental performance of the vehicle. Countries tax motor vehicles, for example, via one-off registration or sales taxes and via recurrent taxes on vehicle use or ownership. These taxes usually take the form of specific taxes or ad valorem taxes on the price (OECD, 2022[9]).

Drivers in Andalusia are liable for the national registration tax when they first register a motorised vehicle in Spain, based on the car price. The tax rate is a function of CO_2 emissions with exemptions applying, for example, to industrial, commercial and agricultural two-seat vehicles and rental cars. Rates vary from 0% for vehicles emitting up to 120 g CO_2 per km, to 14.75% for vehicles emitting 200 g CO_2 per km and more (OECD, 2022[9]). Since 2002, tax revenue collected through the vehicle tax is distributed to the autonomous communities. In addition, Andalusian drivers are liable annually for the municipality-specific circulation tax that remains regulated by the central government. Tax rates depend on multiple criteria, such as the vehicle category, the engine power, and the number of seats, but do not consider environmental aspects explicitly.

Revenue can also be derived from road use (R_M). Country c may tax the number of kilometres driven by driving type, k. The type of driving can relate to specific vehicle characteristics, specific roads (tolled vs non-tolled), a specific time of the day, or can depend on an area's population exposure or congestion level. Typical road-pricing systems take the form of distance charges (e.g. motorway or city tolls) or congestion charges. Alternatively, taxation can take an access charge approach, for example in form of a vignette or some types of congestion charging (e.g. cordon fees). Such systems require fees to be paid to access the public road network for a specific period of time but have no direct link to the amount of kilometres driven.

Drivers in Andalusia currently do not face road tolls, although pricing driving on national roads as of 2024 was announced in 2021 as part of the Spanish recovery plan under the European Recovery and Resilience Facility (see Section 2).

The three tax bases in road transport are connected. For example, driving a given distance with a specific vehicle technology determines the amount of energy that a vehicle uses; e.g. driving 100 km with an efficient internal combustion engine requires less energy than driving the same amount of kilometres with an inefficient engine. In this context, improving the fuel-efficiency of internal combustion engines will reduce the amount of fuel used to drive a given distance everything else equal. Technology shifts towards electric vehicles will affect the vehicle and energy tax base. Finally, tax reform in one of the three areas will necessarily affect the other tax bases. For example, increasing fuel taxes will have the immediate effect of increasing tax revenues via the impact on R_E. However, drivers may reduce fuel consumption as a consequence of higher fuel taxes, either by driving less or by driving more efficiently (e.g. through shifting towards more fuel-efficient vehicles or alternative fuel vehicles) or both. Taking the public transport or sharing private vehicles can contribute to less driving. (OECD/ITF, 2019[10]) develops scenario analysis for the penetration of alternative fuel vehicles and models the effects of tax reform on CO_2 emissions and revenues for the case of road transport in Slovenia.

The tax framework discussed in this section does not apply in isolation but operates in the context of other tax and non-tax instruments that likely impact the different tax bases. Important regulatory elements include the European CO_2 emission performance standards for new passenger cars and light commercial vehicles (Regulation (EU) 2019/631). The European Parliament and Council recently agreed (October 2022) to further increase the stringency of these standards and to include a zero-emission target for all new vans and cars sold in the EU in 2035 (European Parliament, 2022[13]). Complementary to the regulation of CO_2 emissions from cars and vans, is the regulation of air pollution via EURO standards. In November 2022, the European Commission presented a proposal of a new Euro 7 standard to further reduce air pollution from vehicles and to improve air quality (European Commission, 2022[11]).

Finally, other tax and non-tax policy provisions may affect decisions of vehicle use and driving. For example, the personal tax treatment of company cars may encourage increased vehicle ownership, more driving and the use of larger and heavier vehicles (Crawford and Smith, 1995[2]). This risks to be the case, when employee compensation related to company cars used for private purposes is taxed more lightly than cash wages and was estimated for Spain to result in significant revenue forgone and environmental costs (Harding, 2014[11]).

4.4.2. What tax types are suited to cover different external costs?

Some tax types account better for external costs than others (Figure 4.5), which should be considered in the tax instrument choice beyond political and administrative feasibility (Van Dender, 2019[4]; OECD/ITF, 2019[10]). Some of the external costs are driven by characteristics of vehicles and fuel types, others relate to the location of driving, the time of day, or the intensity of their harmful effects through population and ecosystem exposure. None of the traditional tax bases alone (fuels or vehicles) may be enough to reflect the full estimate of costs (Crawford and Smith, 1995[2]). Including other tax bases, such as distances driven, into the policy consideration may be useful to cover external costs more comprehensively, as well as using alternative policy instruments, such as subsidies, fuel and emissions standards (Crawford and Smith, 1995[2]).

This section first shows that some tax types are better suited to fit specific tax bases in general, and then discusses the Andalusia case specifically. Additional aspects to be considered for a sustainable tax policy in road transport include behavioural reactions, revenue stability and distributional consequences and will be discussed in Section 4.4.3.

Figure 4.5. Some tax types account better for specific external costs than others

	Fuel or carbon tax	Vehicle tax	Distance-based charges
Long-run revenue stability	☹	☺	☺
External cost management			
• CO_2 emissions	☺	😐	☺
• Air pollution	😐	😐	☺
• Driving-related external costs (e.g. accidents, congestion, noise and air pollution exposure, road damage, use of public space)	☹	☹	☺
Administrative and implementation costs	☺	☺	☹

Source: Adapted from OECD/ITF (2019[10]).

Fuel and carbon taxes are a very well suited to account for the external costs related to CO_2 emissions, because CO_2 emissions are proportional to fuel consumption and can be reflected in the tax rate applicable to the consumption of the fuel. Fuel taxes account less well for air pollution and congestion costs, which depend heavily on aspects that cannot be covered by the fuel tax: vehicle technology (e.g. engine type), driving behaviour (e.g. acceleration, risk taking), the specific driving location and pollution exposure that varies across geographic areas – although some of these dimensions can influence air pollutant levels through variation in fuel consumption and thus the tax may relate to health impacts.

Fuel excise and carbon taxes are relatively easy to administer, as the number of fuel producers or importers is low. Compliance costs for taxpayers are usually low too, in particular for passenger vehicles. Compliance costs increase when fuel used for commercial purposes obtains benefits in form of reduced rates, as it requires truck companies to either file refund claims in all countries where fuel was purchased or to adjust tax returns and respond to audits.

Vehicle taxes on the other hand are less efficient in targeting external costs in road transport, particularly driving-related external costs, such as accidents, congestion, noise and air pollution or road damage. A reason for this is that they can only account for average vehicle characteristics (e.g. average level of pollutant emissions per km) but not the externalities related to driving behaviour, the amount and the place of driving, which affects pollution levels and exposure. Despite their lack of covering driving-related costs, vehicle taxes can reflect the average emissions profile of a vehicle. The vehicle purchase tax in Israel, for example, accounts for a large set of emissions, including CO_2, PM and NOx (Box 4.2).

Box 4.2. Israel's vehicle purchase tax

In 2009, Israel's Green Tax Reform modified the existing vehicle purchase tax to better align with external costs related to climate change and air pollution. Cars were associated to one of 15 categories each with a specific tax rate, ranging from category 1 (no emissions) to category 15 (the most polluting cars). The maximum tax rate was set for category 15 at 83% of the purchase price. Lower rates applied to cars in the other categories.

The tax rates for categories 1-14 were calculated by deducting a certain amount from the maximum rate. The amount of the deduction depended on a specific formula that associated the pollution profile of the car (in particular, NOx, CO, PM, HC and CO_2 emissions) to estimates of the negative impact those pollutants have on society. The goal was to create a differentiation between various levels of emission.

Note: More details in Annex Table 4.A.5.
Source: Roshal and Tovias (2016[12]) and OECD (2016[13]; 2020[14])

Distance-based charges, if carefully designed, have the potential to deliver more efficient road transport because they can reflect external costs related to driving (Parry and Small, 2005[15]). For example, distance-based charges can vary depending on the average pollution profile and weight of a vehicle and as such reflect the costs related to air pollution and road damage. They can also mirror spatial and temporal variation in driving; thereby reflecting population exposure to external costs, such as noise and air pollution, and integrating costs from congestion during peak hours and locations (Van Dender, 2019[4]). However, if the introduction of distance-based charges displaces traffic from highways in less populated areas towards roads that pass close to or inside of highly populated areas, the exposure of population to pollution may increase and lead to a detrimental effect on health. Congestion charges, on the other hand, if they deter access to congested and highly populated areas have the potential to improve pollution exposure and therefore health effects. The Belgian region of Brussels has recently launched the plan to introduce a road pricing system based on distances driven, as well as on the location and time of driving (Box 4.3). Congestion charges are successfully used in a number of different cities, including Stockholm (Box 4.4). Both case studies show that political economy considerations need to be managed well for successful reform.

Benefits from distance-based charging are also evident in terms of their revenue stability, as driving likely adjusts less quickly to pricing and taxation than energy use. A downside of distance-based charging involves the high administrative and implementation costs. Van Dender (2019[4]) reviews the literature on the costs of electronic tolling systems, concluding that distance-based charging systems have historically been expensive, but that historical data may not be a good indicator to predict costs in the future. Technological progress in charging techniques allows systems to become fine-tuned to particular circumstances and be run efficiently to bring costs down compared to historical estimates. In countries where tolling infrastructure already exists, benefits to extending distance-based charges to a wider tax base may outweigh additional costs.

Privacy concerns about data collection through distance-based charging system should be addressed. For example, simple odometer readings can assess distances travelled by a vehicle without collecting detailed information on the driving. Odometer reading cannot implement rates that vary with location and congestion levels, it can only be directed to cars registered in a country and not cover tourist driving (except if visitor's use cars rented from local agencies) and drive-through vehicles entering a country from neighboring regions. A minimum charge levied on non-residents could be discussed in this respect. GPS-based pricing systems, which track a vehicle's position and driving, can accommodate differentiated rates.

Carefully designing GPS-systems may reduce potential privacy concerns. In some existing programmes, driving-related data are destroyed as soon as drivers paid their road user charge (e.g. in Oregon's experimental distance-based charging programme and the German truck tolling system, see (Kirk and Levinson, 2016[16]; Langer, Maheshri and Winston, 2017[17]). The The White Book for Tax Reform in Spain (Comité de personas expertas, 2022[18]) offers a medium term proposal in this sense, namely a tax on the actual use of vehicles, taking account of location, time and type of vehicle.[3]

Box 4.3. The Region of Brussels' plan to implement road pricing

A road tax reform was proposed by the government of the Belgian Region of Brussels with the objective to replace an existing tax on the ownership of vehicles by a road user tax that charges the use of vehicles below 3.5 tonnes based on the amount and the time of distances driven. Heavier vehicles are not covered as already included in the Belgian-wide road user charge for heavy vehicles. If coming to life, the scheme would be the first European system that charges light vehicles based on distances.

The approach chosen seeks to combine road pricing with some form of congestion charging, by including rates that vary by distance, by time of day and location. No exemptions are foreseen at this stage. The charge would include electric vehicles on the grounds that they contribute to congestion. All the information needed to inform the amount of tax due would be recorded through the SmartMove app and rely on camera technology that is currently being used for the region's low emission zone.

The policy was announced in July 2020. The announcement has led to strong opposition from the other Belgian regions (Flanders and Wallonia) as the scheme would also cover all vehicles entering Brussels, including drivers from those regions, and not focus only on vehicles registered in Brussels. In August 2022, the start of a test phase has been announced as the region is looking for volunteers to test the technology SmartMove, including Brussels' inhabitants but also commuters from Wallonia and Flanders.

Note: See Annex 4.A
Source: L'Echo (2022[22]), SmartMove (2022[23]) and Wilson (2020[24])

Box 4.4. The Stockholm congestion charge

Congestion charges are implemented in many different cities, including in Europe (e.g. Gothenburg, London, Milan, Stockholm, Valetta) and beyond (Singapore). These schemes differ in their designs but typically serve the main objective to cover external costs from congestion and improve traffic management.

The Stockholm congestion charge was implemented in 2007, with a trial period in 2006. A vehicle is liable when passing tolling stations installed in the city centre during charging hours. The tax applies to vehicles registered in and outside of Sweden. Exemptions apply to specific vehicle types, including electric vehicles and hybrids.

The rate of the charge varies according to the time of day when the vehicle passes toll stations located in the city and whether driving occurs during peak or off-peak season. Rates are higher during peak commuting times (6:30-9:00 and 15:30-18:00) and from 1 March to the day before Midsummer Eve as well as between 15 August and 30 November.

Some public and political resistance with the introduction of the charge had to be manged ahead of the introduction of the scheme. Opposition was based on privacy and equity concerns. For example. it was argued that lower-income households living outside the city would have to pay for the benefits of wealthier inner city residents. Other concerns related to the risk of errors from misidentifying licence plates.

Source: See Annex Table 4.A.6.

4.4.3. Additional implications from transport taxation: behavioural effects, revenue stability and distributional consequences

Covering external costs is one of different potential tax policy objectives in road transport, as discussed in Section 4.2. Other policy considerations may concern questions about the extent to which taxation can contribute to reaching environmental goals, such as specific air pollution levels, and the net zero transition, or how instruments perform with respect to revenue stability or distributional concerns.

Behavioural effects: reaching environmental goals and revenue stability

Whether countries reach their specific air pollution or GHG reduction goals will depend on the responsiveness of tax bases to changes in tax rates, and this varies with specific design features of a tax and the broader economic context. For example, consumer responsiveness to taxation varies with the original price level (i.e., are prices high or low before the tax reform), household characteristics (such as income and geographical location), and the availability of substitutes, e.g. public transport and car-pooling (Douenne, 2018[25]; Gillingham and Munk-Nielsen, 2019[26]; Spiller, Stephens and Chen, 2017[27]). If substitutes are expensive or unavailable, behavioural responses may only occur at very high price levels. In contrast, where substitutes are readily available at small additional cost, price response can be substantial even at relatively low prices.

Increasing the **price of fuel or fuel taxes** typically lowers the demand for it, for example because drivers reduce travel or change the means of transport (e.g. switch from car to train or bus). Meta-analyses of backward-looking elasticities by Graham and Glaister (2002[28]) and Labandeira, Labeaga and López-Otero (2017[19]) find that gasoline consumption typically reduces around 2-3% in the short run when fuel prices rise by 10%, while the responsiveness is larger in the long run. Recent research shows an average long-run price elasticity of -0.44 in road transport, indicating that a EUR 10 increase in the fuel or carbon tax, decreases emissions from road transport by 4% (D'Arcangelo et al., 2022[20]). A 4% responsiveness to price increases may not be enough to fulfil the national objective of reaching a zero GHG emissions fleet of passenger cars and light commercial vehicles by 2050 (Spanish Climate Change and Energy Transition law) or the regional objective of reducing GHG emissions by 30-43% in the transport and mobility sector in 2030 compared to 2008 levels (PAAC) and achieving air pollution targets. Additional instruments, including non-tax instruments, can usefully support fuel and carbon taxes in this respect. For example, explicit objectives for the penetration of electric vehicles combined with strong financial support to enhance the substitution of technologies through tax and subsidy policies, has proven successful (but also costly) in the case of Norway (Box 4.7 and Section 4.6).

Few caveats apply to the interpretation and transferability of fuel tax elasticity estimates: First, the percentage estimates typically derive from a cross-country analysis and may not be transferable to single countries. Some countries in the sample of D'Arcangelo et al., (2022[20]) tax transport fuels at a relatively low level, which risk pushing the estimates upwards. In a region like Andalusia where taxes on transport fuels are relatively high compared to some of the non-EU countries in the sample, the response to rate increases may be much more limited. Second, evaluating the responsiveness to price or tax rate increases via backward-looking elasticities is less informative in the present situation, where the likelihood of deep change is significant. The applied techniques infer consumer behaviour that is associated with the circumstances prevalent at the sample time, such as income levels and available substitutes, and can evaluate future trends only to a limited extent. For example, they cannot consider the take-up of electric vehicles, when such variation is not yet observed in the data used in the analysis. Yet, the electrification of road transport is supposed to gradually raise the responsiveness of fossil fuel demand to energy and carbon pricing as clean electricity generation increases. The remarkable lack of reaction to the important jump of transport fuel prices due to the recent energy crisis is worth mentioning.

The responsiveness of consumers to price and tax rate changes will also impact the governments' capacity to raise revenues. Excise duties on fuel used in road transport represent a significant share of tax revenue in several countries. Under current policy settings, this tax revenue base is projected to shrink as the fuel-efficiency of internal combustion engines improves and the electrification of the transport sector progresses, driven by policy commitments or the declining costs of electric vehicles. Eroding tax bases lead to declining revenues, which puts stress on government budgets in the long run. Policymakers need to anticipate such potential decline. OECD/ITF (2019[10]) analyses the potential tax revenue erosion from reduced fuel use in road transport and investigates alternative tax policy reform scenarios to compensate for the loss Box 4.5.

In Andalusia, revenues from the hydrocarbon tax represents 8.9% of the region's total tax revenue in 2020 (see Section 2), indicating the size of potential revenue loss if fuel taxes are not replaced by alternative instruments. Revenues may be sustained in the long run by gradually increasing fuel or carbon taxes that cover the external costs closely related with fossil fuel use in vehicles and by phasing-in distance-based charges for cars to reflect external costs closely related with distances driven.

Limitations of behavioural responses to pricing also exist in the context of public transport services. Recent evidence for Spain suggests that a reduction in the price of public transport (as introduced in 2022 across Spanish cities) did not affect air quality – whereas better provision of public transport can improve their uptake (Albalate, Borsati and Gragera, 2022[31]). Comfort and availability seem to matter more for the use of public transport than price.

Box 4.5. Tax Revenue Implications of Decarbonising Road Transport

The OECD jointly with the International Transport Forum (ITF) has developed a model to analyse countries' potential revenue erosion in road transport due to reduced fuel consumption through decarbonisation and to estimate orders of magnitude of alternative tax policy response to compensate for the revenue loss.

"Tax Revenue Implications of Decarbonising Road Transport: Scenarios for Slovenia" provides an in-depth assessment of the transport tax system in Slovenia, where 14.6% of total tax revenue in 2016 was based on fuel and carbon taxes levied on diesel and gasoline used in road transport. It explores revenue impacts under current policy settings and for different scenarios on the take-up of new vehicle technologies. Against this background, the study analyses tax policy reforms with a 2050 horizon. The tax reforms include changes to fuel and carbon taxes, vehicle taxes and distance-based charges, and consider the potential behavioural responses to tax reform.

Key findings

- Under current policies, tax revenues from diesel and gasoline use in private cars is likely to decline substantially in the coming decades in Slovenia. The decline is driven by fuel-efficiency improvements of internal combustion engines and the penetration of alternative fuel technologies in the private car fleet.
- Gradually reforming the tax system, starting now, allows for a smooth adaptation to technological changes in the vehicle fleet and the timely implementation of accompanying measures. Fuel tax revenues from private cars erode only gradually over time as technological changes take time to percolate through the fleet, which leaves leeway to adapt tax policy.
- Shifting from taxes on fuels to taxes on distances driven can contribute to more sustainable tax policy over the long term, improving environmental and mobility outcomes. Such a tax system would gradually shift revenues to an alternative and likely more stable tax base, distance driven.

While the analysis focuses on the specific case of Slovenia, most of the recommendations are transferable to countries with comparable economic and tax system structure.

Source: OECD/ITF (2019[10])

Vehicle taxes can inadvertently set up a tension between revenue raising and environmental objectives. For example, one-off registration taxes for polluting vehicles may reduce fleet turnover and thereby technology take-up, because it can push households to delay buying a new car. Ad valorem taxes, which apply as a percentage on the sales price of a vehicle, that do not vary with environmental criteria tend to incentivise price-sensitive households to choose relatively cheaper or second-hand cars instead of cars with new and expensive technologies. Recurrent annual taxes, on the other hand, can provide regular incentives to invest in clean cars, either by providing lower rates or full exemptions. It has been argued though that vehicle purchase or registration taxes may be more effective in steering consumers' purchasing decisions than annual taxes, due to the myopia of consumers, who tend to attach greater importance to the immediate costs incurred through the registration tax and due to uncertainty about future circulation taxes (Comité de personas expertas, 2022[18]).

When vehicle taxes are used to reflect average emissions profiles of cars, they should include both local air pollution and carbon emissions. Focusing vehicle taxes only on fuel efficiency or CO_2 emissions can stimulate the sale of diesel cars despite their negative impact on health and the environment through air pollution, as has been experienced in the French feebate case (Box 4.6). The vehicle purchase tax in

Israel, for example, accounts for different emissions, including carbon dioxide, particulate matter and nitrogen oxides (Box 4.2).

These characteristics also affect the stability of revenue from vehicle taxation. A recurrent tax on vehicle ownership or circulation induces a stable revenue flow, whereas the one-off character of a registration tax renders revenue dependent on fleet turnover and the business cycle.

Box 4.6. The French feebate programme for CO_2-efficient motor vehicles

In 2008, France introduced a *Bonus/Malus* écologique on passenger vehicles registered in France. It levies a fee (malus) on the purchase or registration of highly polluting vehicles based on the CO_2 emission rate per kilometre as indicated on the vehicle registration certificate, while the revenues collected from the fee are used to support (bonus) clean vehicles.

In 2022, the bonus for new passenger cars with official CO_2 emissions of 0g per kilometre is EUR 6 000, while it is EU 1 000 for cars with emissions up to 50g of CO_2 per kilometre The malus applies to vehicles emitting 128 g of CO_2 per kilometre or more and ranges from EUR 50 to EUR 40 000. In 2022, a new weight component has been introduced to the fee to account for importance of external costs related to heavy vehicles, in particular sport utility vehicles (SUVs).

Feebates can be an effective tool to stimulate purchases of zero- and low-emission vehicles in a revenue neutral way. For example, Durrmeyer (2022[21]) finds that the French policy reduces average CO_2 emissions but came at the cost of creating more emissions of local pollutants. When designed carefully, feebates can be revenue neutral and not add pressure to the public budget. For example, the French feebate is achieving a positive balance since 2014, thanks to a careful design and regular revisions to account for recent developments.

One downside of such systems is that manufacturers can design vehicles in ways that lead to bunching, i.e. characteristics to situate just below the step increase (an alternative is to choose continuous and not a step function to define the rates of the fee) or the pivot point, which transforms the malus into a bonus. In addition, the French system is said to have led to increases in non-climate effects such as air pollution. Singapore's feebate includes air pollution in addition to CO_2 emissions.

Source: D'Haultfœuille, Givord and Boutin (2014[22]), Durrmeyer (2022[21]), ICCT (2022[34]) and Teusch and Braathen (2019[23]).

Current vehicle taxes do generally not account for the observed gap between real-life emissions and the advertised emission profiles derived from test cycles. For given tax rates, the increasing gap between test and real-world emission values leads to substantial amounts of tax revenue forgone in 11 European member states (Forum Ökologisch-Soziale Marktwirtschaft and Green Budget Europe, 2018[24])

Distributional consequences

The potential distributional consequences stemming from transport taxation need to be considered. Estimating and presenting the distributional effects along income and spatial dimensions can form a basis for designing policy measures to accompany tax reform. Accompanying measures may support households that are affected disproportionally by the reform in the short run but cannot easily adjust to the reform due to financial constraints or lack of alternatives. Another way could be to advertise and encourage the use and development of alternative travel modes (such as public transport or car-pooling). Bento (2009[37]) shows that different support measures (flat transfers, income-based transfers or distance-based transfers) can have important and different impacts on the distributional impacts of gasoline taxes.

In the context of the recent energy price crisis, governments use different types of support to help vulnerable populations deal with raising prices, where well targeted support preferrable over price support. Supporting households is costly, but necessary for equity and energy affordability reasons and to sustain political support for the transition towards net-zero emissions. Many countries use price support, which seeks to reduce energy prices paid by consumers (e.g. through price controls, broad energy tax or VAT cuts). Price support seems easier to implement and avoids some of the political economy discussions but puts important pressure on government resources, in particular when it is not targeted but supports all types of households even those that can deal with the crisis by their own means. Price support would better be transformed into well-targeted income support that can contain expenditures and distributional concerns, although it may be administratively more costly. Targeting would best consider several dimensions, such as income, location, patterns of consumption (OECD, 2022[25]; OECD, 2022[26]).

Distributional effects of **fuel taxes** differ across countries, income levels (Sterner, 2012[27]; Flues and Thomas, 2015[28]) and geographic areas within a country, because differences in work distances play an important role in driving patterns. For example, fuel taxes may place a disproportionally high burden on households living in rural areas (see simulation in Bureau (2011[29]) and Spiller, Stephens and Chen (2017[30])), who cannot reduce driving needs in the short run by substituting towards public transport, moving location or changing jobs. Similarly, in the absence of revenue recycling, **distance-based charges** may have adverse effects on households with long commutes and that cannot easily adjust driving patterns in the short run. (Levinson (2010[31]) reviews the equity effects of road pricing.)

Recent studies by Alonso-Epelde et al. (2022[32]), for Spain, and Gore (2022[33]), across EU member states provide insights into distributional patterns of energy consumption across income, location and other socio-economic characteristics of households based on information collected through the EU Household Budget Surveys. They show that energy expenditures in total constitute a significantly larger share of total expenditure for lower-income compared to higher-income households. Looking at the expenditures for transport fuels specifically, they represent the largest share of total expenditure for middle-income households, because low-income households tend to own private cars relatively less often (Figure 4.6). When looking at the distribution across EU member states, household location (rural, intermediate, urban) has stronger implications for energy expenditure than household income. Middle-income households in rural areas spend a much larger share of their total expenditures on energy (notably on transport fuels) than low-income households in an urban context, due to commuting needs and less availability of good public transport alternatives. Similarly, expenditures can vary importantly with household size and composition.

Vehicle taxes may also have distributional consequences. Ad-valorem vehicle taxes may be progressive if low-income households purchase less expensive cars more often. If vehicle taxes are differentiated by emission bands, providing lower rates for more efficient vehicles, and if high-income households predominantly drive fuel-efficient cars, the tax could be regressive. Fully exempting electric vehicles from taxation likely benefits predominantly high-income households that can afford purchasing these vehicle types.

Figure 4.6. Structure of household energy consumption, Spain (% of total expenditure)

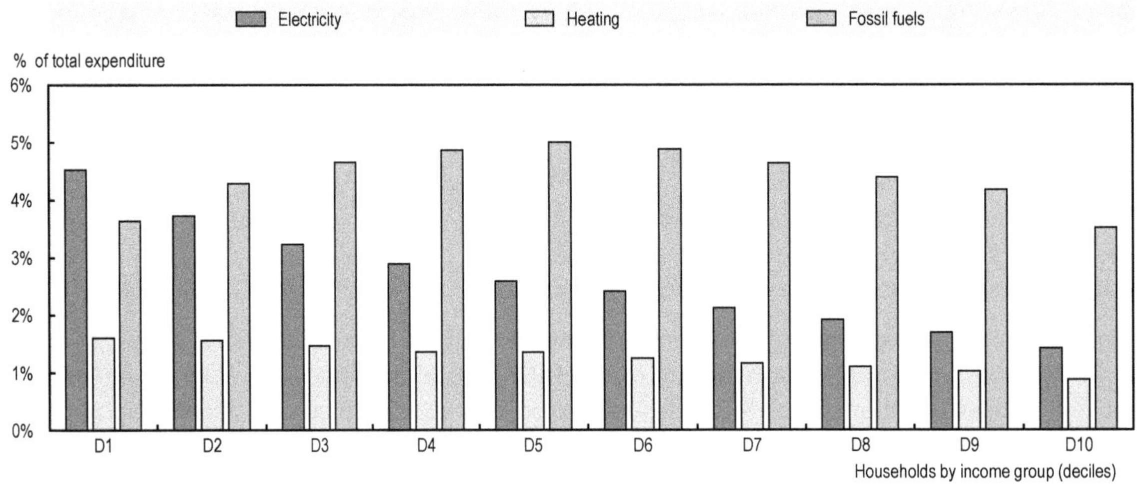

Note: Average household expenditure on different energy products (electricity, heating, fossil fuels – mainly private transport) as a share of total expenditure from the lowest-income decile (1) to the highest-income decile (10).
Source: Alonso-Epelde et al. (2022[32]).

StatLink https://stat.link/dc1b5j

Providing tax exemptions and benefits for electric vehicles are not only expensive in terms of government revenue foregone but are likely to be regressive too. Borenstein and Davis (2016[34]) show that an income tax credit in the United States for plug-in electric vehicles disproportionately benefits the top income quintile, receiving 90% of all credits. The authors explain this strong regressive effect by the fact that low-income households may not invest in expensive electric vehicles and by the non-refundability of the credit. A study by Muehlegger and Rapson (2018[35]) shows that means-tested subsidies directed towards low- and middle-income buyers in California achieve electric vehicle take-up in this segment of the market, but that the revenue cost is large.

Leroutier et Quirion (2022[36]) show that policies targeting local pollution of vehicles (e.g. low-emission zones) may be more regressive than policies that regulate CO_2 emissions of vehicles, such as CO_2 emission standards, in an urban context. This finding is likely driven by the observation that low-income households use pollution-intensive cars more often in the sample, while high-income households typically use car types with attributes that are associated with higher CO_2 emission factors, e.g. heavier, larger and more powerful vehicles.

Distance-related charges may have regressive impacts, as they typically do not consider the economic capacity of different drivers or their access to alternative travel modes (Labandeira, 2022[49]). Such risks could be mitigated by using part of the revenue for public transport improvements or direct transfers to low-income drivers that have no other choice in the short-term than using their car to reach work or school.

4.5. Alignment of Andalusia framework with sound environmental tax principles and strategic reform options

Based on the considerations of a good environmental tax policy framework outlined throughout Sections 4.2 to 4.4, this section discusses how the tax framework applicable in Andalusia aligns with such sound environmental tax policy principles. This section first assesses the alignment of the different tax types with external costs estimates. It then considers how the current framework compares against other important tax policy principles and policy objectives to finally develop strategic reform recommendations.

*4.5.1. **Fuel excise and external cost management***

As discussed in Section 4.4.2 fuel and carbon taxes are well suited to account for the external costs related to CO_2 emissions, because CO_2 emissions are proportional to fuel consumption. The current fuel excise rates that apply in Andalusia exceed a low-end estimates of climate costs (EUR 30 per tonne of CO_2). Translating this EUR 30 benchmark into fuel tax rates requires a diesel tax at 7.99 eurocent per litre and a gasoline tax at 6.86 eurocent per litre. The rates that a driver in Andalusia currently pays settle at 37.9 eurocent per litre for diesel and 50.4 eurocent per litre for gasoline, much above the low-end benchmark of climate costs.

This does not mean the Spanish fuel tax rates are necessarily too high. First, the EUR 30 per tonne of CO_2 reflects a low cost benchmark. Climate cost estimates are uncertain, many studies suggest higher costs already today or costs that increase in the future (Box 3.5). Second, external costs from fuel use are broader than climate costs and include air pollution and noise. Higher fuel taxes may be justified to cover these additional costs. As noted above though, the fuel tax is not an optimal instrument to cover non-climate costs from air pollution, noise or congestion. Such categories depend heavily on aspects that a fuel tax cannot map, such as vehicle technology (e.g. engine type), driving behaviour (e.g. acceleration, risk taking), or the specific driving location and pollution exposure that varies across geographic areas and the time of the day. Countries may use vehicle taxes to reflect average vehicle characteristics and cover location and time specific characteristics through distance-related fees or congestion charges.

Fuel excise rates as they apply in Andalusia vary across fuels and users, which leads to an unequal treatment of taxpayers, potential distortions, and inefficiencies relating to the objective of aligning the tax with external costs. Current rates reveal a common problem: the diesel-gasoline differential describing a situation where diesel use is subject to a much lower rate than gasoline, which cannot be justified on a pure energy basis, nor on a km driven basis, nor on an external costs basis. Equalising the rates is therefore a key policy recommendation (Crawford and Smith, 1995[2]; OECD, 2019[37]). In addition, agricultural fuels are taxed at lower levels and biofuels are exempt although they emit CO_2 emissions at the combustion level.

The unequal treatment of fuel taxation is common across countries (OECD, 2019[37]; OECD, 2022[8]) and Andalusia has little room for manoeuvre to adjust these rates since the fuel excise is regulated at the national level and in the framework of minimum rates set out in the EU Energy Tax Directive (EU ETD). The White Book for Tax Reform in Spain (Comité de personas expertas, 2022[18]) also suggests equalising the taxation on diesel with the one on automotive gasoline at the national level, in line with the currently proposed revision of the EU ETD.

4.5.2. Vehicle taxation and external cost management

Current vehicle taxes applicable on cars registered in Andalusia could be improved to better cover external costs. The annual ownership taxes levied by Andalusia's municipalities do not vary with CO_2 emissions and air pollution. The registration tax at the national level varies by vehicle type and CO_2 profile of car, but it does not differ according to air pollution profiles and is not updated regularly to account for technological advances. As noted in Section 4.4.2, vehicle taxes that only vary with CO_2 emissions, but not with other pollutants have been found to stimulate the sale of diesel cars despite their negative impact on health and the environment (see the French example, Box 4.6). Extending the vehicle taxes to cover *average* and updated emissions profiles of vehicles (CO_2 and air pollutants), may be useful to reflect environmental costs in the decision making of Andalusian car owners and encourage the purchase of less polluting vehicles or electric vehicles. Assuming that the number of trips will be the same, a vehicle with lower emissions would mean an improvement in terms of CO2 and air pollution levels. The use of electric vehicles would also improve noise pollution levels. Phasing-out the current exemption for rental cars, would allow to also incentivise tourist drivers using those cars.

However, such a reform would not cover spatial and time variation of air pollution and population exposure to it (Box 4.1). Similarly, vehicle tax reform will not be able to cover driving-related external costs like congestion and accidents that can be substantial. Distance-based charges would be a better option to cover such variation in external costs (as discussed below).

The possibility mapping in Section 2.4.2 points towards a potential opportunity to create a regional tax in Andalusia on vehicle emissions. If such a possibility is pursued, it is advisable to consider both CO_2 emissions and air pollution at the same time. The vehicle registration tax in Israel, for example, accounts for different emissions, including carbon dioxide, particulate matter and nitrogen oxides (Box 4.2) and could serve as example. A downside of this approach is the administrative complexity related to fine-tuning tax rates to many different emissions and related external costs. Varying tax rates according to environmental indicators such as the vehicle Euro standard could be an alternative approach.

The creation of an additional vehicle tax at the regional level, as implemented in Catalonia (Annex Table 4.A.2), requires careful thinking. From a taxpayer perspective the compliance burden may increase significantly when vehicle owners are subject to three different, but similar taxes (the existing national registration tax, the potentially new regional tax and the existing municipal tax on circulation). In addition, the White Book for Tax Reform in Spain suggested amending the existing national vehicle tax to include environmental criteria. If such reform took place, the potentially new national tax would make a regional tax on vehicle emissions redundant and replace existing regional taxes covering the same taxable event. If the national reform is not pursued, several of the experts' recommendations may be useful to advance action at the regional level.

The White Book for Tax Reform in Spain suggests different reform options of the registration tax at the national level: either to modify the existing national registration tax or to change the entire tax structure. In the *modification scenario*, the experts suggest adapting the rate of the tax to reflect recent technological advances in the sector. Specifically, they suggest extending the number of brackets and to raise tax rates in order to provide a stronger incentive for the purchase of low-emission vehicles. As the tax has not been modified in a long time, its capacity to provide incentives to invest in clean vehicles is limited. In addition, they suggest adding a surcharge based on vehicle weight, which increases external costs, like in the recent reform of the French feebate (Box 4.6). In the *reform scenario*, the experts suggest replacing the ad-valorem levy applicable on the vehicle price by a unitary tax levied on the reported emissions of the vehicle to avoid low tax rates driven mainly by low vehicle prices. Similarly, for the annual circulation tax, the experts suggest incorporating environmental considerations in the calculation of the rates, including a vehicle's category under the Euro standards for example or other environmental impact indicators.

An interesting alternative approach to provide strong incentives to buy and drive clean vehicles through a revenue-neutral approach, is the feebate structure discussed above. While a fee would penalize the use of polluting vehicles, the revenues collected from the fee would be used to support the purchase of clean vehicles. If such an approach is chosen it would best cover external costs from both CO_2 and air pollution and provide support to zero-emissions vehicles only. (Current feebates typically apply support to clean vehicles and efficient internal combustion engines that still emit CO_2 and air pollution (ICCT, 2022[39]).)

4.5.3. Distance-based charges and congestion charging

Currently no distance-based charge or congestion charging exist in Andalusia, although distance-based charges are a superior tool to manage road transport thanks to their ability to reflect external costs related to driving as mentioned throughout the analysis (Section 4.4.2). Potential reasons for the lack of distance-related pricing may relate to political expediency, as such measures may be seen as unpopular, in particular when the public transport system is perceived as a suboptimal alternative to car travel due to concerns about quality of service and travel time. Other reasons may be related to efforts to develop certain activities (tourism for example) or geographic areas. When introduced, such charges can vary depending on the average pollution profile and weight of a vehicle and as such reflect the costs related to air pollution and road damage. They can also mirror spatial and temporal variation in driving; thereby reflecting population exposure to external costs, like air pollution and noise, and integrating additional costs related to congestion during peak hours and in specific locations.

Benefits from distance-based charging are also evident in terms of their revenue stability. The tax base driving likely erodes less quickly than the tax base energy use (Box 4.5). Shifting from taxes on fuels to taxes on distances driven can therefore contribute to more sustainable tax policy over the long term, improving environmental and mobility outcomes.

A downside of distance-based charging is their complexity, high administrative and implementation costs, as well as privacy concerns. Technological progress in charging techniques allows systems to become fine-tuned to particular circumstances, drive costs down in the future and remedy privacy concerns (see Section 4.4.2). An alternative means to charge distances driven is to apply simple odometer readings which assess distances travelled by a vehicle without collecting detailed information on when and where the driving took place. However, charging based on odometer reading cannot implement rates that vary with location and congestion levels and will only be able to cover cars registered in Andalusia. Nevertheless distance-related charging based on odometer reading aligns better with external costs than not having distance-related charging at all.

Emissions in road transport that relate to tourist driving or vehicles entering the region from outside and drive through are currently not covered by the ownership tax, for example because rental cars are exempt. They will also not be covered by distance-based charges that are implemented through odometer reading, although a fixed fee could be applied to non-resident drivers. Removing the vehicle tax exemption of rental cars would be a first step in aligning tourist activity with external costs. Tourist drivers can also be included in distance-based systems or congestion pricing if carefully designed. For example, the road tolling system on expressways and motorways in France (péage) charges all vehicles that pass toll gates independently of their origin, including tourist drivers. In France, most of the roads are operated by commercial companies, who manage the network and set the rates.

In addition, local congestion charging should be considered to reflect the external costs of road transport in urban areas, related to congestion and air pollution. Adame et al. (2014[7]) show that air pollution from CO, NO, NO_2 and PM10 is highest in Andalusian urban centres – where in addition, population density and as such exposure to pollution is highest (Box 4.1). Well-designed congestion pricing that charges vehicles for circulating in particular areas or cities can translate the health-related costs from air pollution and congestion costs in charges for drivers to affect their behaviour. Congestion charges can be usefully combined with the implementation of low emission zones, as is the case of London for example, or other regulatory measures to decongest urban areas (e.g. pedestrianisation of spaces, creation of green areas, densification and improvement in the accessibility of public transport, improvements of infrastructure for alternative means of transport – such as bicycles).

Because air pollution is a very local effect, congestion charging may best be implemented at the regional or local level, where opportunities exist to follow the spatial variation of pollution and their impacts relatively better, and to implement such charges in those cities or areas where negative effects from driving are most important. Congestion charges can usefully be combined with air pollution standards, such as Euro standards, that increase in stringency over the years.

Congestion charges are also recommended in the White Book for Tax Reform in Spain to charge for the external costs related to congestion and air pollution in urban centres. To mitigate the potentially regressive effects of such measures, part of the revenue is suggested to be invested in public transport improvements or transferred to the most affected population through direct compensatory solutions (e.g. transfers, aid for changing vehicles).

Political and social resistance to tax reform that focuses on implementing and improving distance-based charges and congestion pricing requires well-designed policies supported by a tailored and effective communication campaign, which is essential for the success of comprehensive tax reform in the road sector, given the involvement of numerous stakeholders. Early and careful preparation and implementation is required to manage resistance and opposition. Discussions with stakeholders and a gradual implementation approach (e.g. through trial periods, roundtables) will help reduce the risk of disruptions. (See experiences with discussing road pricing in Brussels (Box 4.3) and implementing congestion charging in Stockholm (Box 4.4).) It will also create room for carefully designing policies, developing the necessary accompanying measures and tailoring communication.

Accompanying measures could encourage the development of alternative travel modes, such as public transport, or take the form of support to those households that are affected disproportionally by the reform in the short run but cannot easily adapt to the reform due to budget constraints or missing public transport opportunities. A good understanding of the potential negative consequences needs to be developed (e.g. how changes in tax liability from reform are distributed along income and spatial dimensions) to design appropriate policy the short run, but cannot easily adapt to the reform due to budget constraints.

4.6. Key findings and strategic recommendations

On a pure external cost basis, the alignment of the current Andalusia tax framework could be improved. The design of existing tax instruments does not align well with external costs considerations. Fuel taxes, for example, apply heterogeneously and are not based on carbon content. Also, no specific tax instrument applies to incorporate costs from air pollution. And although external costs related to congestion and accidents risk to be substantial – in particular at the local level, they are not considered explicitly in the current tax system applicable in Andalusia. Distance-based charges or congestion pricing is currently not considered for passenger cars although they align better with most of the external costs in road transport than fuel and vehicle taxes do.

While fuel excise rates are regulated at the national and EU level, the sub-national level is well-placed to manage pricing of air pollution and congestion. Taxes (or feebates) targeted to the emissions of vehicles or congestion pricing in urban centres can help manage local congestion problems and improve local air quality.

Focusing tax reform on trying to mimic external costs perfectly provides only little answers to additional policy considerations, such as the transition to a zero-emission transport sector. It also risks binding administrative resources that are scarce and may better be spent on focusing on key aspects of the most pressing policy needs. Policy considerations going beyond external costs management are discussed in Section 4.2.

The policy objective to implement the zero-emissions transition in the road transport sector is likely the most pressing need. This aligns with countries' commitments made under the Paris Agreement to limit increases in the global average temperature to well below 2 degrees Celsius above pre-industrial levels and to pursue efforts to limit the temperature increase to 1.5 degree Celsius. National climate neutrality, i.e. reaching net zero-emissions by 2050 country-wide, is also a key objective for the Spanish national government. In this context, the 2021 Andalusia Climate Action Plan aims for a 30-43% reduction of GHG emissions by 2030 (compared to 2008 levels) in the transport and mobility sector but misses precise objectives for 2050. Accelerating progress towards achieving the planned reductions by 2030 is necessary, as well as a plan for 2050 objectives. Spain has also committed to phase-out the sales of passenger cars and light commercial vehicles that do not qualify zero-emissions vehicles by 2040.

The Andalusia government is encouraged to decide on a ranking of policy objectives before starting a comprehensive tax reform process. If the main goal is to be serious about climate policy and decarbonising the road transport sector, administrative resources may better be spent in designing a reform that pushes zero-carbon vehicles on the street, instead of engaging in marginal but burdensome reform to align the vehicle tax better with external cost estimates. To tackle air pollution relying on emissions standards of cars and regulating low emission zones for example can prove administratively less burdensome than aligning tax rates with external costs

Different policy levers need to be pulled to achieve zero-carbon in the necessary timeline, which come with different costs and benefits. For example, through a combination of policy instruments, Norway became a leader with respect to electric vehicle penetration in the fleet. Their toolbox comprises clear objectives for the penetration of zero-emissions vehicles combined with well-designed carbon and energy taxation and technology support for zero-emission vehicles through subsidies and tax incentives (Box 4.7). The instruments have been reviewed regularly to avoid excessive spending and align support. Nevertheless, important fiscal resources are behind these efforts.

A combination of tax elements may help Spain and Andalusia to push for a clean passenger vehicle fleet. It will be important to expand and adapt the use of tax approaches while the future policy mix evolves including changes in regulatory approaches, e.g. implementation of low emission zones, upcoming Euro 7 standard, etc.

First, consistent fuel excise and carbon pricing will align climate incentives across the economy and provide strong signals that fossil fuels are not the future. These, considerations are relevant for the national or even the EU level.

Second, vehicle taxes could be reformed to favour only zero-emission vehicles (and not internal combustion engines, even the very efficient ones, as they are still responsible for external costs, which may delay a shift towards full decarbonisation) and to reflect air pollution costs. Relying on Euro standards that increase in stringency over time may be preferable to chasing the nitty gritty of air pollution costs. The vehicle tax could also be transformed into a feebate that penalises ICEs and large vehicles and subsidises zero-emissions vehicles of regular size. Whether such reform would best happen at the national or regional level depends on advances of vehicle tax reform at the Spanish level.

Otherwise, tax incentives provided through the CIT or PIT system can further push the adoption of clean vehicles. Tax incentives can help overcome consumer myopia, financial constraints and other constraints that prevent households from making the relevant investments. But they come with costs that need to be considered. Tax incentives involve forgone tax revenue - akin to expenditures from direct subsidies - but with the downside to be less transparent to the broader public. They also risk to predominantly benefit rich households. Means-tested subsidies that are directed towards low- and middle-income buyers may overcome such shortcomings.

Third, implementing congestion pricing at the regional or local level will help manage local congestion problems while improving local air quality.

Finally, preparing for increased use of distance-based charges. If not pursued at the national level, local level action in this area can bring local benefits (better traffic management, reduced congestion, fewer accidents, lower air pollution, revenue).[4]

While the principal objective of such tax policy choices may be to direct drivers and car-owners towards buying and using zero-emission vehicles to expedite the transition to net zero, they have budgetary impacts that should be considered when designing environmental policy, as should be their potential distributional effects. For example, if the zero-emissions objective in transport is pursued successfully, revenues from energy taxes are likely to erode.

Finally, taxation is not the only instrument in the climate policy toolkit. Synergies and coordination with other policy instruments will be needed to reach success. Several other policy instruments are set at the national level or at the level of the European Union. These include the European regulation on GHG emissions from vehicles, regulations on emissions of air pollutants, Euro standards, the EU Energy Tax Directive, the EU Emissions Trading System, including the potential extension to road transport, and the Spanish National Fund for the Sustainability of the Electricity System (FNSSE). These policies are currently being reviewed with the intention of increasing environmental policy-stringency, particularly in relation to carbon neutrality. It national level policies become more ambitious, the scope for regional level activity declines.

Electrification of the car fleet can only be successful, if accompanied by significant investment in charging infrastructure for electric vehicles, and will contribute to the net-zero transition only if electricity production is decarbonised.

Box 4.7. Decarbonising road transport in Norway

Norway is one of the leading countries to adopt electric vehicles. Over the last 10 years, the country increased the share of electric vehicles in vehicle sales from 1% to 65% in 2021. A national objective asks new passenger and light-duty vehicles sold in 2025 to be zero-emission (electric or hydrogen).

The ambitious policy objective is supported by an extensive instrument package that explains the success of electric vehicles (EVs) in the country. These include but are note restricted to

- Excise taxes and an explicit carbon tax on fuel used in road transport;
- Registration tax based on CO_2 and NO_x emissions; exemptions for EVs;
- Annual motor vehicle tax based on fuel and particulate filter use; exemptions for EVs;
- VAT exemptions and reduced VAT rates: since 2001, electric vehicles are VAT exempt. Fossil fuel vehicles pay the standard 25% rate. As of 2023, the VAT exemption for EVs will be amended. The VAT rate increases with the price of the car, including EVs;
- Exemption and reduced charges on toll roads and ferries: until 2017, EVs were exempt from paying for their use of roads or ferries. The Parliament has agreed on implementing a 50% rule: counties and municipalities cannot charge more than 50% of the price that fossil fuel cars pay on ferries, public parking and toll roads. A rule of maximum 50% parking fee at public parking for zero-emission cars is not yet implemented; and
- Additional preferential treatment of EVs has applied or still applies in the context of the taxation of company cars and for charges at the local level, e.g. use of bus lanes and free municipal parking.

These instruments have been combined with policies that promote investments in charging infrastructure and public procurement that favours zero-emission vehicles. For instance, between 2017 and 2021, legislation established a “charging right” for people living in apartment building. A well-organised charging network has been established to ensure feasibility of long-distance trips with fast charging stations on all main roads in Norway. As of February 2022, more than 470 000 EVs and 4 600 cars can fast-charge at the same time in Norway.

Source: Annex 4.A

References

Adame, J. et al. (2014), "Weekend-Weekday Effect Assessment for O3, NOx, CO and PM10 in Andalusia, Spain (2003–2008)", *Aerosol and Air Quality Research*, Vol. 14, pp. 1862-1874, https://doi.org/10.4209/aaqr.2014.02.0026. [7]

Agencia Tributaria de Catalunya (2022), *Base imponible, cuota tributaria y bonificación*. [43]

Albalate, D., M. Borsati and A. Gragera (2022), *Preliminary results of the impact of massive discounts in public transport on air quality*, https://nadaesgratis.es/admin/resultados-preliminares-del-impacto-de-los-descuentos-masivos-en-transporte-publico-sobre-la-calidad-del-aire. [74]

Alonso-Epelde, E. et al. (2022), *Modelling the direct socio-economic impacts of the New Energy Taxation Directive (ETD) and the extension of the ETS on transport and building sectors*, OTEA (Observatorio de la Transición Energética y la Acción Climática), https://api.otea.info/storage/2022/06/06/ff2c5dc97f2cda8d3ea2812cb0e34e576100920a.pdf. [32]

Bento, A. et al. (2009), "Distributional and Efficiency Impacts of Increased US Gasoline Taxes", *American Economic Review*, Vol. 99/3, pp. 667-699, https://doi.org/DOI: 10.1257/aer.99.3.667. [61]

Borenstein, S. and L. Davis (2016), "The Distributional Effects of US Clean Energy Tax Credits", *ax Policy and the Economy*, Vol. 30, pp. 191-234, https://doi.org/10.1086/685597. [34]

Bruzz (2022), *Consultation committee not even about SmartMove, but 'road pricing not buried'*. [60]

Bureau, B. (2011), "Distributional effects of a carbon tax on car fuels in France", *Energy Economics*, Vol. 33/1, pp. 121-130, https://doi.org/10.1016/j.eneco.2010.07.011. [29]

Comité de personas expertas (2022), *Libro Blanco Sobre la Reforma Tributaria*. [18]

Crawford, I. and S. Smith (1995), "Fiscal Instruments for Air Pollution Abatement in Road Transport", *Journal of Transport Economics and Policy*, Vol. 29/1, pp. pp. 33-51, https://www.jstor.org/stable/20053059. [2]

D'Arcangelo, F. et al. (2022), "Estimating the CO2 emission and revenue effects of carbon pricing: New evidence from a large cross-country dataset", *OECD Economics Department Working Papers*, No. 1732, OECD Publishing, Paris, https://doi.org/10.1787/39aa16d4-en. [20]

D'Haultfœuille, X., P. Givord and X. Boutin (2014), "The Environmental Effect of Green Taxation: The Case of the French Bonus/Malus", *The Economic Journal*, Vol. 124/578, pp. F444–F480, https://doi.org/10.1111/ecoj.12089. [22]

Douenne, T. (2018), "The vertical and horizontal distributive effects of energy taxes: A case study of a French policy", *FAERE Working Paper*, No. 2018.10, http://faere.fr/pub/WorkingPapers/Douenne_FAERE_WP2018.10.pdf. [62]

Durrmeyer, I. (2022), "Winners and Losers: The Distributional Effects of the French Feebate on the Automobile Market", *The Economic Journal*, Vol. 132/644, pp. 1414–1448., https://doi.org/10.1093/ej/ueab084. [21]

Elbil (2022), *Norwegian EV policy*. [53]

European Commission (2022), *Commission proposes new Euro 7 standards to reduce pollutant emissions from vehicles and improve air quality [press release]*, https://ec.europa.eu/commission/presscorner/detail/en/IP_22_6495. [65]

European Parliament (2022), *Deal confirms zero-emissions target for new cars and vans in 2035 [press release]*, https://www.europarl.europa.eu/news/en/press-room/20221024IPR45734/deal-confirms-zero-emissions-target-for-new-cars-and-vans-in-2035. [66]

Flues, F. and A. Thomas (2015), "The distributional effects of energy taxes", *OECD Taxation Working Papers*, No. 23, OECD Publishing, Paris, https://doi.org/10.1787/5js1qwkqqrbv-en. [28]

Forum Ökologisch-Soziale Marktwirtschaft and Green Budget Europe (2018), *Loss of revenues in passenger car taxation due to incorrect CO2 values in 11 EU states*, A report commissioned by the Greens/EFA group in the European Parliament, http://www.foes.de/pdf/2018-03-10_FOES_Taxation_loss_due_incorrect_CO2_values.pdf. [24]

Generalidad de Cataluna (2020), *Decreto Lay 33/2020, de 30 de septiembre, de medidas urgentes en el ámbito del impuesto sobre las emisiones de dióxido de carbono de los vehículos de tracción mecánica y del impuesto sobre las estancias en establecimientos turísticos, y en el ámbito presupuestario y administrativo*. [44]

Gillingham, K. and A. Munk-Nielsen (2019), "A Tale of Two Tails: Commuting and the Fuel Price Response in Driving", *Journal of Urban Economics*, Vol. 109, pp. 27-40, https://doi.org/10.1016/j.jue.2018.09.007. [63]

Gobierno de Espana (2022), *Royal Decree-Law 20/2022 of 27 December on measures in response to the economic and social consequences of the war in Ukraine and in support of the reconstruction of the island of La Palma and other situations of vulnerability, provided in Articles 1 and 72 for changes in VAT and equivalence surcharge rates.*. [72]

Gore, T. (2022), *Can Polluter Pays policies in the buildings and transport sectors be progressive? Assessing the distributional impacts on households of the proposed reform of the Energy*, Research report, Institute for European Environmental Policy, https://ieep.eu/uploads/articles/attachments/7a9ac44a-fa75-4caf-9db5-76d55110217c/Can%20polluter%20pays%20policies%20in%20buildings%20and%20transport%20be%20progressive_IEEP%20(2022).pdf?v=63813977582. [33]

Graham, D. and S. Glaister (2002), "The Demand for Automobile Fuel: A Survey of Elasticities", *Journal of Transport Economics and Policy*, Vol. 36/1, pp. 1-25, https://www.jstor.org/stable/20053890. [64]

Harding, M. (2014), "Personal Tax Treatment of Company Cars and Commuting Expenses: Estimating the Fiscal and Environmental Costs", *OECD Taxation Working Papers*, Vol. 14/OECD Publishing, Paris, https://doi.org/10.1787/5jz14cg1s7vl-en. [11]

Held, M. et al. (2021), "Lifespans of passenger cars in Europe: empirical modelling of fleet turnover dynamics", *European Transport Research Review*, Vol. 13/9, https://doi.org/10.1186/s12544-020-00464-0. [3]

ICCT (2022), *Incentivising zero- and low-emission vehicles: the magic of feebate programs (blog post)*, https://theicct.org/magic-of-feebate-programs-jun22/. [70]

Junta de Andalucia (2022), *Parque distribuido por tipo de vehículo, carburante y año de matriculación.*. [1]

Junta de Andalucia (2022), *Statistics on levels of emission into the atmosphere of pollutants in Andalusia*, https://www.juntadeandalucia.es/medioambiente/portal/landing-page-%C3%ADndice/-/asset_publisher/zX2ouZa4r1Rf/content/estad-c3-adstica-de-niveles-de-emisi-c3-b3n-a-la-atm-c3-b3sfera-de-contaminantes-en-andaluc-c3-ada/20151?categoryVal=. [71]

Kassirer (2020), *France's Feebate for Fuel Efficient Vehicles*. [40]

Kirk, R. and M. Levinson (2016), *Mileage-Based Road User Charges*, Congressional Research Service, https://fas.org/sgp/crs/misc/R44540.pdf. [16]

Labandeira, X. (2022), *Taxation and Ecological Transition during Climate and Energy Crises: the Main Conclusions of the 2022 Spanish White Book on Tax Reform*, Real Instituto Elcano WP 09-2022, https://www.realinstitutoelcano.org/en/work-document/taxation-and-ecological-transition-during-climate-and-energy-crises/. [73]

Labandeira, X., J. Labeaga and X. López-Otero (2017), "A meta-analysis on the price elasticity of energy demand", *Energy Policy*, Vol. 102, pp. 549-568, https://doi.org/10.1016/j.enpol.2017.01.002. [19]

Langer, A., V. Maheshri and C. Winston (2017), "From gallons to miles: A disaggregate analysis of automobile travel and externality taxes", *Journal of Public Economics*, Vol. 152, pp. 34-46, https://doi.org/10.1016/j.jpubeco.2017.05.003. [17]

L'Echo (2022), *L'application SmartMove démarre sa phase de test (press article)*, https://www.lecho.be/economie-politique/belgique/bruxelles/l-application-smartmove-demarre-sa-phase-de-test/10409083.html. [69]

Legifrance (2018), *Articles D251-7 to D251-13*. [41]

Leroutier, M. and P. Quirion (2022), "Air pollution and CO2 from daily mobility: Who emits and Why? Evidence from Paris", *Energy Economics*, Vol. 109, https://doi.org/10.1016/j.eneco.2022.105941. [36]

Levinson, D. (2010), "Equity Effects of Road Pricing: A Review", *Transport Reviews*, Vol. 30/1, pp. 33-57, https://doi.org/10.1080/01441640903189304. [31]

Mottershead, D. et al. (2021), *Green taxation and other economic instruments. Internalising environmental costs to make the polluter pay*, European Commission, Directorate-General for Environment, Brussels, https://environment.ec.europa.eu/publications/green-taxation-and-other-economic-instruments-internalising-environmental-costs-make-polluter-pay_en. [5]

Muehlegger, E. and D. Rapson (2018), "Subsidizing Mass Adoption of Electric Vehicles: Quasi-Experimental Evidence from California", *NBER Working Papers*, No. 25359, National Bureau of Economic Research, https://www.nber.org/papers/w25359. [35]

Norwegian Ministry of Transport and Communications (2017), *National Transport Plan 2018–2029*. [51]

OECD (2022), *Consumption Tax Trends 2022: VAT/GST and Excise, Core Design Features and Trends*, OECD Publishing, Paris, https://doi.org/10.1787/6525a942-en. [9]

OECD (2022), *OECD Environmental Performance Reviews: Norway 2022*, OECD Environmental Performance Reviews, OECD Publishing, Paris, https://doi.org/10.1787/59e71c13-en. [55]

OECD (2022), *Pricing Greenhouse Gas Emissions: Turning Climate Targets into Climate Action, OECD Series on Carbon Pricing and Energy Taxation*, OECD Publishing, Paris, https://doi.org/10.1787/e9778969-en. [8]

OECD (2022), *Pricing Greenhouse Gas Emissions: Turning Climate Targets into Climate Action*, OECD Series on Carbon Pricing and Energy Taxation, OECD Publishing, Paris, https://doi.org/10.1787/e9778969-en. [54]

OECD (2022), *Tax Policy Reforms 2022: OECD and Selected Partner Economies*, OECD Publishing, Paris, https://doi.org/10.1787/067c593d-en. [26]

OECD (2022), "Why governments should target support amidst high energy prices", *OECD Policy Responses on the Impacts of the War in Ukraine*, https://www.oecd.org/ukraine-hub/policy-responses/why-governments-should-target-support-amidst-high-energy-prices-40f44f78/. [25]

OECD (2020), *Accelerating Climate Action in Israel: Refocusing Mitigation Policies for the Electricity, Residential and Transport Sectors*, OECD Publishing, Paris, https://doi.org/10.1787/fb32aabd-en. [14]

OECD (2020), *Accelerating Climate Action in Israel: Refocusing Mitigation Policies for the Electricity, Residential and Transport Sectors*, OECD Publishing, Paris, https://doi.org/10.1787/fb32aabd-en. [46]

OECD (2020), *Consumption Tax Trends 2020: VAT/GST and Excise Rates, Trends and Policy Issues*, OECD Publishing, Paris, https://doi.org/10.1787/152def2d-en. [56]

OECD (2019), *Taxing Energy Use 2019: Using Taxes for Climate Action*, OECD Publishing, Paris, https://doi.org/10.1787/058ca239-en. [37]

OECD (2016), "Israel's Green Tax on Cars: Lessons in Environmental Policy Reform", *OECD Environmental Policy Paper*, No 5, OECD Publishing, Paris,, https://doi.org/10.1787/5jlv5rmnq9wg-en. [13]

OECD (2016), "Israel's Green Tax on Cars: Lessons in Environmental Policy Reform", *OECD Environment Policy Papers*, No. 5, OECD Publishing, Paris, https://doi.org/10.1787/5jlv5rmnq9wg-en. [45]

OECD/ITF (2019), *Tax Revenue Implications of Decarbonising Road Transport: Scenarios for Slovenia*, OECD Publishing, https://doi.org/10.1787/87b39a2f-en. [10]

Parry, I. and K. Small (2005), "Does Britain or the United States Have the Right Gasoline Tax", *The American Economic Review*, Vol. 95/4, pp. 1276-1289, https://www.jstor.org/stable/4132715. [15]

Politico (2020), *Plan to charge Brussels motorists per kilometer driven*. [57]

Regjeringen (2021), *Norway is electric*. [52]

Reveal (2020), *Smart Move: A Breath of Fresh Air for Brussels*. [58]

Road Traffic (2020), *Stockholm Congestion Charge.* [49]

Roshal, V. and A. Tovias (2016), *Car purchase tax: green tax reform in Israel*, OECD, Paris, https://www.oecd.org/israel/OECDWorkingPaper-Green-Tax-Reform-in-Israel.pdf. [12]

SmartMove (2022), *SmartMove sets Brussels on the right track*, https://smartmove.brussels/en. [68]

Spiller, E., H. Stephens and Y. Chen (2017), "Understanding the heterogeneous effects of gasoline taxes across income and location", *Resource and Energy Economics*, Vol. 50, pp. 74-90, https://doi.org/10.1016/j.reseneeco.2017.07.002. [30]

Sterner, T. (2012), *Fuel Taxes and the Poor: The Distributional Effects of Gasoline Taxation and Their Implications for Climate Policy*, RFF Press, New York, https://doi.org/10.4324/9781936331925. [27]

Teusch, J. and N. Braathen (2019), "Are environmental tax policies beneficial?: Learning from programme evaluation studies", *OECD Environment Working Papers*, No. 150, OECD Publishing, Paris, https://doi.org/10.1787/218df62b-en. [23]

Teusch, J. and N. Braathen (2019), "Are environmental tax policies beneficial?: Learning from programme evaluation studies", *OECD Environment Working Papers*, No. 150, OECD Publishing, Paris, https://doi.org/10.1787/218df62b-en. [38]

The Brussels Times (2020), *Brussels eyes half a billion euros per year with new toll plan.* [59]

Time (2022), "What Norway Can Teach the World About Switching to Electric Vehicles". [50]

Tools of Change (2014), *Stockholm's Congestion Pricing.* [48]

Transport Styrelsen (2021), *Hours and amounts in Stockholm.* [47]

Van Dender, K. (2019), "Taxing vehicles, fuels, and road use: Opportunities for improving transport tax practice", *OECD Taxation Working Papers*, No 44, OECD Publishing, Paris, https://doi.org/10.1787/e7f1d771-en. [4]

Wappelhorst (2022), "Incentivizing zero- and low-emission vehicles: the magic of feebate programs". [39]

WHO (2005), *Health effects of transport-related air pollution*, World Health Organization Regional Office for Europe, https://www.euro.who.int/__data/assets/pdf_file/0006/74715/E86650.pdf. [6]

Wilson, S. (2020), *Brussels Government to introduce road pricing across the Brussels region (blog post)*, http://roadpricing.blogspot.com/2020/12/brussels-government-to-introduce-road.html. [67]

Yang (2018), *Practical lessons in vehicle efficiency policy: The 10-year evolution of France's CO2-based bonus-malus (feebate) system.* [42]

Annex 4.A. Detailed case studies: road transport

This section presents selected case studies in the area of greenhouse gas emissions and air pollution across the world with a focus on passenger cars.

France: Vehicle Purchase Feebate

Annex Table 4.A.1. Vehicle Purchase Feebate (France)

Legal bases	Energy code Articles D255-1 to D251-13
Objective	To reduce transport CO_2 emissions by providing consumers with incentives to purchase more fuel-efficient vehicles.
Level of responsibility	Central government (France)
Tax setter(s)	Central government (France)
Revenue beneficiary(ies)	Central government (France)
Tax payer(s)	Vehicle buyers or lessees (with contracts greater than two years)
Tax base (including main exemption(s), credits or deductions)	The buyers of vehicles emitting CO_2 above the fixed threshold pay a fee for their emissions, whilst the purchasers of electric or hybrid vehicles receive a rebate depending on the vehicle price, capped to a maximum.
Tax rate(s) (including their calculation)	The French government establishes vehicles' CO_2 emission thresholds. More specifically, in 2022: • Fee: buyers of vehicles emitting more than 127g/km must pay a fee between EUR 50-40 000 according to the level of emissions. Vehicles over 1 800kg also need to pay an additional EUR 10 per kg exceeding the 1 800kg threshold. • Rebate: purchases of electric or plug-in hybrid vehicles emitting less than 20g/km receive a rebate of 27% up to EUR 6 000 or EUR 1 000, respectively. The thresholds are updated annually and are linked to the values established by European Union.
Governance and implementation	The system is calibrated to be cost-neutral, and all tax income collected from the sales of higher-emission vehicles should be used to subside the purchase of less-emitting cars. Nevertheless, the system ran in deficit until 2014, when it was recalibrated and achieved a constant positive balance since then.

Environmental, social & health impacts	The policy seems to have supported a shift towards sales of cleaner cars. For example, the market share of class B vehicles (lower emitting) increased from 20% to 50% one year after the scheme's implementation. Over the same period, the share of class E vehicles (higher emitting) decreased from 15% to 5%. However, the average emission of new vehicles decreased by only 5% as many buyers made only marginal adjustments in their purchase decisions (Teusch and Braathen, 2019[38]). Between 2008 and 2016, the average CO_2 emission from French vehicles declined from 150g/km to 110g/km (Wappelhorst, 2022[39]) (Wappelhorst, 2022[39]). The feebate also employed multiple marketing strategies that contributed to its effectiveness (Kassirer, 2020[40]).

Source: (Legifrance, 2018[41]; Yang, 2018[42]; Teusch and Braathen, 2019[38]; Kassirer, 2020[40]; Wappelhorst, 2022[39]).

Spain: Catalonian Tax on Emissions from Motor Vehicles

Annex Table 4.A.2. Catalonian tax on emissions from motor vehicles (Spain)

Legal basis	Catalonian Law Decree 33/2020
Objective	To tax the CO_2 emissions from motor vehicles (Agencia Tributaria de Catalunya, 2022[43]).
Level of responsibility	Region (Autonomous Community of Catalonia)
Tax setter	Region (Autonomous Community of Catalonia)
Revenue beneficiary	Region (Autonomous Community of Catalonia)
Tax payer	Vehicle owners that reside in the region
Tax base (including main exemptions, credits or deductions)	The tax base are the CO_2 emissions, measured in grams of CO2 per kilometer, from vehicles of the following categories: (1) Vehicles of categories M1 (vehicles mainly for the transport of people and their luggage) and N1 (vehicles mainly for the transport of goods with a maximum mass not exceeding 3.5 tons); and (2) Vehicles of categories L3e (two-wheel motorcycles), L4e (two-wheel motorcycles with sidecar), L5e (motor tricycles) and L7e (heavy quadricycles). Old classic vehicles (removed from the Vehicle Registry due to the age of the model) and those not generating emissions (e.g. 100% electric vehicles) are not subject to the tax. Exemptions apply to official public vehicles, public service vehicles (e.g., ambulances), vehicles of reduced mobility owners, and consular and international organisations' vehicles.
Tax rate(s) (including their calculation)	The tax is paid annually. The tax rate is progressive and depends on the CO2 emissions profile presented in the certificates issued by the vehicle manufacturer. The Catalonian government publishes the rates annually. The tax rates for 2022 are presented in Annex Table 4.A.3 for vehicle categories M1, L3e, L4e, L5e and L7e and Annex Table 4.A.4 for vehicle category N1. A formula applies when vehicles do not have official pollution data issued by the manufacturers.
Governance and implementation	There was a debate about the constitutionality of this annual tax and its potential overlaps with the Spanish national tax on the first registration of motor vehicles (a one-off registration tax that also varies with the CO_2 emissions profile of the vehicle). Nevertheless, this tax was ruled constitutional by the Spanish Constitutional Court (STC 87/2019, 20 June, FJ 19), which declared: "The regional tax is linked to the "polluter pays" principle. As its periodic nature encourages, insofar as possible in a high-value good such as motor vehicles, not only the first purchase of low-polluting vehicles, but also the replacement of existing vehicles with less polluting ones. It thus aims to change behaviour or at least make people pay for it [STC 53/2014, FJ 6 c)], an aim absent in the State tax at least in this second case of renewal of the vehicle fleet, since, being an instantaneous tax, the State tax is levied at a single stage on the first registration of the vehicle, boat or aircraft, so that its replacement would always involve the creation of a new taxable event that would be avoided, however, if the vehicle already owned and for which its owner has already paid the tax is kept" (STC 87/2019, FJ 19).

Source: (Generalidad de Cataluna, 2020[44]; Agencia Tributaria de Catalunya, 2022[43]).

Annex Table 4.A.3. Tax rates for vehicle categories M1, L3e, L4e, L5e and L7e per (2022)

Level of CO_2 emissions as advertised in the official emissions profile CO_2	Marginal rate (in EUR per g CO_2/km)
Until 95 g/km	0
More than 95 g/km and until 120 g/km	0.70
More than 120 g/km and until 140 g/km	0.85
More than 140 g/km and until 160 g/km	1.00
More than 160 g/km and until 200 g/km	1.20
More than 200 g/km	1.40

Source: (Agencia Tributaria de Catalunya, 2022[43])

Annex Table 4.A.4. Tax rates for vehicle category N1 (2022)

Official emissions of carbon dioxide	Marginal rate (in EUR per g CO2/Km)
Until 140 g/km	0
More than 140 g/km	0.70

Source: (Agencia Tributaria de Catalunya, 2022[43])

Israel: Vehicle Purchase Tax

Annex Table 4.A.5. Vehicle purchase tax (Israel)

Legal basis	Green purchase tax reform from 2009
Objective	To internalise the external costs that vehicles pose to society (OECD, 2016[45]).
Level of responsibility	Central government (Israel)
Tax setter	Central government (Israel)
Revenue beneficiary(ies)	Central government (Israel)
Tax payer(s)	Vehicle owners
Tax base (including main exemption(s), credits or deductions)	The tax base is the vehicle purchase, with rates differing with the pollution grade. In order to incentivise the purchase of hybrid and electric cars, the tax rate was simplified and limited to 30% of the car price for hybrids and 10% for plug-in vehicles in 2009. Tax credits are also given to additional safety mechanisms, such as ABS+4 airbags and emission-lowering devices (e.g. catalytic converters and diesel particulate filters).
Tax rate(s) (including their calculation)	The tax is paid once at the purchase of the vehicle, at a rate of 83% of the price of the vehicle. Deductions (up to NIS 16 629 or EUR 4 722) apply to all vehicles with pollution ratings (reflecting their emission levels) below 15. The pollution grades are calculated using a formula incorporating the five most relevant and harmful pollutants (CO, HC, NOx, PM, HC and CO2), as defined by the central government, with parameters reflecting their estimated social costs. These are updated every two years to reflect the changes resulting from real GDP and population changes. Since the effective purchase tax is a function of the car price and the pollution rating, the cheaper and the less pollutant the car, the lower the tax.
Governance and implementation	In 2006, a pluri-disciplinary commission was established to design the tax. The commission includes different ministries, such as the Ministries of Finance, Transport and Road Safety, National Infrastructure and Environmental Protection, as well as car engineers, industrial engineers and economists. The objective was to formulate a policy to reduce air pollution by estimating and internalising several externalities of transportation. After reviewing alternatives, the commission opted for the tax on vehicle purchases, which differentiated the tax by the pollution level they caused per litre of fuel. Opposition and general uncertainty about the tax were overcome by comprehensive dialogue and public relations campaigns led by the authorities targeting the general public and those particularly against the tax, such as car importers who pushed for a tax based solely on CO2 emissions, like in Europe. To gather support from this group, the governments worked with them to build a full and comprehensive database of all car models and pollution volumes, which took over a year. To effectively generate the intended increase in the share of less polluting vehicles, it became mandatory to mark the green grade and fuel consumption at every advertisement, point of sale and the Ministry of Transport website. Additional measures were taken to renew the fleet, including a scrapping scheme for disposing of vehicles older than 20%. This successful scheme was renewed in 2013.
Environmental, social & health impacts	The share of heavy-polluting cars (bands 13-15) was reported to have been reduced from 23.5% in 2009 to 7.4% in 2014, while the share of low-polluting vehicles increased from 1.8% to 47.2% (OECD, 2016[45]).

	The inclusion of local air pollutants in addition to CO2 seem to have reduced CO2 emissions, while avoiding a shift towards sales of diesel cars, which impose a higher damage to public health through air pollution compared to gasoline cars (OECD, 2020[46]). However, by decreasing the final prices of cars, the tax may have contributed to an increase in car ownership in Israel (17% between 2013 and 2016), exacerbating related issues like congestion, infrastructure erosion, noise, and pollution.

Source: (OECD, 2016[45]; OECD, 2020[46])

Sweden: Stockholm Congestion Tax

Annex Table 4.A.6. Stockholm congestion tax (Sweden)

Legal bases	Act (2004:629) on the congestion tax
Objective	To reduce congestion and improve accessibility (Transport Styrelsen, 2021[47])
Level of responsibility	Central government (Sweden)
Tax setter(s)	Central government (Sweden)
Revenue beneficiary(ies)	Central government (Sweden) and municipality (Stockholm)
Tax payer(s)	Vehicle owners
Tax base (including main exemption(s), credits or deductions)	All vehicles passing tolling stations in the city centre during charging hours are liable. The tax applies to vehicles registered in and outside of Sweden. The system works with automatic number plate recognition technology. Payment gates are equipped with number plate recognition cameras which photograph vehicles' plates and cross-references with Sweden's National Vehicle Registry to record the charge. Vehicles that are exempt from the tax include emergency vehicles, buses, diplomatic vehicles, disabled person vehicles, military vehicles, hybrid or electric cars, motorcycles and mopeds, and foreign-registered vehicles. Furthermore, traffic to and from Lidingö connected is exempt from the tax, provided that the vehicle passes the Ropsten payment station and some other payment station within 30 minutes of each other. The reason is that vehicles can reach the area of Lidingö from Stockholm only through roads that are part of the congestion charge zone. The Essingeleden motorway (E4) that passes through the charging area is also exempt because it is the main road when travelling past central Stockholm.
Tax rate(s) (including their calculation)	The rate of the charge varies according to the time of day when the vehicle passes toll stations and according to driving during peak or off-peak season (see Annex Table 4.A.7). The peak season is from 1 March to the day before Midsummer Eve and between 15 August and 30 November. The rest of the year is off-peak season. The tax is not charged on public holidays, as well as certain days before a public holiday and during July (except the first week of July). The maximum amount per day and vehicle is SEK 135 (EUR 9.6).
Governance and implementation	The Stockholm congestion tax was introduced as a trial in early 2006. The trial was followed by a period without taxation and a referendum, where Stockholm residents voted for the permanent implementation of the tax. Before the trial, the main barriers to congestion charging were public and political opposition which feared that license plate numbers would be misidentified (resulting in cour appeals or refusals to pay) or that the system would favour wealthier inner-city residents and punish lower-income people living outside the city (Tools of Change, 2014[48]). None of the public objections materialised: false plate identifications were kept to a minimum (97% accuracy) and calculations showed that the wealthier population in the inner city paid more charges than residents. The Royal Institute of Technology conducts repeated surveys to evaluate the public support for the tax, which was lowest just before the trial, increased after it began, and reached about 70% of public support in 2011. All driver categories demonstrate support for the charge, with "have no

	car" and "pays often" demonstrating the highest and lowest support rates, respectively (Tools of Change, 2014[48]).
Environmental, social & health impacts	It is reported that the congestion charge has come with traffic reductions of around 20% that have held constant over time (Annex Table 4.A.8). The highest decline was reported in the afternoon (-23% between 4pm and 6pm) followed by the morning (-18% between 7am and 9am), which demonstrates that a larger share of travels is made in the afternoon and that there is a higher flexibility in the departure from working hours. Additionally, the net social benefit of the congestion tax was estimated at approximately EUR 65 million per year, with the main drivers being the shorter and more reliable commutes, lower GHG emissions (between 10-15% compared to 2005 levels), improved traffic safety, public transit revenues and health and environmental impact (airborne pollutants reduced between 10 and 14%) (Tools of Change, 2014[48]). Finally, the system is said to have led to many drivers switching from private to public transport, which increased the number of passengers in the public transit system by approximately 4-5%. Due to investments in public transport capacity, this increase was reported to not result in a general crowding in the public transport (Tools of Change, 2014[48]).

Source: (Tools of Change, 2014[48]; Road Traffic, 2020[49]; Transport Styrelsen, 2021[47])

Annex Table 4.A.7. The Stockholm congestion charge, amount of the charge per time interval and peak vs. off-peak season in SEK and EUR

Hours	Off-peak season tax amount in SEK (EUR)	Peak season tax amount in SEK (EUR)
6:00-6:29	15 (1.37)	15 (1.37)15
6:30-6:59	25 (2.28)	30 (2.74)
7:00-8:29	35 (3.19)	45 (4.11)
8:30-8:59	25 (2.28)	30 (2.74)30
9:00-9:29	15 (1.37)15	20 (1.83)
9:30-14:59	11 (1.00)	11 (1.00)
15:00-15:29	15 (1.37)15	20 (1.83)20
15:30-15:59	25 (2.28)25	30 (2.74)30
16:00-17:29	35 (3.19)35	45 (4.11)
17:30-17-59	25 (2.28)	30 (2.74)30
18:00-18:29	15 (1.37)15	20 (1.83)20

Source: (Transport Styrelsen, 2021[47]).

Annex Table 4.A.8. Estimated yearly reduction of traffic (in vehicle kilometers) compared to 2005 levels

	2006a	2007b	2008	2009	2010	2011	2012	2013
Traffic reduction from charges, compared to 2005	-21.0%	-18.7%	-18.1%	-18.2%	-18.7%	-20.5%	-21.4%	-22.1%

Note: Charged weekdays are 6am to 7pm. Calculations are not available for 2012-2013 for the second row.
2006a is the trial period from January to July 2006 and 2007b is after the tax was introduced in August.
Source: (Tools of Change, 2014[48])

Norway: Electric Vehicle Support in Norway

Norway is leading the world in electric car adoption. Over the last 10 years, the country increased the share of electric vehicles in automobile sales from 1% to 65% in 2021 (Time, 2022[50]) (Time, 2022[50]). Additionally, in the National Transport Plan for 2018-2029, the Norwegian government presented three main goals to achieve "a transport system that is safe, enhances value creation and contributes to a low-carbon society": (i) improving the mobility within the country, (ii) reducing accidents in line with the Vision Zero plan, and (iii) reducing climate emissions towards a low-carbon economy and decreasing other negative environmental impacts (Norwegian Ministry of Transport and Communications, 2017[51]) (Norwegian Ministry of Transport and Communications, 2017[51]). Among this plan, the government set several targets to decarbonise the national fleet (Regjeringen, 2021[52]) (Regjeringen, 2021[52]):

- New cars and light vans must be zero-emission vehicles from 2025 (including electric cars and hydrogen cars),
- New city buses must be zero-emission vehicles or use biogas in 2025,
- New heavy vans, 75% of new long-distance buses and 50% of new trucks must be zero-emission vehicles by 2030,
- The distribution of goods in the largest city centres must be close to zero emissions by 2030.

The ambitious policy objective is supported by multiple policy instruments, including tax incentives and perks over fossil fuel car owners, which explain the success behind the increasing electric vehicle fleet in the country (Elbil, 2022[53]) (Elbil, 2022[53]). These include:

- Excise taxes and an explicit carbon tax on fuel used in road transport: explicit carbon prices, which include both ETS permit prices and carbon tax, covered 80.8% of greenhouse gas emissions in CO_2 eq in 2021 (51.2% and 54.3% respectively), while the fuel excise taxes, which are an implicit type of carbon pricing, covered 26.5% of emissions in 2021 in Norway. The explicit carbon price represented EUR 60.3 per tonne of CO_2 eq on average in 2021 (EUR 33.2 for the carbon tax) and EUR 33.2 per tonne of CO_2 eq for the fuel excise taxes (OECD, 2022[54]; OECD, 2022[55]).
- Registration tax: rates vary according to weight, CO_2 and NOx emissions. When information about emissions is unavailable, the tax is calculated based on cylinder capacity. Exemptions apply to electric vehicles, while plug-in hybrid and flexifuel (i.e. that can use fuel with at least 85% ethanol) benefit from rebates (OECD, 2020[56]).
- Traffic insurance tax (replaced the annual motor vehicle tax since 2018): the tax is based on fuel and particulate filter use. The daily charges are NOK 9.47 (EUR 0.92) for diesel cars without factory-fitted particle filters and NOK 8.12 (EUR 0.79) for other cars, NOK 5.65 (0.55) for motorbikes and NOK 1.31 (EUR 0.12) for mopeds and tractors. Electric vehicles are exempt (OECD, 2020[56]) (OECD, 2020[56]).
- Exemptions and reduced VAT rates: since 2001, electric vehicles are VAT exempt in contrast to fossil fuel vehicles, which pay a standard 25% VAT. As of 2023, the VAT exemption for electric vehicles will be replaced by a new VAT scheme where the VAT rate increases with the price of the car (i.e. the more expensive the car, the higher the VAT rate).
- Exemption and reduced charges on toll roads and ferries: between 1997 and 2017, electric vehicles did not have charges to pay for their use of roads or ferries. The Parliament has agreed on implementing a 50 % rule: counties and municipalities cannot charge more than 50 % of the price that fossil fuel cars pay on ferries, public parking and toll roads.

These instruments have been combined with policies that promote investments in charging infrastructure. For instance, between 2017 and 2021, legislation established a "charging right" for people living in an apartment building. A well-organised charging network has been established to ensure the feasibility of long-distance trips with fast charging stations on all main roads in Norway. Since 2015, the state has invested over NOK 136 million (EUR 13.2 million) in charging infrastructure, and as of February 2022, more than 470 000 of electric vehicles and 4 600 cars can fast-charge at the same time in Norway (Elbil, 2022[53]) (Elbil, 2022[53]).

Belgium: Brussels Road Pricing

SmartMove is a tax project, which aims to improve mobility in the Brussels Region. The goal is to reduce the externalities that traffic imposes on the environment, health, economy and quality of life (time lost in traffic jams), and the government hopes to reduce individual car trips by 25%. To this end, it aims to substitute the current "ownership tax" levied on vehicle possession for a tax based on actual kilometres driven. The tax is planned to be based on the distance driven, time of driving and engine capacity of the vehicle to better capture the negative environmental impact of driving (Politico, 2020[57]) (Politico, 2020[57]). Reduced rates would be offered outside peak hours. All the information needed to inform the amount of tax due would be recorded through the SmartMove app and rely on camera technology currently used in the region's low emission zone (Reveal, 2020[58]) (Reveal, 2020[58]).

The basic levy is planned to be EUR 1 in off-peak hours and EUR 2 during peak hours, which would be corrected for engine power (from 20 fiscal horsepower onwards, the multiplication becomes six, the highest rate and for those who drive an average car the rates become two or three) and added to an additional charge per kilometre driven (17 cents per kilometre during peak hour and 9 cents during off-peak hours). Driving at night or over weekends and holidays would be free of charge, and all income would be reinvested in mobility measures (The Brussels Times, 2020[59]) (The Brussels Times, 2020[59]).

This tax would substitute the vehicle ownership tax in the Belgian Region of Brussels. Drivers from the other two Regions (Flanders and Wallonia) risk being subject to both taxes, which has led to strong opposition in these Regions. The implementation of the project will depend on finding an agreement with the other Regions and is currently stalled. In recent months, the Region of Brussels has been considering a large-scale test phase (Bruzz, 2022[60]).

Notes

[1] The entire Section 4.4 draws importantly from OECD/ITF (2019[10]).

[2] During the ongoing energy price crisis in 2021/22, Spain has adjusted downward some of these rates. For example, a reduced VAT rate of 5% applies to electricity use. Additionally, the Royal Decree-Law 20/2022 of 27 December on measures to respond to the economic and social consequences of the war in Ukraine and to support the reconstruction of the island of La Palma and other situations of vulnerability was approved (Gobierno de Espana, 2022[72]).

[3] A non-official English summary of these suggestions were recently published (Labandeira, 2022[73]).

[4] The White Book for Tax Reform in Spain (Comité de personas expertas, 2022[18]) offers a medium term proposal in this sense, namely a tax on the actual use of vehicles that varies according to location, time and type of vehicle. Such a charge would replace most of the existing taxes in road transport (fuel, vehicles) and also those on congestion and infrastructure should they be introduced. Introducing such taxation would best be implemented gradually and considering potential distributional impacts – likely through the help of pilot evaluations (Labandeira, 2022[73]).

Part III Water pollution and usage

5 Legal stocktake: Water pollution and usage

Water is a strategic resource of economic, ecological and social importance in Andalusia (Junta de Andalucia, 2019[1]). Water consumption per capita is particularly high in Spain and Andalusia. The consumption is mainly driven by the agricultural sector. Regarding households, water consumption was 133 litres per inhabitant per day on average in 2020 in Andalusia, close to the national average (Estadística sobre el Suministro y Saneamiento del Agua, 2020[2]), but higher than the EU average of 124 litres per inhabitant per day (EurEau, 2021[3]). Although 89.7% of the regional population is connected to wastewater treatment, only 4.8% of treated wastewater is reused in Andalusia, below the national average of 11.2% (Official Association of Biologists of Andalusia, 2021[4]).

Regarding water pollution, the quality of surface water and groundwater is of great significance and is closely monitored in Andalusia. The quality is measured through the biological, hydromorphological, chemical and physical-chemical of water. The level of nitrates is one of the most relevant parameters to control water quality as it is closely related to the presence of fertilisers and wastewater discharges. Based on this parameter, the quality of surface water in Andalusia improved in 2019 as compared to 2018 as nitrate levels decreased in all hydraulic basins of the region, except in the Guadiana basin where levels remained stable and in the Guadalquivir basin where they increased (Junta de Andalucia, 2019[1]).

This chapter proposes possible opportunities for reform to Andalusia's existing environmental tax system governing water pollution and usage. The proposed opportunities are derived from the analysis of the legal and policy framework governing water pollution and usage at the EU, national, and regional government levels and the analysis on the distribution of responsibilities in policy areas relevant to reducing water pollution and usage between the different levels of government (EU, national, regional and local). The key possibilities will be assessed against environmental tax policy principles in the economic analysis.

5.1. Legal framework on water pollution and usage

This section outlines the legal and policy instruments governing water pollution and usage at the EU, national, and regional levels. In doing so, it provides context on the policies, targets, and strategies in place for this environmental domain. This then serves as the basis for the subsequent section, 5.2, on the responsibilities across levels of government relating to water pollution and usage.

5.1.1. At the EU level

As part of the EU Green Deal (see Part II, Section 2.1.1), the EU Commission adopted the EU Action Plan "Towards Zero Pollution for Air, Water and Soil" (European Commission, 2021[5]). The Plan sets a zero pollution vision for 2050, with the aim to reduce air, water and soil pollution to levels no longer considered harmful to health and natural ecosystems. The Action Plan also sets 2030 targets to reduce pollution at source, of which the improvement of water quality by reducing waste, plastic litter at sea by 50% and microplastics released into the environment by 30%. The Plan includes several actions, which comprise the revision of the standards for the quality of water, including in EU rivers and seas. It also proposes a zero pollution hierarchy, integrating the precautionary principle and the polluter payer principle.

The EU Commission also adopted the **European Water Framework Directive (2000/60/EC),** on 23 October 2000, which determines Spain's national water regulatory framework (European Commission, 2000[6]). The Directive was developed based on multiple international conventions on water protection and management, notably the United Nations Convention on the Protection and Use of Transboundary Watercourses and International Lakes, the Convention for the Protection of the Marine Environment of the North-East Atlantic, the Convention for the Protection of the Mediterranean Sea Against Pollution and its Protocol for the Protection of the Mediterranean Sea Against Pollution. It currently forms the backbone of water management in Europe. The Directive was established to overcome fragmented water policies tackling specific water-related domains and to provide an integrated framework for the protection and sustainable use of water within the EU (art. 1). To this end, it proposes a new water governance framework, which prescribes river basin districts as the managerial units for water management, defines water quality levels to be pursued and specifies limits to water abstraction. The Directive also determines that Member States shall follow the "*principle of recovery of the costs of water services, including environmental and resource costs associated with damage or negative impact on the aquatic environment should be taken into account in accordance with, in particular, the polluter-pays principle* (art. 1)" (European Commission, 2000[6]). The EU Directives that are integrated into the Water Framework Directive are presented in the Box 5.1.The Water Framework Directive was transposed into national law via the Spanish Water law (Royal Legislative Decree 62/2003) (Gobierno de Espana, 2003[7]).

The requirements under the European Water Framework Directive have been completed by other EU regulations transposed (or to be transposed) into Spanish law:

- The **EU Groundwater Directive (2006/118/EC)** sets the quality standards required by art. 17 of the Water Framework Directive (European Commission, 2006[8]). This Directive establishes a regime that sets standards for groundwater quality and introduces measures to prevent and limit pollutants inputs into groundwater. The Directive has been transposed into Spanish law by Royal Decree 1514/2009 (Gobierno de Espana, 2009[9]).
- The **EU Environmental Quality Standards Directive (2008/105/EC)** was developed to respond to art. 16 of the Water Framework Directive (Commission, 2008[10]). It also provides the list of priority (hazardous) substances described in the Water Framework Directive (updated again in Directive 2013/39/EC). The Directive has been transposed into Spanish law by Royal Decree 60/2011 (European Commission, 2021[11]).
- The **EU Water Reuse Regulation (2020/741)** sets minimum requirements for water reuse for agricultural irrigation from 2023 in accordance with the Directive 91/271/EC (European Commission, 2020[12]). The Regulation was adopted as part of the new Circular Economy Action Plan (CEAP) adopted on 11 March 2020 (European Commission, 2020[13]). It sets (i) harmonised minimum water quality requirements for the safe re-usage of treated urban wastewaters in agricultural irrigation, (ii) harmonised minimum monitoring requirements, (iii) risk management provisions to assess and address health and environmental risks, (iv) permit requirements, and (v) transparency on water re-usage projects. As an EU regulation, the objectives set are directly legally binding for EU Member States and thus Spain, without the need to be transposed into national legislation (see Part II, Section 2).

Box 5.1. Relevant European Union Regulation integrated into the Water Framework Directive

The **EU Nitrates Directive (91/676/EEC)** forms an integral part of the Water Framework Directive, serving as one of the most important instruments for safeguarding water quality against agricultural pressures by preventing nitrate pollution from reaching water bodies and by promoting sustainable farming practices. The Directive requires Member States to establish agricultural action programme measures to: (1) limit inorganic N fertiliser application; (2) limit organic manure application; (3) promote seasonal restriction on the application of slurry, manure sand sludge on sandy and shallow soils; (4) maintain farm records on cropping, livestock and fertiliser application. The Directive was transposed into Spanish law by Royal Decree 261/1996, which entitles autonomous communities to develop programs to prevent and mitigate nitrogen contamination.

The **EU Plant Protection Products Directive (91/414/EEC)** - repealed by Regulation 1107/2009/EC – lays down rules for the authorisation, placing on the market, use and control of plant protection products (safeners and synergists). The precautionary principle underpins the provisions of the Regulation. Regarding water quality, it determines that authorisations can only be granted to products that prove not to be harmful to the environment, particularly to groundwater quality. It was transposed into Spanish law by Royal Decree 2163/1994.

The **EU Biocides Directive (98/8/EC)** – repealed by Regulation 528/2012/EC – concerns authorisation and placing on the market of biocidal products such as pesticides, herbicides, or fungicides. The precautionary principle underpins the provisions of the Regulation. In respect to water quality, it determines that authorisations can only be granted to products that prove not to be harmful to the environment, particularly to groundwater quality. The Directive was transposed into Spanish law by Royal Decree 1054/2002.

The **EU Bathing Water Directive (76/160/EC)** – repealed by the Directive 2006/7/EC – requires the Member States to monitor and assess the bathing water quality for faecal bacteria. It was transposed into Spanish law by Royal Decree 734/1988, and Directive 2006/7/EC incorporated into Royal Decree 1341/2007.

The **EU Drinking Water Directive (98/83/EC)** – recast in Directive 2020/2184 – sets the minimum quality standards for water intended for human consumption. These should be assessed using microbiological parameters and chemical parameters. It was transposed into Spanish law by Royal Decree 140/2003 - amended by Royal Decree 902/2018.

The **EU Sewage Sludge Directive (86/278/EEC)** concerns the use of sludge in agriculture. It aims to increase the amount of sewage sludge in agriculture whilst protecting the environment from heavy metals in soil and sludge. To this end, it sets limits for the concentration of heavy metals in sewage sludge and bans the use of sewage sludge that exceed these limits. It was transposed into Spanish law by Royal Decree 1310/1990.

Source: Author's own elaboration.

5.1.2. At the national level

In accordance with the European Water Framework Directive, the Spanish Water law (Royal Legislative Decree 1/2001, of July 20, 2001) determines river basin districts as the basic managerial units of Spanish water resources (Gobierno de Espana, 2003[7]). Each river basin district is managed by a River Basin Authority. These authorities are responsible for establishing water management plans, which include the assessment of the water resources, an order of preference between the different uses of water, specific environmental objectives and measures to achieve the objectives therein (see below). The plans are articulated through adaptive processes carried out through the continuous monitoring and reviewing of the current hydrological plan, which is updated every six years. These plans are currently on their third cycle, with the newest set of plans adopted for the 2022-2027 period (Ministerio para la Transicion Ecologica y el Reto Demografico, 2021[14]).

In July 2021, the Spanish government adopted the National Plan for Wastewater Treatment, Sanitation, Efficiency, Savings and Reuse (DSEAR Plan) as a governance tool for the third cycle of the river basin management plans. The objective of the DSEAR Plan is to incorporate improved procedures and working methodologies, aligned with the principles of the environmental transition and the demographic challenge, into the updated river basin management plans. The Plan highlights seven areas for improvement based on lessons learnt from the two previous river basin management planning cycles, including for the co-ordination and co-operation mechanisms between the different administrations involved in river basin districts' management plans and to the economic and financial regime of water to adapt it to new challenges (Ministerio para la Transicion Ecologica y el Reto Demografico, 2021[15]).

The plans are co-ordinated at the national level following these steps: (i) the objectives and criteria for hydrological planning are set out in the Spanish Water Law and the Hydrological Planning Regulation (Decree 907/2007) (Andalucia, 2015[16]), and (ii) all hydrological plans are co-ordinated through the National Hydrological Plan, which harmonises all discrepancies and differences between basin districts. The National Hydrological Plan is elaborated through a participatory process involving the public sector and civil society. At the EU level, the hydrological plans are shared with the EU Commission, which publishes reports on the progress of implementation of the plans. This water management system implies that, although the central government establishes water-related levies (see below), they are managed and regulated by the river basin district authorities, who are ultimately responsible for determining their values in compliance with the Spanish Water law.

The Royal Decree 47/2022 on protecting waters against diffuse pollution produced by nitrates from agricultural sources was adopted in January 2022 and repealed Royal Decree 261/1996. It aimed to provide a stronger response to the problem of water diffuse pollution. The Royal Decree is based on the same instruments as the previous regulation, which include the designation of vulnerable areas, performance programs, monitoring programs and status reports. The Royal Decree also provides the possibility for River Basin Authorities to establish limits on new water concessions and other activities that may result in nitrate contamination (Gobierno de Espana, 2022[17]).

5.1.3. At the regional level

In line with the Spanish Water law, Andalusia adopted the Andalusian Water law on 30 July 2010 (law 9/2010), which establishes a set of environmental objectives and principles on the treatment of water as an exclusively economic resource (Junta de Andalucia, 2010[18]). The law regulates the responsibilities between the Autonomous Community of Andalusia and local governments with the aim to achieve water protection and sustainable water usage. More specifically, it regulates (i) the organisation of the river basin district authorities and their management plans, (i) water works of interest of the Community, (iii) the supply, sanitation and purification system of urban water use, (iv) the assessment and management of flood risks, (v) the revenue earmarked for infrastructure of the integral water cycle and public service provisions, as well as (vi) the system of penalties for non-compliance with the rules governing water use. The law applies to continental, transitional, coastal and ground water integrated in intraregional and interregional basin districts that pass in the Andalusian territory.

In addition, and in accordance with the Spanish Water Law, the Royal Decree 1620/2007 establishes the legal framework for the reuse of treated water in Andalusia. The Royal Decree includes a list of permitted uses according to specific quality criteria in its Annex I.A (e.g. watering of private gardens, irrigation of urban green areas, street cleaning, irrigation of crops with certain water application system), and of prohibited uses (e.g. human consumption, food industry, hospital facilities, fountains, or any other uses that may pose a risk on health or environment). It also guarantees quality control for the reuse of water (Gobierno de Espana, 2007[19]).

In 2020, in anticipation of the third cycle of hydrological plans covering the period from 2022 to 2027, the Andalusian government also launched the Andalusian Water Pact in 2020 (Junta de Andalucia, 2020[20]). The Pact established a participatory process in which public and private agents may discuss water-related issues to identify possible solutions. The approach focused on the investment priorities, the financial mechanisms and the governance systems of the autonomous communities.

5.2. Responsibilities related to water usage and pollution across levels of government

5.2.1. At the EU level

The EU's environmental responsibilities, as described above, are shared between the EU and Member States (art. 4) (European Union, 2012[21]). In the area of water, the EU has the ability to establish environmental policies, notably on water pollution. Art. 192 of the TFEU however stipulates that policies relative to the quantitative management of water resources or affecting the availability of these resources shall be adopted unanimously by the Council, after consultation of the European Parliament, the Economic and Social Committee and the Committee of the Regions (European Union, 2012[21]).

5.2.2. At the national, regional and local levels

As for the above-mentioned domains, the distribution of responsibilities between the different levels of government in Spain is defined in the Constitution and the Statute of Andalusia. They are listed in Table 5.1. The Constitution grants exclusive responsibilities to the central government on the legislation, management, and concession of hydraulic resources, the public works of general interest and the basic legislation on environmental protection (art. 149) (Gobierno de Espana, 1978[22]). By contrast, the autonomous communities may assume responsibilities over projects of hydraulic uses, canals and irrigation of interest to the autonomous community, mineral and thermal waters, fishing in inland waters, shellfish and aquaculture, hunting and river fishing and the management of environmental protection (art. 148). Additionally, they can assume responsibilities in matters that may indirectly affect water resources, such as agriculture and livestock raising, woodlands and forestry and the promotion and planning of tourism within their territory (art. 148) (Gobierno de Espana, 1978[22]). The responsibilities related to water for local governments are set out in Andalusia's Statute of Autonomy (Junta de Andalucia, 2007[23]).

Table 5.1. Distribution of responsibilities relating to water use and pollution across levels of government in Spain

	Matter: Water
Central government	**Exclusive responsibilities:** Legislation, management, and concession of hydraulic resources and uses when the waters flow through more than one autonomous community (art.149.1.22); Public works of general interest or whose performance affects more than one autonomous community (art.149.1.24); Basic legislation on environmental protection, without prejudice to the responsibilities of the autonomous communities to establish additional protection standards (art. 149.1.23).
Andalusia	**Exclusive responsibilities:** Projects, construction and exploitation of hydraulic uses, canals and irrigation of interest to the autonomous community; Mineral and thermal waters (art.148.1.10); Fishing in inland waters, shellfish and aquaculture, hunting and river fishing (art. 148.1.11); and Management of environmental protection (art. 148.1.09). Managing the participation of the users, the guarantee of supply, parcel regulation and works of transformation, modernisation and consolidation of irrigation systems for the saving and efficient use of water (AS. 50.1). Adopting additional measures for the protection and sanitation of water resources and aquatic ecosystems, execution and exploitation of state-owned works if established by agreement; Managing competences of the hydraulic public domain attributed by national legislation (AS. 50.2). Matters that may indirectly affect water resources, such as: (i) agriculture and livestock raising, in accordance with general economic planning (art. 148.1.07), (ii) woodlands and forestry (art. 148.1.08), and (iii) the promotion and planning of tourism within its territorial domain (art. 148.1.18).
Provinces	**Responsibilities:** Securing co-ordination and provision of municipal services.
Municipalities	**Responsibilities:** Regulation, management and provision of the water supply and wastewater treatment (AS. 92)

Note: AS: Andalusian Statute.
Source: Author's own elaboration based on (Gobierno de Espana, 1978[22]; Junta de Andalucia, 2007[23]).

As previously seen, river basins districts are the base unit of the institutional framework for terrestrial water management in Spain. The management structure of a river basin district depends on whether the basin falls entirely within the borders of an autonomous community (intraregional) or whether it crosses regional boundaries (interregional). An intraregional river basin is managed by an Autonomous Water Agency, which reports to the autonomous community's regional government. By contrast, an interregional river basin is managed by a River Basin Authority that reports to the national Ministry of Environmental, Rural and Marine Affairs. The different bodies involved in terrestrial water management in Spain are described in Box 5.2.

Box 5.2. The institutional framework for terrestrial water management in Spain

There are several bodies involved in the Spanish terrestrial water management system, which are described below (Fuentes, 2011[84]).

- The **central government** is responsible for all water policy on resources shared by more than one autonomous community (Table 5.1). It hence oversees the river basin authorities, which monitor interregional river basin districts. The central government finances transport and supply infrastructure, partially through the river basin authorities. It also sets policy priorities, which are subject to EU Directives, in the national hydrological plans covering multi-year periods.
- The **National Water Council** is responsible for drafting the National Hydrological Plan and for providing recommendations on all policies affecting water resources nationally. It is made up of representatives from national, regional and local governments, as well as representatives from user associations (e.g. agriculture), scientists and non-governmental organisations. Central and regional governments usually hold a majority. Its composition, organisational structure and operations are regulated by the Royal Decree 1383/2009, adopted on 28 August 2009.
- The **River Basin Authorities** are responsible for the management of water resources, including groundwater, wastewater releases into natural water streams, public storage and long-distance transport infrastructure for interregional river basin districts. They develop and monitor river basin districts' management plans and administer water resources at the interregional level, including by granting concessions. They have a high degree of organisational, functional and budgetary autonomy. A **Competent Authorities Committee** coordinates all water-relevant policies of the governments and of the river district authorities. For By contrast, intraregional river basin districts are administered by an **Autonomous Water Authority**.
- **The autonomous communities** are responsible for natural resources, agricultural policies, subject to EU Directives and central government's guidelines, and the responsibilities listed in the Table 5.1. They fund transport and supply infrastructure for water resources that are not shared across regions, although some recent regional Statutes specify the creation of reserves from shared river flows. Through the Autonomous Water Authority, they have complete oversight over intraregional river basin districts.
- **Municipalities** are responsible for the supply and collection and treatment of wastewater (Table 5.1). They may provide these public services themselves or through licensed public or private enterprises.
- Water users (e.g. agricultural producers) are required to create **user associations** when they share a common concession or the same outlet. The associations establish norms for distribution and control, regulate the use and maintenance of shared hydraulic systems, organise the shared payments and resolve problems among members. They play an important role in the River Basin Authorities, in which they have assigned members in the users' assembly.

Source: (Fuentes, 2011[84]).

There are currently six river basin districts in Andalusia (Figure 5.1), which may be changed by Royal Decree. The river basin districts of Tinto-Odiel-Piedras, Guadalete y Barbate (both referred to as "Antlantica Andalucia" in the Figure), and Mediterráneo, fall entirely within the territory of Andalusia and are managed by the Andalusian Water Administration under the Andalusian Water law (9/2010) (art. 97 to 99) (Junta de Andalucia, 2010[18]). The river basin districts of Guadalquivir, Guadiana, and Segura are interregional (or international, i.e. shared with Portugal in the case of Guadiana) basins and are thus managed by a River Basin Authority (Gobierno de Espana, 2007[24]).

Figure 5.1. River basin districts in Andalusia

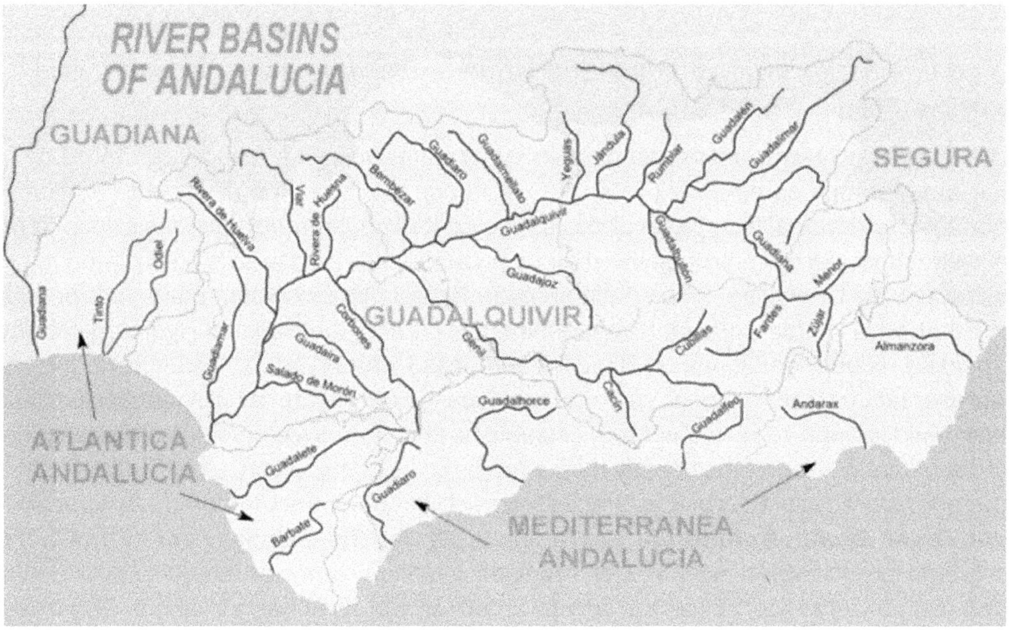

Source: (Absolute Axarquia, n.d.[25]).

5.3. Current levies related to water across levels of government in Spain

The Spanish Water law (29/1985), repealed by Royal Decree 1/2001 and consolidated in law 62/2003, established five levies related to water at the national level in Spain (Gobierno de Espana, 1985[26]; Gobierno de Espana, 2001[27]; Gobierno de Espana, 2003[7]). In the case of interregional river basin districts, the below levies are managed and collected by the competent river basin authority or by the central government's tax administration in case of a previous agreement. In the latter, the river basin authority will share the pertinent data and information with the State Tax Administration Agency, which will collect the tax and make it available to the river basin authority.

The **fee for the use of a public hydraulic domain** (art. 112): concessionaires and authorised persons are charged a fee for the use or occupation of a public hydraulic domain that requires administrative authorisation[1]. The fee corresponds to: (i) 5% of the market value of the land in case of occupation of land of the public hydraulic domain, (ii) 5% of the value of the use or benefit from the use of the hydraulic domain in case of use of the domain, or (iii) 100% of the value of the materials consumed in case of use of the assets of the hydraulic domain. It aims to finance the protection and improvement of the public hydraulic domain (Gobierno de Espana, 2001[28]).

The **hydroelectric development fee** (art. 112 bis): the holders of a hydroelectric exploitation are charged a fee for the use and exploitation of the public hydraulic domain for hydroelectric development purposes. The fee corresponds to the economic value of the hydroelectric energy produced by the holder of a hydroelectric exploitation for the use and exploitation of the public hydraulic domain, measured in plant bars, and declared by the holder in a self-assessment. It aims to finance the protection and improvement of the public hydraulic domain (Gobierno de Espana, 2001[28]).

The **pollution control fee** (art 113): the persons who carry out discharges into the public hydraulic domain are charged a fee. The fee corresponds to the volume of discharge authorised multiplied by the unit price for discharge control. The unit price is the basic price (i.e. EUR 0.01751 per m3 for urban waste-water and EUR 0.04377 per m3 for industrial wastewater) multiplied by a coefficient of increase or reduction, which is established by regulation depending of the nature of the pollution from the discharge. The fee aims to finance the assessment, control, protection and enhancement of the river basin district where pollution is emitted. When a taxable person of the pollution control fee also has to pay other taxes related to the protection, improvement and control of the river basin district, the amount of these taxes may be deducted from the amount of the pollution control fee (Gobierno de Espana, 2001[28]).

The **regulation fee** (art. 114): the persons benefitting from surface water or groundwater regulation works[2], financed wholly or partially by the central government, are charged a fee to compensate the costs related to the building, operations and maintenance of these works. The fee corresponds to the sum of the expected operation and maintenance costs of the works carried out, the administrative costs related to the works and 4% of the value of the investments made by the central government (Gobierno de Espana, 2001[28]).

The **water use tariff** (art. 114): the persons benefitting from other hydraulic works than those falling under the regulation fee (e.g. works to correct the deterioration of the public hydraulic domain) are charged a tariff to finance the investment, operation and maintenance costs of these works. The fee is calculated in the same way as the regulation fee (see above) (Gobierno de Espana, 2001[28]).

National regulation also allows for the possibility of establishing the below levies:

The **irrigation charges**: irrigation water users of the same water concession are charged to finance the construction, maintenance and improvement of irrigation infrastructure in the Community[3] (Royal Decree 849/1986) (art. 198 to 231) (Gobierno de Espana, 1986[29]). These charges are regulated by each Irrigation Community (Gobierno de Espana, 2003[7]).

The **amortisation rate and operating rate of water companies**: the rates are applied to compensate water companies for their costs associated with the investment, operation, and maintenance of hydraulic infrastructures. The contributions are paid by the River Basin Authority under the terms defined in an agreement between the water company and the River Basin Authority, which is regulated under the Spanish Water law (art. 126) (Gobierno de Espana, 2003[7]).

The **fee for the occupation and use of the public maritime-terrestrial domain:** fee established to compensate the costs associated with the protection and enhancement of the maritime-terrestrial domain (Coastal Law 22/1988) (art. 84) (Gobierno de Espana, 1988[30]). The fee is regulated under the Order of 30 October 1992.

The Autonomous Community of Andalusia created additional water-related levies. The Water law of Andalusia (9/2010) (Junta de Andalucia, 2010[18]) established the following:

- **Improvement fee** (art. 72-78): the taxable matter is the urban use of water from any sources, whether supplied by public or private supply networks. The fee applies to water users (i.e. the holder of water supply contracts). If the users hold different water supply contracts, the fee shall apply on each of the contracts. The tax base is the volume of water invoiced by the water supply companies (expressed in cubic metres). The fee is levied through two modalities: (i) a regional fee (art.79 to 90) and (ii) a local fee (art. 91 to 96) It aims to finance hydraulic infrastructures for the provision of water supply, sewage, and wastewater treatment services. (Junta de Andalucia, 2010[18]). The fee accounted for a small share of tax revenue in Andalusia in 2020 (close to 1%) (Ministerio de Hacienda y Funcion Publica, 2022[31]). Under the Andalusian Decree-law 7/2022, this fee has been suspended temporarily from January 1st to December 31st 2023 in order to mitigate the effects of inflation on households and industries (Junta de Andalucia, 2022[32]).
- **General services fee** (art. 114): the aim of the fee is to cover administrative expenses of the Andalusian Water Administration to guarantee the proper use and conservation of water (Junta de Andalucia, 2010[18]). This fee still remain to be implemented (Adame Martínez, 2020[33]).

Andalusia established another levy, which is regulated under the Andalusian Law 18/2003 on fiscal and administrative measures (Junta de Andalucia, 2003[34]):

- **Tax on discharges into coastal waters** (art. 39 to 55): the taxable matter is the discharge to coastal waters, which is carried out from land to any maritime-terrestrial public domain or to its area of protection. The tax is levied on the persons who carry out discharges into the maritime-terrestrial public domain. The tax base is the amount of the pollutant load of the discharge, which is equal to the sum of the polluting units. The polluting units of each parameter of the discharge (established in Annex I of the law) are the discharge flow (in thousand m^3 per year) multiplied by the value of a parameter, divided by a reference value in accordance with the Annex I of the law. The aim of the tax is to promote the good chemical and ecological status of coastal waters (Junta de Andalucia, 2003[34]). The tax represented a negligible amount of revenue in Andalusia in 2020.

Finally, municipalities may establish fees under the Urban Water Supply Regulation (Decree 120/1991) (art. 94 to 104) (Junta de Andalucia, 1991[35]):

- **Municipal fees for the provision of water supply, sewage, and wastewater treatment services**: municipalities may establish fees related to the provision of water supply, sewage and wastewater treatment services. The fees are levied on users of drinking water, sewerage and wastewater treatment services in a given municipality. They may consist of a fixed part per user and/or a variable part depending on the volume of water invoiced (in m^3). Several types of use may be identified (i.e. domestic, commercial, industrial, official bodies and other uses). The fees aim to compensate the local water supply company for the operating costs associated with the provision of urban water services (e.g. supply of drinking water, sewage and wastewater treatment). The municipality is responsible for developing the specific regulation of each fee. Special surcharges may also be established for operations other than water supply, connection supply, service connection charges, contracting fees, charges for the financing of infrastructure, deposits and specific services (Junta de Andalucia, 1991[35]).

A list of existing water-related levies in Spain, including those levied by the autonomous communities, is provided in Table 5.2.[4] The levies are structured according to their domain (i.e. freshwater or maritime) and category (i.e. water abstraction, water usage and water pollution).

Table 5.2. Existing levies related to water in Spain

Water domain	Category	Levy	Competence	Taxable matter	Payer
Freshwater domain	Water abstraction[1]	Fee for the use of public hydraulic goods	National	The occupation, use, or exploitation of hydraulic public domain assets in the channels of natural currents, continuous or discontinuous, and in the beds of lakes and lagoons and those of surface reservoirs in public channels.	Concessionaires and authorised persons for the use or occupation of public hydraulic domain.
		Hydroelectric development fee	National	The use for purposes of hydroelectric exploitation of the dams of the reservoirs or the channels built with funds from the Public Administration. Said use must be foreseen in the corresponding District Hydrological Plan.	Holders of a hydroelectric exploitation.
	Water usage[2]	Regulation fee	National	The availability or use of water flows for irrigation, population supplies, industrial uses or uses and installations of any kind, which are benefited or improved by	Persons benefitting from surface water or groundwater regulation works.

				regulation hydraulic works.	
		Water use tariff	National	The availability or use of water flows for irrigation, population supplies, industrial uses or uses and installations of any kind, which are benefited or improved by specific hydraulic works.	Persons benefitting from other hydraulic works than those falling under the regulation fee.
		Irrigation charges	National	The use of hydraulic infrastructure from the Irrigation Community.	Irrigation water users of the same water concession.
		Amortisation rate and operating rate of water companies	National	Availability and use of the water resources generated from the hydraulic infrastructures built by the State Water Companies	Contributions are paid by the River Basin Authority.
		Improvement fee	Regional	The availability and urban use of drinking water from any source, supplied by public or private supply networks. Water losses in the supply networks will be assimilated to urban use.	Water users (i.e. the holder of water supply contracts).
		General services fee	Regional	Performance of activities and the provision of general administration services of the Public Administration, which directly or indirectly affect the conservation and exploitation of hydraulic works, as well as the different uses and exploitation of groundwater and surface water.	Water users.
		Municipal fees for the provision of water supply, sewage, and wastewater treatment services	Local	Obtaining management services for the urban water cycle, which include drinking water supply, sewerage and wastewater treatment.	Users of drinking waters, sewerage and wastewater treatment services in a given municipality.
	Water pollution	Pollution control fee	National	The realisation of discharges to the hydraulic public domain.	Persons who carry out discharges into the public hydraulic domain.
Maritime domain	**Water usage**	Fee for occupation and use of the public maritime-terrestrial domain	National	The occupation or use of maritime-terrestrial public domain assets.	Concessionaires and authorised persons for the use or occupation of maritime-terrestrial public domain.
	Water pollution	Tax on discharges into coastal waters	Regional	Ecological tax to be paid by those who discharge into the maritime-terrestrial public domain in order to promote the good chemical and ecological status of coastal waters.	Persons who carry out discharges into the maritime-terrestrial public domain.

Notes:

1. Water abstraction refers to the process of taking or extracting water from a natural source.
2. Water usage refers to various uses of water, including drinking, irrigation, treatment and industrial applications.

Source: Author's own elaboration.

The below table presents different water-related levies that exist among the autonomous communities in Spain (at the exception of Navarra and the Basque Country) (Table 5.3).

Table 5.3. Levies on water in the Spanish Autonomous Communities

AND	CAT	GAL	AST	CANT	RIO	MUR	VAL	ARA	CLM	CAN	EXT	BAL	MAD	CYL
Improvement fee (*Canon de Mejora*) on urban water user	Water fee (*Canon del Agua*)	Water fee (*Canon del Agua*)	Tax on environmental effects of water use (*Impuesto sobre las afecciones ambientales del uso del agua*) on water use (agriculture exempt)	Wastewater fee (*Canon del agua residual)* on wastewater generation (agriculture exempt)	Sanitation fee (*Canon de saneamiento*) on wastewater generation (measured through water usage; agriculture exempt)	Sanitation fee (*Canon de saneamiento*) on wastewater generation (measured through water usage)	Sanitation fee (*Canon de saneamiento*) on wastewater generation (measured through water usage)	Water pollution tax (*Impuesto sobre la contaminación de las Aguas*) on wastewater generation (measured through water usage; agriculture exempt)	Adduction fee (Canon de aducción) on the water supply services	Desalinization fees (*Tarifas desalinizadoras*) – Charges the use of desalinated water	Sanitation fee (*Canon de saneamiento*) on wastewater generation (measured through water usage; agriculture exempt)	Sanitation fee (*Canon de saneamiento*) on wastewater generation (measured through water usage; agriculture exempt)	Sanitation fee (*Canon de saneamiento*) on wastewater generation (measured through water	Adduction fee (*Canon de aducción*) on the water supply services
General services fee (*Canon de servicios generales*) to cover administrative costs (to be implemented)	Charges water use (agriculture exempt)	Charges water use (agriculture exempt)				Tax on discharges into coastal waters (*Impuesto sobre vertidos a las aguas litorales*) on maritime water pollutanters not connected to public infrastructure		Environmental tax on certain uses and exploitation of water reservoirs (*Impuesto medioambiental sobre determinados usos y aprovechamientos de agua embalsada*) on hydropower generation	Water treatment fee (*Canon de depuración*) on water treatment services (measured in outflow of water treatment facilities)	Isnsular council fee (*Tarifas del Consejo Insular*) on the use of reuse water (including agriculture)			Reuse water fee (*Tarifas de reutilizacion*) on the use of reuse water (including agriculture)	Water treatment fee (*Canon de depuración*) on water treatment services (measured in outflow of water treatment facilities)
Tax on discharges into coastal waters (*Impuesto sobre vertidos a las aguas litorales*) –		Tax on the environmental damage caused by the construction and use of water reservoirs (*Impuesto sobre el daño medioambiental agua embalsada)* on hydropower generation												
Charges maritime on		Discharge coefficient												

water polluters		(*Coeficiente de vertido*)												
		Charges wastewater generation												

Note: AND: Andalusia; CAT: Catalunia, GAL: Galicia, AST: Asturia; CANT: Cantabria; RIO: La Rioja; MUR: Murcia; VAL: Valencia; ARA: Aragon; CLM: Castilla-La-Mancha; CAN: Canaria; EXT: Extremadura; BAL: Baleares Islands; MAD: Madrid; CYL: Castilla-y-Leon.
Source: Author's own elaboration.

5.4. Possibilities for improvements to water-related taxation in Andalusia

This section identifies some opportunities to reform the environmental taxation related to fresh water pollution and usage in Andalusia. It is structured as follow: (i) water abstraction, with a focus on the agricultural, industrial and tourism sectors, (ii) water usage, with a focus on the agricultural, industrial and tourism sectors, and (iii) water pollution, focusing on taxes related to the use of fertilisers and pesticides. As for the chapter on GHG emissions and air pollution, it also includes opportunities at the national level, which may improve environmental outcomes in Andalusia. The opportunities presented result from the analysis of the legal framework, the responsibilities mapping and the existing levies in Spain as discussed in the precious section. A selection of these opportunities will be further analysed from an economic perspective in Activity 1.3 of the report. Case studies on the use of such instruments in other countries and Spanish regions will also be included. Where relevant this discussion looks at related aspects such as distributional consequences and health.

5.4.1. Water abstraction and usage: agriculture, industry and tourism

Improvements in taxation related to water abstraction for agricultural and industrial purposes represent important possibilities of reform in Andalusia. The main possibility identified at the regional level is the creation of an Andalusian levy for water abstraction. Additional opportunities exist at the national level, which are presented below. The White Book for Tax Reform in Spain also provides recommendations on taxation related to water abstraction in Spain (Box 5.3).

Possibility 1 (regional or national): creating a regional or national levy on water abstraction

The current levies related to water abstraction do not reflect the environmental costs of this activity. The entities abstracting water only pay for water abstraction concessions granted by River Basin Authorities, regardless of the volume of water abstracted (Greenpeace, 2019[36]). The levies are associated to the use of exploitation sources of water (e.g. fee for the use of public hydraulic goods) and hydraulic infrastructure (e.g. irrigation charges).

Possibility 2 (national): developing incentive mechanisms on sustainable groundwater abstraction for Irrigation Communities

Irrigation charges determined by Communities often do not consider the environmental costs of water abstraction. Most Communities only consider the compensation required for developing and maintaining irrigation infrastructure in their charges (Fuentes, 2011[37]).

The central government has the opportunity to create a legal provision for enabling River Basin Authorities to develop mechanisms able to incentivise sustainable groundwater abstraction for Communities (e.g. charge on groundwater abstraction to users in Communities abstracting water persistently above a sustainable level) (Fuentes, 2011[37]).

Possibility 3 (regional): creating a tourism tax with an environmental criteria

There is no tourism tax that integrates the environmental cost of touristic activities. Such a tax at the regional level could cover all water-related environmental costs emerging from touristic activities. The tourism tax could be considered horizontal and incorporate several criteria related to water and waste for example.

5.4.2. Water pollution: pesticides and fertilisers

Most of the current mechanisms for limiting the use of fertilisers and pesticides are associated with (i) water quality standards, (ii) the limits to the use of pesticides and fertilisers and (iii) the bans on specific chemicals. The Spanish Constitution allows Andalusia to implement stricter water quality standards within its territory. The White Book for Tax Reform in Spain also made a recommendation on taxation related to water pollution (Box 5.3).

Possibility 4 (national): creating a tax to disincentivise the use of pesticides and fertilisers

The EU provides a wide range of water quality standards regulations, which have been translated into Spanish law, granting the autonomous communities the power to develop programs to prevent and mitigate water pollution. Many of these standards concern pollutants strictly associated with agricultural activities (e.g. nitrogen and phosphorus).

A national tax on the manufacturing or importation of pesticides and fertilisers could be established. This tax could be accompanied by public campaigns to raise awareness on the risks associated with these chemicals and the benefits of adopting more sustainable agricultural practices. The revenue from the tax could be used to compensate farmers switching to more sustainable agricultural practices (Adame Martínez, 2020[33]).

Other experts have suggested be to link water concessions for water abstraction by Irrigation Communities to water quality standards by amending the Spanish Water law (Greenpeace, 2019[36]; Ministerio para la Transición Ecológica y el Reto Demográfico, 2020[38]). Successful cases of taxes on the manufacturing and importation of fertilisers and pesticides exist in Norway and Sweden (Adame Martínez, 2020[33]; Gago et al., 2006[39]).

Box 5.3. Recommendations from the White Book on taxes related to water

The White Book for Tax Reform in Spain provides several recommendations related to water taxation to help improve water quality and fight against water scarcity (Table 1.2).

1. Introduction of coordination and cooperation measures to improve the design and effectiveness of regional taxes on environmental damage to water

Some autonomous communities have established taxes on environmental damage caused by the use and exploitation of water reservoir (e.g. Galicia, Castilla y León and Aragón), which led to litigation and, according to the Committee, raises doubts about their environmental effectiveness. The Committee recommends to intensify the environmental characteristics of these taxes.

Regarding taxes on discharges into coastal waters, Andalusia and Murcia have been pioneers in their establishment. The Committee of experts recommends to exempt from these taxes the direct reuse of reclaimed water that have a concession or authorisation, as there is no discharge in these cases, and the discharges into coastal waters of desalinated waters.

2. Reform of fees associated with coverage of hydraulic infrastructure costs

The national fees associated with the coverage of the costs of hydraulic infrastructures (e.g. the regulation fee and the water use tariff) price the benefit on the use of water (i.e. availability of the resource, improvement) resulting from hydraulic works financed by the central government. The Committee considers these fees as deficient with a high degree of litigation since it is difficult to identify the scope of the beneficiaries for each work and there is a lack of clarity in the liquidation of the rate. The Committee recommends to review these fees to improve both qualitative and quantitative elements of their design to shift from a "quota levy" to a levy able to recover water environmental and resource costs.

3. Creation of a tax on the extraction of water resources

The Committee proposes the creation of a national tax on water extraction to incentivise the proper use of a scarce resource. The tax would be carried out by the central government and not transferred to the autonomous communities. The taxable matter would be the extraction of water for any use, in order to tax the use of a resource belonging to the public domain. The tax base would be the volume of water extracted and the fee would be proportional to it. The Committee also underlines that a use factor could be applied, as well as a territorial factor, depending on the difficulties of extraction.

4. Creation of a national tax on the nitrogen content of fertilisers used in agriculture

The Committee also suggests the creation of a national tax on the nitrogen content of fertilisers used in agriculture, combined with a VAT increase for these products to reduce diffuse nitrate pollution in Spanish water bodies.

Source: (Comité de personas expertas, 2022[40]).

References

Absolute Axarquia (n.d.), *Rivers of Andalucia*, https://www.absoluteaxarquia.com/andalucia/rivers-andalucia/. [25]

Adame Martínez, F. (2020), *Análisis desde la perspectiva ambiental de la tributación de la Comunidad Autónoma de Andalucía en el entorno de las restantes Comunidades Autónomas y la Unión Europea*, Fundación Pública Andaluza Instituto de Estudios sobre la Hacienda, https://www.iehpa.es/assets/files/repositorios_digitales/-1618503944_es-res.pdf (accessed on 15 February 2022). [33]

Andalucia, J. (2015), *Demarcación Hidrográfica del Guadalete-Barbate: Revisión de tercer ciclo (2021-2027).* [16]

Comité de personas expertas (2022), *Libro Blanco Sobre la Reforma Tributaria*. [40]

Commission, E. (2008), *Directive 2008/105/EC of the European Parliament and of the Council of 16 December 2008 on environmental quality standards in the field of water policy, amending and subsequently repealing Council Directives 82/176/EEC, 83/513/EEC, 84/156/EEC, 84/491/EEC, 86/280/EEC and amending Directive 2000/60/EC of the European Parliament and of the Council.* [10]

Estadística sobre el Suministro y Saneamiento del Agua (2020), *El consumo medio de agua de los hogares fue de 133 litros por habitante y día, un 2,2% menos que en 2016*. [2]

EurEau (2021), *Europe's Water in Figures*. [3]

European Commission (2021), *Pathway to a Healthy Planet for All EU Action Plan: 'Towards Zero Pollution for Air, Water and Soil'*. [5]

European Commission (2021), *Regulation (EU) 2019/631 of the European Parliament and of the Council of 17 April 2019 setting CO2 emission performance standards for new passenger cars and for new light commercial vehicles, and repealing Regulations (EC) No 443/2009 and (EU) No 510/2011*. [11]

European Commission (2020), *Circular Economy Action Plan*. [13]

European Commission (2020), *Regulation (EU) 2020/741 of the European Parliament and of the council of 25 May 2020 on minimum requirements for water reuse*. [12]

European Commission (2006), *Directive 2006/118/EC of the European Parliament and of the Council of 12 December 2006 on the protection of groundwater against pollution and deterioration*. [8]

European Commission (2000), *Directive 2000/60/EC of the European Parliament and of the Council of 23 October 2000 establishing a framework for Community action in the field of water policy*. [6]

European Union (2012), *Consolidated version of the Treaty on the Functioning of the European Union*. [21]

Fuentes, A. (2011), "Policies Towards a Sustainable Use of Water in Spain", *OECD Economics Department Working Papers*, No. 840, OECD Publishing, Paris, https://doi.org/10.1787/5kgj3l0ggczt-en. [37]

Gago, A. et al. (2006), *Environmental Taxes in Spain: A Missed Opportunity*. [39]

Gobierno de Espana (2022), *Real Decreto 47/2022, de 18 de enero, sobre protección de las aguas contra la contaminación difusa producida por los nitratos procedentes de fuentes agrarias*. [17]

Gobierno de Espana (2009), *Real Decreto 1514/2009, de 2 de octubre, por el que se regula la protección de las aguas subterráneas contra la contaminación y el deterioro*. [9]

Gobierno de Espana (2007), *Real Decreto 125/2007, de 2 de febrero, por el que se fija el ámbito territorial de las demarcaciones hidrográficas*. [24]

Gobierno de Espana (2007), *Real Decreto 1620/2007, de 7 de diciembre, por el que se establece el régimen jurídico de la reutilización de las aguas depuradas*. [19]

Gobierno de Espana (2003), *Ley 62/2003, de 30 de diciembre, de medidas fiscales, administrativas y del orden social*. [7]

Gobierno de Espana (2001), *Real Decreto Legislativo 1/2001, de 20 de julio, por el que se aprueba el texto refundido de la Ley de Aguas*. [27]

Gobierno de Espana (2001), *Real Decreto Legislativo 1/2001, de 20 de julio, por el que se aprueba el texto refundido de la Ley de Aguas*. [28]

Gobierno de Espana (1988), *Ley 22/1988, de 28 de julio, de Costas*. [30]

Gobierno de Espana (1986), *Real Decreto 849/1986, de 11 de abril, por el que se aprueba el Reglamento del Dominio Público Hidráulico, que desarrolla los títulos preliminar I, IV, V, VI y VII de la Ley 29/1985, de 2 de agosto, de Aguas*. [29]

Gobierno de Espana (1985), *Ley 29/1985, de 2 de agosto, de Aguas*. [26]

Gobierno de Espana (1978), *The Spanish Constitution*. [22]

Greenpeace (2019), *Propuestas de fiscalidad ambiental: avanzando hacia un mundo más justo y sostenible*. [36]

Junta de Andalucia (2022), *Decreto-ley 7/2022, de 20 de septiembre, por el que se modifica la Ley 5/2021, de 20 de octubre, de Tributos Cedidos de la Comunidad Autónoma de Andalucía*. [32]

Junta de Andalucia (2020), *Pacto Andaluz por el Agua*. [20]

Junta de Andalucia (2019), *Informe de Medio Ambiente en Andalucía*. [1]

Junta de Andalucia (2010), *Ley 9/2010, de 30 de julio, de Aguas para Andalucía*. [18]

Junta de Andalucia (2007), *Organic law 2/2007 dated 19 March 2007 on Reform of the Statute of Autonomy for Andalusia*. [23]

Junta de Andalucia (2003), *Ley 18/2003, de 29 de diciembre, por la que se aprueban medidas fiscales y administrativas*. [34]

Junta de Andalucia (1991), *Decreto 120/1991, de 11 de junio, por el que se aprueba el Reglamento del Suministro Domiciliario de Agua*. [35]

Ministerio de Hacienda y Funcion Publica (2022), *Autonomous Community Funding*. [31]

Ministerio para la Transicion Ecologica y el Reto Demografico (2021), *National Plan for Wastewater Treatment, Sanitation, Efficiency, Savings and Reuse*. [15]

Ministerio para la Transicion Ecologica y el Reto Demografico (2021), *Sintesis de los borradores de planes hidrologicos de las demarcaciones hidrograficas intercomunitarias (2022-2027)*. [14]

Ministerio para la Transición Ecológica y el Reto Demográfico (2020), *Libro Verde de la Gobernanza del Agua en España*, https://www.miteco.gob.es/es/agua/temas/sistema-espaniol-gestion-agua/libro-verde-gobernanza-agua_tcm30-517206.pdf (accessed on 15 February 2022). [38]

Official Association of Biologists of Andalusia (2021), *Andalucía reutiliza 4,8% de su agua residual tratada, por debajo de la media*. [4]

Notes

[1] The hydraulic public domain *(Dominio Publico Hidraulico, DPH)* refers to inland surface and underground waters, the channel of continuous or discontinuous natural currents, beds of reservoirs and lakes. Marine waters are not included in the hydraulic public domain. Administrative authorisation is required for the use of the hydraulic public domain when the objective is to carry out works, to plant or cut vegetation, to navigate, to extract aggregates, to establish bridges, piers, etc. (Gobierno de Espana, 2001[28]).

[2] Regulation works or canalisations are systems of water conduction, channels and natural or artificial reservoirs.

[3] A Community refers to all the users of the water and other goods of public hydraulic domains from the same intake or concession. It has been established and regulated by art. 81 of the Spanish Water law (Gobierno de Espana, 2003[7]). Their establishment is compulsory. Communities are governing bodies with their own Statutes and Ordinances, drafted and approved by themselves and then by the River Basin Authority, which can only refuse their approval or introduce variants. There are different categories of Communities based on their use (e.g. irrigation, water supply, industrial). They can be surface or groundwater. If the concession of the waters involves several intakes, the Basin River Authority shall determine whether all the users are to be integrated in one Community or several independent Communities.

[4] The full description of water-related taxes (e.g. taxable matter, payers, destination) in Andalusia is available at the following link: https://www.juntadeandalucia.es/medioambiente/portal/areas-tematicas/agua/gestion-del-agua/recuperacion-de-costes/tarifas-y-canones-uso-agua-dominio-publico-hidraulico-dph-y-dominio-publico-maritimo-terrestre-dpmt.

6 Assessment: Water use and water pollution

This section considers the pricing of water usage and water pollution in Andalusia. Regarding water usage, the focus is on the design of current Andalusia-specific taxes, and deals with all water users: agriculture, households and industry. The assessment includes, but is not restricted to, the general service fee on water usage that is under discussion at the time of writing. The analysis is on fees in place in 2022.[1] For water pollution, the focus is on the introduction of new taxes in agriculture, particularly through taxing the use of pesticides and fertilizers. Industry and households already face a price on water pollution, which is set at a national level. The section deals with some of the main issues related to water pollution and usage but does not aim at being exhaustive about these complex and constantly evolving issues.

Water use and water pollution are analysed separately, even though both are interlinked. Indeed, on the one hand, water use may engender water pollution. The reasons for this include groundwater extractions beyond the capacity of aquifers deteriorating their quality and favouring the seeping of seawater into the aquifer (Aznar-Sánchez et al., 2019[1]); excessive irrigation causing erosion and transporting nutrients, pesticides, and heavy metals to surface water;[2] or reduced capacity for dilution when point sources are discharged to surface waters (OECD, 2017[2]). On the other hand, water pollution increases the costs of water use by increasing treatment requirements and increases water scarcity by reducing the quantity of water that is safe to use (FAO & IWMI, 2017[3]).

The focus is on market-based instruments (mainly taxes, fees and charges) even though other non-market instruments, such and command-and-control (CAC) type instruments (e.g. regulation and technology standards) are briefly discussed at the end of the section. Theoretically, along the same arguments as in the case of climate mitigation instruments, market-based instruments are generally more cost-efficient. Prices allow to decentralise the decision to abate by leaving it up to the users to determine where it is most efficient to reduce water consumption and pollutant use. However, in the case of water, if the price cannot align closely with the source of pollution (due to many diffuse pollution sources) or if, for water use, supply and demand curves are not known, hard to measure and vary too frequently, command-and-control type instruments may be better suited or used as a complement. As a reminder, a distinction is made between taxes on the one hand, and fees and charges on the other hand. The term "levy" covers taxes, fees and charges (see Box 1.2 in Part I for additional details).

Existing levies and the possible introduction of new taxes will be assessed in the context of European legislation, national laws and existing taxes. As exposed in Section 5, the Water Framework Directive (WFD) was introduced in 2000 at the European Union (EU) level and determines all EU member states' water regulatory framework. It incorporates or is complemented by eight other directives (see Box 5.1). From 26 June 2023, the European Union Water Reuse Regulation will set minimum requirements for water reuse for agricultural irrigation. At the Spanish national level, the Water Law was introduced in 1985 and has since been adjusted to the various EU-level directives. The Basin Plan was introduced in 1998, the 2001 National Hydrological Plan proposed large scale projects to transfer water from basins with excess water supply to basins in deficit and the 2005 A.G.U.A. Programme brought forward desalinisation and water reuse as national priorities. Finally, Spain – as all OECD members – is an Adherent to the OECD

Council Recommendation on Water, which puts forward four key principles in water management matters (see Box 6.1.).

Box 6.1. The OECD Council Recommendation on Water

In 2016, OECD member countries unanimously adopted a Council Recommendation on Water. The adoption marked the outcome of a two-year consultation process.

The Recommendations fall into five categories: (i) managing water quantity, (ii) improving water quality, (iii) managing water risks and disasters, (iv) ensuring good water governance, (v) ensuring sustainable finance, investment and pricing for water and water services.

Water management services (for both usage and pollution) are recommended to be financed with the following four key principles in mind, which might apply to either pollution, usage or both:

The Polluter Pays principle – which applies to pollution – to serve the following purposes:

Influence behaviour to reduce pollution,

or generate revenues to alleviate pollution and compensate for social costs.

The Beneficiary Pays principle – which applies to usage:

Aims at sharing the costs of water management between different water users such as industry, households and agriculture.

The Equity principle – which applies to both pollution and usage:

Focuses on who, within a group of users, bears the costs and benefits of water management.

Aims at ensuring equity in the access to water services and protection against water-related risks.

The Policy Coherence principle – which applies to both pollution and usage:

Ensure that different policy areas (agriculture, energy, land use, urban development or trade) do not have negative impacts on water availability, quality and freshwater ecosystems, or increase the cost of water management.

Source: Adapted from OECD (2016[4]).

In Spain, water use and water pollution policies take place at the river basin level (see Section 5). They fully fall under Andalusian authority for river basins that are entirely within the Autonomous Community. Only urban water management takes place at the municipality level.

Andalusia is part of six river basins, three of which are entirely within Andalusia. These are the Andalusian Mediterranean Basins, which make up 20.6% of the territory and the Guadalete-Barbate and Tinto-Odiel-Piedras river basins, which taken together make up 15.4% of it. Even though it does not fall entirely within the region, Guadalquivir is the greatest river basin in Andalusia. It makes up 60% of the Andalusian territory. The other river basins make up 3.8% of the territory for the Guadiana river basin and 0.2% for the Segura river basin.[3]

6.1. Pricing water usage

After a brief discussion on the reasons why pricing water usage properly is becoming an increasingly pressing issue worldwide and in Andalusia more specifically, this subsection first provides a general description of water users in Andalusia. This is followed by an analysis of water costs in general along with the criteria water pricing should address – in particular environmental economics principles. A description of water pricing instruments in Andalusia is then provided. This enables an analysis of the alignment of water pricing in Andalusia with sound economic principles. The analysis highlights that a better setting of objectives along with a better knowledge of demand patterns and service costs for different users would improve supply cost-recovery, sustainable use and equity. Finally, the introduction of water abstraction charges or taxes are discussed.

6.1.1. Water use and scarcity in Andalusia

In the coming decades, freshwater availability is projected to decrease and drought cycles to increase, including in Andalusia. Climate change models project warming temperatures, increased variability in precipitation patterns, and more frequent and extreme weather events (OECD, 2020[5]). Assuming no efficiency gains, some research finds that the world is to face a global freshwater deficit[4] of 40% by 2030 (2030 Water Ressources Group, 2009[6]). Andalusia will be particularly affected by these issues, as in the southern part of Spain drought cycles will very likely become more frequent as the 21st century progresses. Decreases in rainfall adversely affect groundwater recharge as well as the availability of surface water reservoirs. Moreover, higher temperatures increase evaporation, so that lower water inflows into the ground are complemented by greater water exits from the ground due to the phenomenon of evapotranspiration (Luis Caparrós-Martínez et al., 2020[7]). General decreases in precipitations are to be felt in certain river basins in particular, among which the Guadalquivir river basin,[5] which in recent years, has been experiencing increasing drought events.[6]

Spain is characterised by a high temporal and spatial variability in water resources, with certain regions – especially in the South – experiencing water scarcity and long periods of droughts. Mean annual precipitation varies from 2,200 mm in northern areas to 120 mm in the South-East. Consistent with this observation, mean annual runoff[7] varies from 50 mm/year (in particular in South-Eastern areas of Spain) to more than 800 mm/year (Northern areas and some mountainous areas) (Estrela and Sancho, 2016[8]). The important heterogeneity in water resources has resulted in the construction of numerous hydraulic works, such as dams[8], reservoirs and inter-basin water transfers (e.g., the Tagus-Segura Water Transfer), and the intensive use of groundwater through the drilling of wells. These supply-side strategies have helped deal with water scarcity in Spain so far, but the increasing risks linked to climate change and increasing water scarcity over the world call for a focus on demand-side instruments (such as taxes and levies, or certain non-market-based instruments) – even if used in parallel with other supply-side strategies such as desalination and reuse.

Water use is generally divided into agricultural, industrial and urban use. Energy and recreational uses are relatively less important and are not always documented. Urban use refers to grid use, as opposed to non-grid use which is taking water directly from rivers, sources, etc. More precisely, the Draft Hydrological Plan for the Mediterranean river basin defines urban water use as uses by households, regulated accommodation (e.g., hotels, rural tourism, campsites), non-regulated accommodation, industry connected to the urban grid, commercial and institutional uses, losses and uncontrolled uses. The definition is very similar in the other Hydrological Plans. Industry is understood as "industry not connected to urban grids".

In Andalusia, water is principally used for agriculture and urban supply. Depending on the river basin, between 65 and 87% of water use is for agriculture purposes, and between 9 and 25% is for urban supply (see Figure 6.1).[9] Agricultural use is mainly for irrigation: where data is available, use for feedstock is limited to between 0.2 and 2.2% of all agricultural water use. Urban use includes industrial, business and residential users connected to the grid, but the figures do not enable differentiating between these. Industrial use is generally limited (below 3%) except in the Tinto-Odiel-Piedras river basin, where it makes up about 16% of water use. The figure for recreational water use, including swimming pools and golf courses does not always exist, but where it does it hovers around 1% of water use except in the Mediterranean river basin where it goes up to 2.6%. Finally, energy production makes up 1.4% of water demand in the Guadalquivir river basin and 3.9% in the Guadalete-Barbete river basin. Projections to 2027 and 2039 from the six Draft Hydrological Plans indicate the shares of users in water demand are not expected to change much.

Hovering around 80%, the share of water demand for irrigation in Andalusia is above the worldwide average (around 70%), and reflects the fact that Spain is a country where irrigation as opposed to rain plays a major role in agricultural practices. Agriculture holds an important part in the Andalusian economy. It made up 6.7 of Andalusian GVA in 2021[10] and made up 30.8% of Spanish agricultural GVA (INE, 2023[9]).

Figure 6.1. Water demand by user

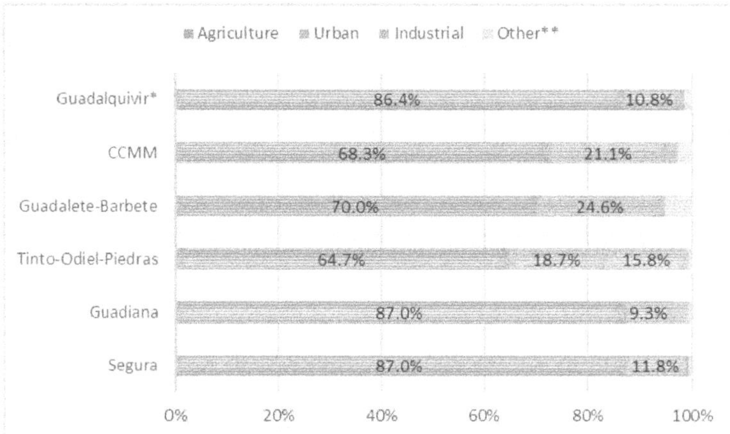

Note: The data presented is for 2021. Only percentages above 9% are indicated in the graph.
* The Guadalquivir Draft Hydrological Plan presents the demand for principal uses.
** Other stands for energy production use in the Gudalquivir river basin, for recreational use in the Mediterranean river basin, is composed of 3.9% for energy production and 1.4% for recreational use in Guadalete-Barbete, stands for recreational use in the Tinto-Odiel-Piedras river basin and recreational use (golf) in the Segura river basin.
Source: Table 56 in https://www.juntadeandalucia.es/medioambiente/portal/documents/20151/497870/DI_MEMORIA_GB.pdf/caaf7b70-5ec4-61ee-795a-5377b35f8d73?t=1582033431000 for the Guadalete-Barbete river basin, Table 30 in https://www.juntadeandalucia.es/medioambiente/portal/documents/20151/773714/DI_MEMORIA_TOP.pdf/5a437aad-1f91-6268-308d-7434e69f21c0?t=1582039745000 for the Tinto-Odiel-Piedras river basin, Table 161 in https://www.juntadeandalucia.es/medioambiente/portal/documents/20151/1152494/DI_MEMORIA_CMA.pdf/fd981d2a-5d61-ea14-8788-c52a0d3f03ad?t=1582032688000 for the Mediterranean river basin, Table 34 in https://www.chsegura.es/export/sites/chs/descargas/planificacionydma/planificacion21-27/docsdescarga/docplan2227Consolidado/01_MEMORIA/Memoria_PHDS_2022-27VCAD.pdf for the Segura river basin, Table 25 in https://www.chguadalquivir.es/documents/10182/2322527/PHGuadalquivirCAD_Memoria.pdf/b0a2577a-ac09-073e-29e6-12f2d3e37e9e for the Guadalquivir river basin and Table 38 in https://www.chguadiana.es/sites/default/files/2022-04/Memoria_1.pdf for the Guadiana river basin.

StatLink https://stat.link/f2ciuw

Water used for irrigation in agriculture is mainly abstracted from surface water and groundwater. For all river basins of which Andalusia is part except for Segura, in the period of 2011-2014, between 71 and 74% of water abstraction for irrigation is from surface water, while between 26 and 28% is from groundwater; a very low share is not freshwater (Figure 6.2). In a similar period (2012-2015), overall use of unconventional water resources (i.e., desalination and reuse) was highest in the Andalusian Mediterranean Basins and the Segura river basin.

Using groundwater instead of surface water for irrigation has several advantages, among which its availability even in times of drought, a lower need for investment at first and its immediate accessibility. In Spain, until 1985, a landowner owned the groundwater underneath their land. This situation changed with the Spanish Water Law of 1985, which established the public nature of all water resources as a general rule and the priority status of Hydrological Planning. Still, many users who extracted groundwater before 1986 were able to maintain their private water rights, and as a consequence a large amount of water used for irrigation has remained under private ownership, which in many cases has led to overexploitation (Luis Caparrós-Martínez et al., 2020[7]).

According to the Hydrological Plan of the second EU WFD cycle,[11] groundwater bodies in the six river basins districts in Andalusia present heterogeneous quantitative statuses. In five of the river basin districts, at least one fifth of groundwater bodies had poor quantitative status. Quantitative statuses ranged between the Tinto-Odiel-Piedras river basin, where all four associated groundwater bodies had good quantitative status and the Segura river basin, where 63% of its 63 groundwater bodies had poor quantitative status.

Figure 6.2. Origin of water used for irrigation in agriculture, 2011-2014

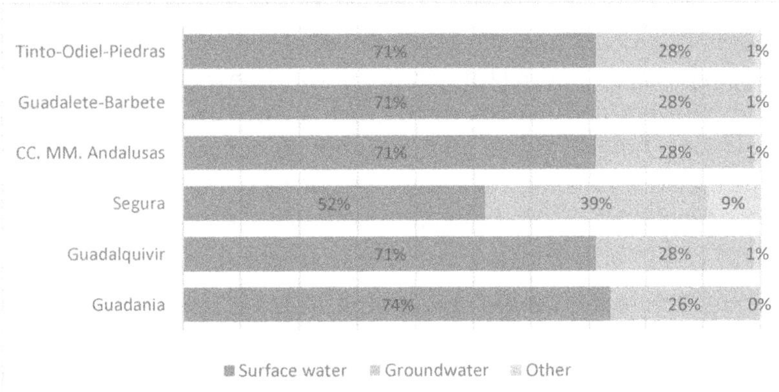

Note: "Other" includes desalination and reuse.
Source: Table 62 in the Draft Hydrological Plan of the Guadalete-Berbete river basin, https://www.juntadeandalucia.es/medioambiente/portal/documents/20151/497870/DI_MEMORIA_GB.pdf/caaf7b70-5ec4-61ee-795a-5377b35f8d73?t=1582033431000.

StatLink https://stat.link/qfsa42

The White Book for Tax Reform in Spain (Comité de personas expertas, 2022[10]) defines two phases of the water cycle: upstream and downstream. The 'upstream' phase, which the Hydrographic Confederations regulate, directly supplies consumers of non-potable water (agriculture, energy sector, large industries). The 'downstream' phase begins with the transfer of water for treatment to become drinking water. It also covers distribution to consumers by the municipalities, as well as its collection and treatment before its return to the natural environment, in which the Autonomous Regions play an important regulatory role.[12] Table 6.1 presents the different services and users they concern.

Table 6.1. Water services and users concerned

Service	User concerned
Upstream surface water services	Agriculture (irrigation, livestock farming) / Industry / Hydroelectric industry
Upstream groundwater services	Agriculture (irrigation, livestock farming) / Industry / Hydroelectric industry
Downstream distribution of water for irrigation	Agriculture (irrigation)
Downstream urban supply	Households / Industry
Self-provision	Agriculture (irrigation, livestock farming)
Reuse	Urban / Agriculture (irrigation, livestock farming) / Industry
Desalination	Urban / Agriculture (irrigation, livestock farming) / Industry

Source: Authors.

6.1.2. The costs of water use

If the water market is properly managed,[13] the costs related to water use include: (i) supply costs, (ii) administrative and governance costs and (iii) environmental externalities[14] (OECD, 2010[11]; Rogers, Bhatia and Huber, 1998[12]; Rogers, 2002[13]; Cardone and Fonseca, 2003[14]). Supply costs may be understood as operation, network and maintenance costs as well as capital costs – they can include a fixed and variable component. Administrative and governance costs include those incurred in regulating the service, institutional capacity building, and the cost of devising and implementing the policy and enabling environment for the sector. Environmental externalities arise because harm is imposed on the ecosystem.

A recent European Commission report (Mottershead et al., 2021[15]) discusses analyses of environmental externality costs in the water use context (which are referred to as scarcity costs in the report), with a focus on the externalities associated to ecosystems. They provide various examples of such externalities. These can relate to depriving fish of water as their habitat, but also to a lack of water as a support to wetlands, as a necessity for healthy vegetation or again as a carbon sequestration provider.

Putting a number on the different cost dimensions of water use is not straightforward. While supply costs along with administrative and governance costs may be measured in a similar fashion to other services (with certain difficulties), externalities can prove harder to measure. Some difficulties are discussed in (OECD, 2010[11]). For example, relating to capital costs, it is not clear whether capital charges relating to "stranded assets" that no longer provide a useful service should be included.

Regarding the valuation of externalities linked to water use, difficulties arise from the various parts of the ecosystems concerned. For now, no single estimate incorporating all externalities exists, but some do for specific externalities to ecosystems. Mottershead et al. (2021[15]) for the European Commission review Australian and Spanish studies and retain an environmental cost value of EUR 0.30 per cubic metre of extracted water. The recently released White Book for Tax Reform in Spain presents total environmental costs by river basin. According to those calculations, environmental costs represent between 6 and 10% of costs in all river basins other than Segura, and 23% for the Segura river basin district.

Water is generally not properly managed through a market mechanism, and allocation between users is not always well defined. In this case, opportunity costs can arise, whereby for example, the upstream user of a river may deprive the downstream user from getting enough water. Moreover, the fact that water is generally not managed through a well-functioning market engenders a lack of consideration for the scarcity this can cause.

6.1.3. Criteria for pricing water

The overall pricing of water usage seeks to satisfy several criteria, and addressing one does not guarantee that the others are addressed as well – in fact there might even be trade-offs between them (Grafton, Chu and Wyrwoll, 2020[16]; OECD, 2010[11]). The five main economic and environmental criteria are the following, in the order of the discussion to follow (not their importance):

1) Cost of service recovery (i.e., water prices cover the full current and future supply, administrative and governance costs of water use and guaranty financial sustainability);
2) Universal access and affordability;
3) Promotion of sustainable water use for human populations;
4) Internalisation of externalities caused to ecosystems;
5) Equity.

Cost of water service recovery (criterion 1) should theoretically be guaranteed through long-run marginal cost of supply (LRMC) pricing (Olmstead and Stavins, 2009[17]), even though in practice, water prices lie well below these costs. LRMC pricing is supposed to reflect the full economic cost of water supply (since fixed costs are taken over the long term, so allowed to vary). This includes the sum of the transmission, treatment and distribution cost, as well as some portion of the capital cost of current reservoirs and treatment systems. In practice, LRMC pricing reveals complicated to measure (it requires full metering of consumption as well as full information on capital costs in the long run) and implement (e.g., prices should vary over the course of the year, and with the source from which water is abstracted). Moreover, it might not fit all criteria of financial sustainability (e.g., it does not ensure that utilities can accumulate sufficient funds for investment) (OECD, 2017[18]). Alternatives to LRMC pricing include, but are not restricted to, average cost pricing, short-run marginal cost pricing combined with public provisions guided by cost-benefit analysis, etc. Australia is one of the only countries to use LRMC pricing in practice (Tooth, 2014[19]).

Most countries price water using a fixed charge, which covers connection costs to the public water supply and sewerage systems, and a volumetric rating system, which covers the volume of water supplied at a fixed specific rate per cubic meter. This approach is closer to *average cost pricing*, but may leave out an important component of cost of water recovery, i.e., the inclusion of a forward-looking aspect in the service cost recovery. Indeed, anticipating future costs by investing ahead of time is key to ensure financial sustainability.

Affordability concerns (criterion 2) are generally addressed through reduced rates on low levels of water consumption or preferential rates for certain groups. It is worth stressing here that affordability and distributional concerns do not only arise for households, but may also arise for small farmers, for instance. In the face of increased prices on water, this population could also face strong adaptation and hence affordability issues.

For now, many countries address affordability concerns by charging lower rates to vulnerable households, but as in the case of energy prices, higher efficiency would be achieved through targeted support policies (Arbués and García-Valiñas, 2020[20]; Van Dender et al., 2022[21]). Reduced water tariffs may be relatively simple to introduce and to communicate in general, but are not necessarily well targeted to the most vulnerable consumers and weaken incentives to reduce water use when supply is tight. Affordability can be addressed through targeted support. Public transfers can be justified by the need for public policies to guarantee access to a minimum level of water for everyone. Indeed, on 28 July 2010, the United Nations General Assembly adopted a historical resolution that recognised "the right to safe and clean drinking water and sanitation as a human right that is essential for the full enjoyment of life and all human rights".[15] In order for this basic need to be satisfied, two-part water tariffs may be used that charge less or nothing for the amount of cubic meters per person which are deemed to be the necessary minimum level. Public transfers to water companies could help cover the foregone cost-recovery from the lower rates on small volumes of water.

Cost of service recovery and hence financial sustainability on the one hand and affordability and social concerns on the other, face trade-offs but also go hand-in-hand. Financial sustainability, by increased costs, can jeopardise affordability. However, if financial sustainability is not guaranteed, this increases the risk of underinvestment (e.g. to renew urban water infrastructures) and more remote and poorer areas not being furnished in water. For now, the whole Spanish population is connected to water provision networks,[16] but care should be taken for this not to become a concern in the future.

Given the lack of a water market, sustainable use for human societies (criterion 3) is threatened through potential overuse by households, firms or farmers, which should be addressed through some form of management – ideally through pricing. Instruments or mechanisms used to ensure sustainable water use will depend on administrative capacity and formality or informality of water use. On top of potential overuse by households, firms or farmers, it is also important to stress another source of inefficiency in this regard: water losses in networks. In Spain, this is an issue for importance, given that leakage rate for public water supply is estimated at 30%.[17]

Demand and supply curve estimates may provide a better framework set the right prices to encourage sustainable water use. However, whether sustainable water use is managed through pricing or non-pricing policies, it is important that alternatives to certain levels of water use are promoted which enable demand elasticities to be high enough so as to ensure effectiveness of the policy and avoid affordability, productivity (especially in the agricultural sector) and political feasibility issues. Moreover, salience of water prices is also key to ensure significant responsiveness (García-Valiñas, Martínez-Espiñeira and Suárez-Varela Maciá, 2021[22]).

In the medium-term, supply-side solutions to sustainable use also exist, such as reuse and desalination. However, their high costs can jeopardise financial sustainability and desalination is very energy-intensive (even though it is becoming increasingly efficient (OECD, 2020[23])), which can cause other environmental, in particular climate-related, issues.

Regarding externalities to ecosystems, theoretically, when externalities can be measured, a form of Pigouvian taxation can be introduced. Pigouvian taxation aims to reflect the external costs that individual water consumption puts on society through the ecosystem in the decision making of the individual water user and hence to steer behaviour accordingly.

In line with the OECD Council Recommendation on Water, pricing should be such that the burden falls in an equitable way on users (criterion 5). The water pricing system as a whole should avoid one type of user generating the greatest costs for the system, while the other users bear the price. However, it is important to bear in mind the interlinkages between the various water users, and in particular general equilibrium effects. Indeed, while households are at the very end of the supply chain, firms and farmers use the water to produce goods for consumption, ultimately, by households in Andalusia or Spain more generally or for export. Hence, cost pass-through should also be accounted for, especially for the Equity principle. In the following, equity is only assessed in a partial equilibrium setting. Its evaluation in a general equilibrium context would require a much deeper analysis.

Equity can positively impact sustainable use. Indeed, sharing the burden in an equitable manner can have an effect on sustainable use by making users responsible for their water consumption.

6.1.4. Policy instruments for pricing water in Andalusia

Spain and Andalusia more specifically currently recover water-related costs through user charges and through public spending. At the Spanish national level, slightly above half of water management is funded through the public budget as opposed to water tariffs charged to individual users (OECD, 2021[24]). Most national and Andalusian water-use tariffs include a mix of cost-recovery, sustainable use and affordability criteria.

Table 6.2 presents the pricing mechanisms that apply in Andalusia in 2022 (i.e., Spanish, Andalusian and more local levies) and maps them to the different pricing criteria they address (intentionally or not), and the phases of the water cycle they concern. As exposed in Section 5, on top of multiple national levies, there are two Andalusia-specific levies, one local levy and one municipal levy. These four levies are discussed in more detail below.

Table 6.2. Criteria addressed by water-use levies in Andalusia

Levy	Service concerned	Brief description	Criteria effectively addressed
Regulation fee (National)	Upstream surface water services /	The persons benefitting from surface water or groundwater regulation works, financed wholly or partially by the central government, are charged a fee to cover the costs related to the building, operation and maintenance of these works. The fee is to be distributed among the different hydrological exploitation systems according to criteria of rationalisation of water use, fairness in the distribution of obligations and self-financing of the service	Cost of service recovery (Sustainable use and equity)
Water use tariff (National)	Upstream surface water services /	The persons benefitting from other hydraulic works than those falling under the regulation fee (e.g., works to correct the deterioration of the public hydraulic domain) are charged a tariff to finance the investment, operation and maintenance costs of these works.	Cost of service recovery (Sustainable use and equity)
Fee for the use of public hydraulic goods (National)	Upstream surface water services	Concessionaires and authorised persons are charged a fee for the use or occupation of a public hydraulic domain that requires administrative authorization. It aims to finance the protection and improvement of the public hydraulic domain.	Environmental cost
Hydroelectric development fee (National)	Upstream surface water services	The holders of a hydroelectric exploitation are charged a fee for the use and exploitation of the public hydraulic domain for hydroelectric development purposes. It aims to finance the protection and improvement of the public hydraulic domain.	Cost of service recovery
Irrigation charges (National)	Downstream distribution of water for irrigation	Irrigation water users of the same water concession are charged to finance the construction, maintenance and improvement of irrigation infrastructure in the Community. These charges are regulated by each Irrigation Community.	Cost of service recovery / Equity
Amortisation rate and operating rate of water companies (National)	Downstream urban supply	The rates are applied to compensate water companies for their costs associated with the investment, operation, and maintenance of hydraulic infrastructures. The contributions are paid by the River Basin Authority under the terms defined in an agreement between the water company and the River Basin Authority, which is regulated under the Spanish Water law.	Cost of service recovery
Fee for occupation and use of the public maritime-terrestrial domain (National)	Desalination	Concessionaires and authorised persons are charged for the use or occupation of maritime-terrestrial public domain. This fee was established to cover the costs associated with the protection and enhancement of the maritime-terrestrial domain.	Environmental cost
Improvement fee* (Andalusia & local)	Downstream urban supply	The fee applies to water users. The base is the volume of water invoiced by the water supply companies. The fee is levied through two modalities: (i) a regional fee and (ii) a local fee. It aims to finance hydraulic infrastructures for the provision of water supply, sewage, and wastewater treatment services.	Cost of service recovery / Sustainable use / Affordability
General service fee** (Andalusia)	Upstream surface water services / Upstream groundwater services	The aim of the fee would be to cover administrative expenses of the Andalusian Water Administration to guarantee the proper use and conservation of water.	Cost of service recovery / (Sustainable use and equity)
Municipal fees for the provision of water supply, sewage, and	Downstream urban supply / Reuse	Municipalities may establish fees related to the provision of water supply, sewage and wastewater treatment services. The fees are levied on users of drinking water, sewerage and	Cost of service recovery / Sustainable use

wastewater treatment services		wastewater treatment services in a given municipality. The fees aim to compensate the local water supply company for the operating costs associated with the provision of urban water services (including supply of drinking water, sewage and wastewater treatment).	/ Affordability

Note: * In the Fall 2022, the Andalusian government has just set (Decreto-Ley 7/2022, de 20 de Septiembre) the temporary suppression of the Improvement fee from 1 January to 31 December 2023.
**As indicated in Section 5, the general service fee has not been implemented yet and is currently under discussion.
Source: Authors, based on the detailed description of the fees provided in Section 5.

The Andalusian **improvement fee** concerns the availability for urban use of drinking water from any source, supplied by public or private supply networks. It is meant to provide compensation for the costs of investment in hydraulic infrastructures of any nature corresponding to the integral cycle of water for urban use[18] borne at the Autonomous Community level and declared to be of general interest. The base and rates of the fee are further described in Box 6.2. . Water supplying entities and the natural or legal persons who own other supply networks are liable for water losses in supply networks. In addition, local improvement fees exist with very similar features to the Andalusian-level one and are meant to cover the costs of investment in water infrastructure borne by *local authorities*.

Box 6.2. Base and rate of the Andalusia improvement fee

Base

The base is the volume of water invoiced to urban users[19] by the supplying entities, expressed in m^3. In the event of water losses in the supply networks, the base is calculated as the difference between the volume supplied to the supply entity and the volume invoiced by the same.

Rate

The total charge is the result of adding a variable charge for consumption and, where applicable, a fixed charge for availability.

The fixed charge is at EUR 1 per month per user for domestic use. No fixed amount is charged to non-domestic users.

The variable charge is exposed in Table 6.3.

Table 6.3. Variable charge for the improvement fee

Type of use	Bracket	Rate (EUR/m3)
Domestic use	Consumption between 0 and 2 m³/hh/month	0
	Consumption between 2 and 10 m³/hh/month	0.10
	Consumption between 10 and 18 m³/hh/month	0.20
	Consumption over 18 m³/hh/month	0.60
Non-domestic use	n.a.	0.25
Losses in supply network	n.a.	0.25

Note: hh stands for household.

If the number of members in a household exceeds four, the upper limit of each of the progressive rate brackets may be increased by 3 cubic metres for each additional person living in the dwelling.

In the first 5 years of application of the fee, the application of the variable fee is progressive over time with the following percentages: 30, 45, 60, 80, 100%.

Note: Note that the improvement fee has been temporarily suspended by the Andalusian government from 1 January to 31 December 2023.
Source: Based on https://www.juntadeandalucia.es/medioambiente/portal/web/guest/areas-tematicas/agua/gestion-del-agua/recuperacion-de-costes/tarifas-y-canones-uso-agua-dominio-publico-hidraulico-dph-y-dominio-publico-maritimo-terrestre-dpmt, as accessed on 14 June 2022.

The **municipal fees for the provision of water supply, sewage, and wastewater treatment services**, are by definition set at the municipal level, and may widely differ across municipalities (Arbués and García-Valiñas, 2020[20]). As their name indicates, they are meant to cover the operating costs of the supplying entity for the provision of urban water cycle services. The fee may consist of a fixed part per user (which often depends on the metre calibre) and a variable part (generally progressive with the amount of water use, for households) depending on the cubic metres of water billed within the settlement period considered. A distinction is made between the types of domestic, commercial, industrial, official bodies and other uses, with no explicit mentioning as to why differential treatment applies. The tariff system can be complemented with a system of bonuses, linked either to the sustainable use of water or to social criteria.[20] The Andalusia or local improvement fee applies in addition to this fee, as it is meant to cover different costs.

The **general services fee** is to be introduced at the Andalusia level with the aim of compensating the administration costs of the Public Administration to ensure the proper use and conservation of water.[21] At the time of writing, the fee is still being discussed and no clear schedule for implementation is available. This fee would come on top of the existing national regulation fee and water use tariff, which are both meant to cover administrative costs as well administrative costs that are covered by the general services fee are to be subtracted from the level of administrative costs covered by the current regulation charges and water use tariffs. Like these two national-level charges, the payable amount determined by the general services fee is to be distributed among the different hydrological exploitation systems according to criteria of rationalisation of water use, fairness in the distribution of obligations and self-financing of the service.

6.1.5. Alignment of the Andalusian water pricing system with economic and environmental criteria

In order to assess the alignment of the Andalusian water pricing system with economic and environmental criteria, it is key to not only consider the Andalusia-specific levies (including local and municipal levies), but also to consider the system as a whole (and not instrument by instrument), including national-level levies. Thus, the following first provides an economic and environmental analysis of Andalusia-specific instruments taken separately. Then, it provides an assessment of the overall water pricing system that applies in the Autonomous Community.

Andalusian fees

The improvement fee addresses the criterion of cost of service recovery. Indeed, the fixed and variable charge for domestic users are a way to address cost recovery, and so is the variable cost for non-domestic users. It is however not clear why the variable cost is progressive for domestic users and not so for non-domestic ones, nor why domestic users are the only ones liable for the fixed cost. This is further discussed on the basis of the other three economic criteria.

The improvement fee's progressive rate structure for households and its differential rates between domestic and non-domestic users can help deal with affordability, can ambiguously affect sustainable use, and may not align well with equity concerns. First, the exemption of the first two cubic metres consumed per household per month can help address affordability issues for the poorest households. However, this same feature might engender sustainable use issues, by creating no disincentive for consumption below

$2m^3$/month. Given that today, the average daily consumption of water per person in Andalusia is around 128 litres,[22] this amounts to about half of a one-person household water consumption being provided for free.[23] A more precise characterisation for this would require defining minimum levels of necessary water consumption per person, which could better help set a per person threshold below which water consumption is provided for free.

Second, the progressive rates on water consumption faced by households may help discourage water overconsumption (Olmstead and Stavins, 2009[17]), which contributes to more sustainable water use. However, the unit to which these progressive rates apply being the household and not the user, can create equity issues between households depending on their size. Indeed, as shown in a few examples in Box 6.3 , this creates a penalty the larger the household is, for dwellings of four or less individuals. The adjustment of the upper limit of each of the progressive rate brackets for larger households then advantages them as compared to smaller households.

Third, the differential rate between domestic and non-domestic users may create an equity issue between the two (see Box 6.3). This is reinforced by the exemption from the fixed fee for non-domestic users. However, this could be justified on a sustainable use or cost-recovery basis if water demand of non-domestic users was more elastic than that of domestic users or if service costs were lower for that sector, hence calling for lower rates.

Fourth, base coverage of the fee encourages sustainable use by covering water losses in the supply networks, a novel and important feature of this fee. Indeed, as stressed in section 6.1.3, water losses in networks are important in Spain, and worldwide and generally do not face any cost associated to this. This in turn generally provides no incentive for better management or renewable of obsolete infrastructure, and can help address future service-cost recovery, hence financial sustainability.

Finally, affordability is also tackled through the progressive introduction of the fee in time: this provides time for households to adapt. It also gives them visibility on what costs they will face going forward.

Box 6.3 Equity issues created by the progressive rate structure of the improvement fee

To see how equity issues may arise between households of different sizes, Table 6.4 presents the variable fee (i.e. abstracting from the EUR 1 fixed charge paid by households for infrastructure reasons) paid by user for an individual consumption of $4m^3$ per month, depending on the size of the household. This table highlights how both due to the $2m^3$/month exemption, the progressive rate structure of the fee, and the adjustment in brackets for dwellings of more than 4, households of different sizes end up having different liabilities *per user*.

Table 6.4. Variable fee paid by individual for an individual consumption of $4m^3$ per month

Household size	Total household consumption (m^3)	Total variable fee per individual (EUR)
1	4	0.2
2	8	0.3
3	12	0.4
4	16	0.5
5	20	0.44
6	24	0.4

Note: The calculations were based on hypothetical households that would have been liable for this fee for more than five years.

To see how equity issues may arise between domestic users and non-domestic users based on the variable part and the fixed part of the improvement fee, Table 6.5 presents the improvement fee liability faced by a household and that by a firm for the same volume of water consumption. The amount to be paid would be lower for the business than for the household, who, however, would generally have a smaller monthly budget. Again, however, this different may be justified by other factors such as different demand elasticities or different service costs.

Table 6.5. Total fee paid by a household and by a business for a total consumption of 32m^3 per month

Type of user	Total fee per user (EUR)
Household of 6	8.2
Business	8

Note: The calculations were based on a household liable for the improvement fee for more than 5 years.

The municipal fees address similar criteria to the Andalusia and local improvement fees. Sustainable use is addressed with the additional instrument of a bonus system and affordability with lower tariffs for households considered as vulnerable. However, such measures do not exist in all municipalities. In Malaga and Seville, for example, lower tariffs do not exist. In Cordoba, they exist for certain retired people, families where all members are unemployed or families at risk of social exclusion (Arbués and García-Valiñas, 2020[20]).

The general services fee addresses cost of service recovery and covers an important element of service costs: administrative costs. However, given that administrative costs (albeit possibly different ones) are already covered and in a similar fashion by the existing national regulation fee and water use tariff, one can question the efficiency of introducing yet another levy to the system. While this might cover previously uncovered costs, this introduces additional complexity to the system, as well as potential additional administrative costs. Consolidating fees covering administrative costs might be a more efficient way forward.

It would also seem like the general services fee is meant to address sustainable use as well as equity. Indeed, the payable amount determined by this fee is "to be distributed among the different hydrological exploitation systems according to criteria of rationalisation of water use [and] fairness in the distribution of obligations." The calculation of the distribution of the amount payable between users (proportionally to water use or not) is key to determining whether rationalisation is indeed promoted. Indeed, some research points to higher responsiveness to progressive rates (Olmstead and Stavins, 2009[17]). Rationalisation effects will also depend on responsiveness of hydrological exploitation systems to incentives provided by this calculation. However, and as discussed below, it may not be required to incorporate rationalisation objectives in this levy itself.

The Andalusian water pricing system as a whole

Overall, the water pricing system in Andalusia contains many national-level instruments meant to recover service costs and certain regional and local-level instruments that might help promote affordability, sustainable water use as well as equity.

Despite the multiple cost of service-recovery instruments, the recovery of costs from water use is far from complete across the Andalusian river basins, according to the analysis presented in the 2022 White Book for Tax Reform in Spain (Table 6.6). This indicates that the Andalusian water management cycle is forgoing revenues that can be justified based on the costs related to water use. In the four main Andalusian river basins (Guadalquivir, CC. MM. Andalusas, Guadalate-Barbate and Tinto-Odiel-Piedras), cost recovery rates are around 80% and even reach about 87% for Guadalate-Barbate. While these are amongst the highest in Spain, they still fall short of full recovery rates, i.e. 100% – even with environmental costs left out. Moreover, these estimates do not include all costs mentioned previously, such as administrative and governance costs. It is not clear either whether the costs considered in this table are existing costs, or also include a forward-looking view on future costs. As mentioned in the economic and environmental criteria discussion, this is key to encourage investment.

Table 6.6. Annual cost of water use and revenues at river basin district levels

In Million Euros

River basin	Financial cost (operating and maintenance costs, AEC of investment)	Environmental cost (AEC)	Total cost	Revenues	Cost recovery rate (including environmental cost)
Guadiana	537.81 (91.72%)	48.57 (8.28%)	586.38	353.06	60.21%
Guadalquivir	1032.00 (93.66%)	69.88 (6.34%)	1101.88	870.76	79.02%
Segura	805.70 (77.22%)	237.67 (22.78%)	1043.37	700.02	67.09%
CC. MM. Andalusas	743.90 (90.43%)	78.70 (9.57%)	822.60	659.65	80.91%
Guadalete-Barbate	163.78 (91.89%)	14.46 (8.11%)	178.24	154.11	86.46%
Tinto-Odiel-Piedras	121.15 (92.30%)	10.47 (7.70%)	131.26	109.37	83.01%

Note: AEC stands for annual equivalent cost. Environmental costs represent the following: "Environmental costs are valued at the economic cost of the actions necessary to minimize the environmental cost associated exclusively with the provision of water services." The percentages in parentheses represent the share of costs finance and environmental costs respectively in total costs.
Source: Hydrological Plans (2022-2027) of the river basin districts (in approval process), available at: https://www.miteco.gob.es/es/agua/temas/planificacion-hidrologica/planificacion-hidrologica/PPHH_tercer_ciclo.aspx. The whole table with all river basin districts information is Table 24 in the Libro Blanco.

From a partial equilibrium point of view,[24] the current water pricing system in Andalusia is not well aligned with the equity criterion, both due to coverage and to rates. Figure 6.3 summarises the levies directly faced by the three main water users. Regarding coverage, irrigation is only subject to fixed fees to cover principally the costs of construction, repair and improvement of related works and installations of the Community as well as common operating and administration expenses.[25] Regarding overall rates, the fee depends on the share of each irrigator in the Community but not directly on the volume of water used. For this reason, it may be that the agricultural sector bares a lower cost for water-related services, while it is the principal sector in terms of water demand (about 80% in Andalusia). Regarding rates facing urban users, the design of Andalusian and local taxes often exempts or offers reduced fixed rates to industries for water supply access. Moreover, the progressive rates faced by households are applied at a dwelling level with certain provisions for larger households, so that equity between individuals may not be respected. The rates faced by industry not connected to the grid are not straightforward to infer from the pricing system. A caveat is called for regarding rates on different users, however, as it might be the case that certain users face lower rates because the cost of service linked to their water usage is lower or because their demand elasticity is higher.

Figure 6.3. Direct levies faced by users for their water-use

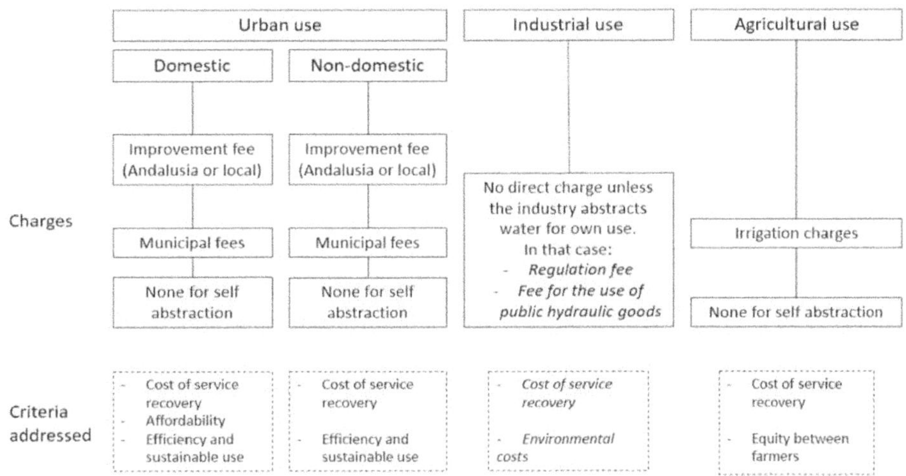

Source: Author's own elaboration.

Setting clear objectives is key to addressing all five criteria for good water management exposed in section 6.1.3. Regarding cost of service recovery, it would be important to know the costs each user type are responsible for. Is it the case that water use for urban purposes is more costly due to higher treatment and control requirements? If this were known, then this could be a rationale for lower rates facing industries not connected to the grid and agriculture. It could help set recovery rates in line with the service costs that should be recovered. Regarding affordability, it is important to know who the most vulnerable households and other water users (certain small farmers for example) are. This would enable directly targeting them. In the case of water, income for households and size of farm for farmers could constitute indicators to work with. Regarding sustainable water use, in the absence of a market, maximal amounts of acceptable water use should be set.[26] Once these three objectives are set, it is more straightforward to ensure equity between the different users. Finally, environmental costs could be accounted for by using as a base, for example, the EUR 0.3/m^3 highlighted in the European Commission study by Mottershead et al. (2021[15]).

Last but not least, the economic and environmental criteria for water pricing discussed above do not cover the last recommendation of the OECD Council Recommendations on Water related to water use, the Policy Coherence principle,[27] which certain regional, national or EU-level policies sometimes negatively affect. According to Box 2 in Fuentes (2011[25]) regional governments are in charge of natural resources, agricultural policies – subject to European Union directives and central government guidelines – as well as of land-use planning. While cereals (such as wheat and rice) are associated with low value added relative to their water needs, they are essential in ensuring national food security. However, the recent take-off of avocado and mango culture (Campos, 2021[26]) in Andalusia, and in particular along the Costa del Sol, questions the Policy Coherence principle, given the very important water needs of these crops as compared to the traditional olive and vine cultures. On the other hand, while probably not sustainable in the long-run, these cultures might have helped promote rural communities' well-being in the short run. These issues are further discussed in the last subsection on non-pricing policies. At the EU-level (and hence not in the resort of Andalusia), a recent report of the European Court of Auditors (2021[27]) highlights that Common Agricultural Policy (CAP) funds are generally not aligned with efficiency water use requirements and instead risk promoting greater water use.

6.1.6. Policy instruments to address environmental concerns: sustainable use for water preservation

While service cost-recovery, affordability, sustainable use, environmental externalities and equity, affordability are all important components of water pricing, sustainable use is currently the most pressing and least accounted for issue at a region-wide level. Indeed, for now, households appear to be the principal bearers of this criterion, while neither water pricing in agriculture nor in industry accounts for this. Given that agriculture represents about 80% of water use in Andalusia, it is important that sustainable use is also promoted in this sector.

Regarding sustainable use objectives, the EU WFD sets certain goals in terms of water preservation. In particular, it establishes the target of "good" quantitative status for all groundwater bodies by 2027 and specifies limits to water abstraction. These could be minimal targets for a potential future environmental component of water pricing in Andalusia. In particular, the Hydrological Plan of the second EU WFD cycle[28] presents the results of analyses of the ecological status of surface water bodies and of the quantitative status of groundwater bodies, classifying them under "Good" or "Poor". Such classifications could serve as a basis for target-setting also in Andalusia.

An abstraction tax on water abstracting entities could be introduced to address water scarcity and help reach water preservation and environmental objectives. If not done at a national level, this could be envisaged at the Andalusian level. In theory, levies on abstraction are designed to reflect externality costs of water use and to discourage low value uses. The rates could differ across water sources (surface or groundwater) and water bodies (depending on their stress level). This tax would then be passed through onto the different water users. Box 6.4. provides examples of abstraction levies in other jurisdictions.

Box 6.4. Abstraction charges

France and Estonia present two examples of jurisdictions which apply abstraction charges.

In Estonia, the tax does not applies to all users seeking to abstract water. Exceptions apply to small volumes of abstracted water, to irrigation and fish farming use as well as to use for energy purposes. Higher rates apply, the deeper the water abstracted and for the Tallin catchment areas as opposed to other areas. Abstracting water from deeper surfaces can indeed have more consequences for the environment, as the deeper it is, the more lengthy aquifer recharge may be. Slow or limited aquifer recharge, in turn, can have important consequences for river flows in the summer, for soil and eventually for desertification phenomena.

France has a tax that applies to all abstraction activities, except for sea water and certain activities including some mining activities, aquaculture, geothermal energy and frost control for perennial crops. Where a person has a borehole for their water supply, they are required to install a metering device that measures the volume of water abstracted. Rates are differentiated according to the intended use, are highest for drinking water supply and lowest for gravity-based irrigation. They may also differ across water basins.

Note: For further information, see Annex Figure 6.A.1.
Source: (Andersen et al., 2006[28]) for Estonia and https://www.legifrance.gouv.fr/codes/section_lc/LEGITEXT000006074220/LEGISCTA000006159222/2008-01-01/#LEGISCTA000006159222 (Article L213-10-9) for France.

An abstraction tax would be legally implementable at the regional or national level (see Section 5) and has been suggested at the national level in the White Book for Tax Reform in Spain. Such a tax could be more effective regarding national water conservation and planning if implemented at a national level, because it would ensure all water potentially abstracted in Spain would be included in the tax. However, different

Spanish regions have different needs in terms of water use and different risks of water shortages, which can be present a justification for policy action at the regional level. Moreover, pushing for abstraction taxes at the regional level first could also support building political momentum for a national-level instrument and public acceptability, similar to how regional carbon taxes in Andalusia and certain other Autonomous Communities have contributed to the promotion of such a tax at the national level (see the recommendations of the White Book for Tax Reform in Spain). Cooperation with river basin authorities will be key, as they are in charge of granting rights for and controlling water abstraction. Tight control should be ensured in order to monitor levels of water abstraction.

Setting the abstraction tax at the required level to reach water conservation objectives, would require a better understanding of demand curves for the different types of users: households, businesses, industry and agriculture. Moreover, for the abstraction tax to fulfil its water conservation objective, alternative means for lower water use should be promoted for all users. This would help effectiveness of the policy as well as help overcome political barriers and affordability concerns for all users as well as productivity and competitiveness issues in the industry and agriculture sectors. Finally, water user responsiveness can be increased by public awareness campaigns. Recent findings point to the importance of a mix of pricing and non-pricing policies measures to better manage water demand (European Environment Agency, 2017[29]; Leflaive, 2022[30]). Measures to increase user responsiveness and price elasticities of water users are further discussed in section 6.3.

Abstraction taxes may be a useful instrument to target formal abstraction activities – and would target most uses – but are more challenging to implement on self-water abstraction, which is a significant, albeit not a majority, practice in the Andalusia agricultural sector. For example, in the Guadalete-Barbete river basin, water self-abstraction represented about 14% of water use in the sector.[29] Indeed, monitoring self-abstraction is not straightforward. Fuentes (2011[25]) suggests a mechanism that would introduce monitoring through user associations (also referred to as Communities) to avoid over-use induced by this highly decentralised water abstraction source. This is a means that has observed in many communities where water use is informal and hence hard to regulate (Ostrom, 1965[31]). While user associations do not typically deal with these issues, there are some examples of successful resource management among Andalusian Communities, which have set up internal mechanisms of abstraction controls and fines, without the need for government intervention. This could be promoted at a larger scale, by supporting user associations in their monitoring effort, organising exchange of best practice amongst associations or by introducing financial incentives (such as fines) for those associations whose users are globally responsible for over-abstraction of aquifers. Indeed, as was seen above, over-abstraction is regularly measured by River Basin Authorities. If the financial incentive imposed on user associations is then passed on to farmers it would represent a type of decentralised abstraction fee. Legally, such provisions would have to be implemented at the national level, however (see Section 5).

Another way of pricing water abstraction is to allocate water usage quotas to introduce "cap-and-trade" water markets. Australia has been a leader in the development of such water markets, especially in the Murray-Darling Basin. This long-term experience of over thirty years now has enabled to gather enough evidence for in-depth analyses (Grafton and Wheeler, 2018[32]). While cap-and-trade systems present usual benefits and drawbacks, in the case of water, the administrative needs for implementation have proven to be extremely complex and require sophisticated administration. More precisely, in the Australia example regulating numerous small farmers and helping them adjust to the complex functioning of a water market was a challenge. Moreover, setting water quotas may prove to induce perverse effects: opportunity costs of not using up or selling assigned quotas are high, which provides an incentive to use up the entirety of available quotas in a given period. This does not incentivise sustainable water use, can lead to over-exploitation and is said to have led to the drying-off of certain riverbeds in Australia. The OECD toolkit for water policies and governance (OECD, 2021[24]) provides additional details on the lessons from the Australian water market experience.

This being said, the management of water resources through robust allocation mechanisms has been extensively analysed in a recent OECD note (Leflaive, 2022[30]) and a report (OECD, 2015[33]). These expose how designed and implemented allocation regimes can perform well under average and extreme conditions and can be adapted to changing conditions at the least cost over time. They can be more effective at managing water scarcity, allocating sufficient shares of water to the ecosystem and addressing equity issues than pricing. This is especially the case because of the relatively low responsiveness of water users to prices (which can be addressed through other policies, as discussed in the following). However, in practice, several issues arise such high degree of path dependency – hence difficulties to effectively adjust the allocation arrangements – or reduced return flows.[30]

Finally, as highlighted in section 6.1.3, ensuring financial sustainability of the water network can help reach goals of sustainable water use and equity. In this regard, a discussion, along with case studies on the United States and on the United Kingdom, on possibilities for improving financing models and public-private partnerships to finance investments is provided OECD (2017[2]). A recent OECD note (Leflaive, 2022[30]) points to findings relating to more efficient of water appliances and networks contributing to decreasing domestic demand for freshwater. Another alternative consists in building dual networks.

Dual water distribution networks would separate the distribution of potable water from that of non-potable water, that would either be untreated or poorly treated. The former would be supplied for drinking purposes and the latter for purposes such as street-cleaning or recreational uses (e.g. private gardening, swimming pools). Dual networks could have the benefit of increasing water reuse (reclaimed wastewater), which is a strategy used in Israel for instance (see (OECD, 2017[2]) for additional details on the surge of water reuse in Israel). They could also enable differential pricing for high and low water uses. The trade-off here stands between the costs of building and maintaining two distinct networks and the costs of water treatment.

6.2. Pricing water pollution

After an exposition of the main sources of water pollution, this subsection discusses the main externalities from water pollution and presents estimates of their costs. It then analyses the water pollution pricing mechanisms in place today in Spain and hence Andalusia, their alignment with the Polluter Pays and Equity principles of the OECD Council Recommendation on Water and exposes additional instruments that could be introduced to deal with water pollution. The Policy Coherence principle is also briefly discussed.

6.2.1. Sources of water pollution

Water pollution comes from urban, industrial and agricultural users and may originate from point sources or diffuse sources. Point sources of pollution refer to "direct discharges to receiving water bodies at a discrete location, such as pipes and ditches from sewage treatment plants, industrial sites and confined intensive livestock operations". Diffuse (or non-point) sources of pollution refer to "indirect discharges to receiving water bodies, via overland flow and subsurface flow to surface waters, and leaching through the soil structure to groundwater" (OECD, 2017[2]).

In the urban and industrial sectors, water pollution is mainly due to wastewater and direct industrial discharges. In the past decade, there has been an increased focus on contaminants of emerging concern (CECs). "Emerging" refers to their recent appearance in water, or to a recent detecting of these contaminants at concentrations significantly higher than expected. Moreover, their risk to human and environmental health may not be fully understood. Examples include pharmaceuticals, industrial and household chemicals, personal care products, manufactured nanomaterials, and their transformation products (OECD, 2020[5]).

The usual water pollution sources from the agricultural sector include sedimentation[31] and pesticides use[32] as well as certain practices of nutrient use (applied in the form of chemical fertilisers, manure, and sludge), animal feeding, livestock grazing and irrigation (EPA, 2005[34]). CECs are also a concern for the agricultural sector and were analysed a decade ago already in Boxall (2012[35]). These include manufactured nanomaterials (e.g., nanopesticides or nanomedicines), veterinary medicine or increased pollution risk from manure and sludge (as feedstock and plants have ingested CECs from other sources).

Moreover, the widespread of antibiotics by humans or in the agriculture sector has been increasing the presence of antibiotic residues in surface and groundwater. This in turn facilitates a permanent exposure of microorganisms that can reinforce resistance to antibiotics (Cycoń, Mrozik and Piotrowska-Seget, 2019[36]), which is a growing threat today.[33]

6.2.2. External costs of water pollution

The main externalities from water pollution relate to health and ecosystems (which shall be referred to as environmental externalities), but are also economic. For example, groundwater provides a non-negligible share of drinking water to both humans and the agricultural sector so the higher its pollution level the higher treatment costs are. In 2021, groundwater accounted for 11.5% of water use in the urban sector in the Guadalquivir river basin, for about 13% in the Guadalete-Barbete river basin and about 5% in the Tinto-Odiel-Piedras river basin. In 2010 already farming was the main source of groundwater pollution in many countries, and increasingly so (OECD, 2010[37]).

Mottershead et al. (2021[15]) calculates external costs from water pollution for the European Commission. They consider water pollution from nitrogen and phosphorus use, which arises primarily from non-point sources, in particular agriculture. Nitrogen and phosphorus are present in most fertilisers currently used in farming (European Commission, 2019[38]). These pollutants have two main environmental externalities: (i) eutrophication,[34] which causes damages to ecosystems and loss of amenity to households with waterfront properties and recreational water users and (ii) human health impacts, and mostly cancers from nitrite pollution of drinking water due to nitrogen as well as osteoporosis due to the presence in food of cadmium contained in mineral phosphorus. Their impact is due to the surplus of nitrogen and phosphorus, which is not absorbed by plants. The costs of pollution from these two inputs vary with geographical area, due to differences in population density (higher density increases the number of people exposed to the risks of eutrophication or health impacts), proximity to surface and coastal waters and sources of drinking water supply.

According to the study, Spain is amongst the median countries in terms of the size of external costs induced from both nitrogen surplus (EUR 0.85/kg/year[35]) and phosphorus (EUR 0.9/kg/year[36]). However, when taking the volume discharged into account and looking at total annual costs (i.e. in EUR/year) and when adding wastewater costs due to households and industries, Spain is amongst the EU member countries with the highest annual costs – both in total (2,113 million euros per year) and as a share of GDP (see Figure 6.4).

Figure 6.4. Annual costs of water pollution as a share of GDP for selected EU countries

In percentage

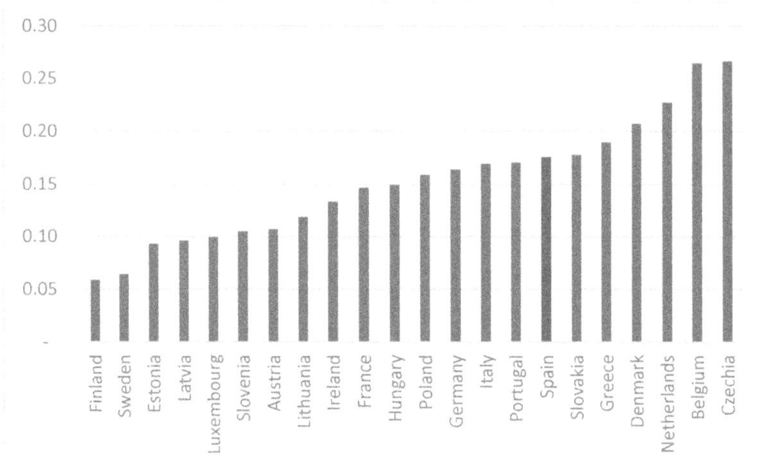

Note: Annual costs of water pollution as a share of 2018 GDP.
Source: Mottershead et al. (2021[15]).

StatLink https://stat.link/4qklvr

6.2.3. Pollution pricing in Andalusia

Taxation is typically used as an instrument to internalise external costs from water pollution but a decision should be made as to whether the tax is targeting pollution or inputs creating pollution. Which instrument is most appropriate will depend mainly on how well polluter and toxicity of pollution can be identified:

Pollution taxes or charges can be designed to cover external costs from point sources. Typically, they apply to industrial users and households.

Input taxes may be more appropriate when facing diffuse pollution, which makes it difficult to identify polluters clearly, such as agriculture.

The state of water pollution pricing in Spain and Andalusia: description and analysis

In Spain, and hence in Andalusia, a **pollution control fee**[37] applies to both urban and industrial[38] wastewater on the persons who carry out discharges into the public hydraulic domain. The national pollution control fee is proportional to the volume of water discharged but also takes into account its polluting impact. Urban wastewater discharge faces a basic volumetric price of EUR 0.01751 per m^3 and industrial wastewater of EUR 0.04377 per m^3. Then, urban and industrial users are charged the product of the volume they discharge by the basic price of the discharge to which a coefficient is applied according to three factors. The first is the nature and characteristics of the discharge, the second is the degree of pollution and the third depends on the environmental quality of the physical environment into which it is discharged.

The pollution control fee addresses the purely economic costs of water pollution. Indeed, by charging a fee proportional to the volume of water discharged, it at least partly covers the monetary costs that pollution imposes on the assessment, control, protection and enhancement of the river basin district where pollution is emitted.

The fee also has an environmental component to it. Indeed, the coefficient applied in the calculation of the charge includes three different factors that account for its environmental impacts. Depending on how these are measured, they may have the feature of accounting for environmental externalities or of discouraging water pollution – this is especially so for the degree of pollution factor (second factor). In the industry sector, the design of taxes discouraging the release of polluting substances into water is complex, as estimating firm-level emissions and their damages may be prohibitively costly, due to the presence of multiple small producers and to chemical and weather factors addressed above (OECD, 2017[18]). The Spanish pollution control fee constitutes an attempt to do so, but the effectiveness of this tax will depend on the level of the three factors. Box 6.5. presents the example of France, which has a tax targeting the pollution level of water discharges.

Box 6.5. The French pollution tax on non-domestic activities

In 2006, France passed a law to tax pollution generated from non-domestic activities, which was implemented in 2008. This tax applies to a yearly average of polluting substances rejected by different industrial and agricultural activities. The rates are set per polluting substance. For agricultural activities, it also applies to livestock farms with more than 90 (or in some cases 150) animals.

For point source pollution, the pollution levels are determined by a regular measurement of industrial discharges. When direct measurement is not possible or when facing diffuse pollution, a theoretical level of pollution based on the activity is computed. The theoretical pollution level of an activity is calculated based on magnitudes and coefficients which characterise that activity. It is determined from general measurement campaigns or studies based on representative samples.

Table 6.7 presents some rates and inclusion thresholds applied since 1 January 2021. Different rates apply depending on whether the pollutants are discharged into surface or ground water and whether they are discharged into the sea or into rivers.

The tax is of EUR 3 per livestock units when there are more than 1.4 livestock units per hectare of utilised agricultural area.

Table 6.7. Tax rates and inclusion thresholds applying to non-domestic water pollution, France

Selected pollutants

Pollution component	Tax rate (in EUR per unit)	Threshold
Suspended solids (per kg)	0.3	5 200 kg
Chemical oxygen demand (per kg)	0.2	9 900 kg
Five-day biochemical oxygen demand (per kg)	0.4	4 400 kg
Reduced nitrogen (per kg)	0.7	880 kg
Oxidised nitrogen, nitrites and nitrates (per kg)	0.3	880 kg
Total phosphorus, organic or mineral (per kg)	2	220 kg
Environmentally hazardous substances discharged to surface water bodies (per kg)	10	9
Environmentally hazardous substances discharged to groundwater bodies (per kg)	16.6	9
Heat discharged into the sea, except in winter (per megatherm)	8.5	100 Mth
Heat discharged to river, except in winter (per megatherm)	85	10 Mth

Note: For further information, see Annex 6.A.
Source: https://www.legifrance.gouv.fr/codes/id/LEGIARTI000041528273/2021-01-01/.

Finally, by differentiating the rates paid by urban users and industrial users, the pollution fee acknowledges that the pollution impact of wastewater from the latter is greater than from the former. This can both address economic costs (treating industrial wastewater is on average more costly than treating urban wastewater) but also environmental costs. Indeed, the environmental costs of industrial wastewater discharges are typically higher than those of urban wastewater discharges.

No such fee applies in the agricultural sector, even though as highlighted earlier it is the main sector responsible for aquifer pollution today. This may be due to the fact that, as discussed above, diffuse pollution is not well targeted by such a tax. Input taxes, which do not exist nor in Spain nor in Andalusia, and other means of addressing water pollution originating from agriculture are discussed in the next part.

In terms of the OECD Council Recommendations on Water, the current system falls short of aligning with both the Polluter Pays and the Equity principles. The Polluter Pays principle is not respected for the agricultural sector, since as of now, water pollution is not priced in that sector – neither through a pollution levy nor through an input tax. The lack of price paid for water pollution for agriculture users is not only misaligned with sound environmental tax policy principles, but it also impacts the Equity principle. Indeed, it results in an important share of water polluters not paying for the pollution they generate.

Many mechanisms aim to contribute to Policy Coherence as defined in the OECD Council Recommendations on Water, especially in the agricultural sector where many practices responsible for water pollution cannot be dealt with through pricing directly. In particular, not only are there limits to input (fertilisers and pesticides mainly) usage but there also exist many policies to limit the polluting impacts of these inputs. The Common Agricultural Policy (CAP) for example forbids farmers from applying nitrogen-based fertilisers on bare land. Farmers should grow temporary crops on these lands that have the sole purpose of absorbing extra nitrogen from the soil, and avoiding the contamination of groundwater. However, direct pricing mechanisms lack. The possible introduction of pricing measures is discussed in the following.

6.2.4. Pricing water pollution in the agricultural sector: proposals

For the agricultural sector, a tax on polluting inputs can be a way of dealing with the environmental impact of water pollution from that sector. While limits on usage of fertilisers and pesticides exist in Andalusia, currently no pricing mechanism to deal with this issue is in place. Limits on polluting inputs may be relatively simpler to introduce and to communicate in general. However, strict implementation in the case of regulation is key, and Spain has recently been referred to the Court of Justice of the European Union by the European Commission for poor implementation of the Nitrates Directive.[39] Moreover, pricing, as opposed to command-and-control policies is a more cost-effective way of dealing with over-use. It decentralises abatement decisions, and leaves it up to the agents for whom marginal abatement costs are lowest to abate first. More concretely, for example, it leaves the decision to use more or less fertilisers to the farmer who knows best about the cost and benefits implied in lower fertiliser use. In that sense, the famer to whom it is least expensive to reduce or avoid fertiliser use (e.g. in terms of forgone yields), will do so first. A further discussion on non-pricing instruments is conducted in the last subsection.

If there is no action at the central government level, Andalusia could introduce a tax on pesticides and fertilisers at the regional level to target water pollution in the agricultural sector. Such a tax would target the quantities purchased of a specific product (that would need to be defined), where rates depend on their respective environmental impact. Polluting substances and water quality are discussed in many EU directives (see Box 6.6).

Box 6.6 EU directives relating to water pollution

As highlighted in Section 5, the EU WFD defines water quality levels to be pursued, which are defined for aquifers in the EU Groundwater Directive. The EU Environmental Quality Standards Directive provides the list of priority hazardous substances. Other EU Directives related to water pollution are integrated into the Water Framework Directive. The EU Nitrates Directive requires EU member states to establish agricultural action programme measures to prevent nitrate pollution from reaching water bodies. The EU Biocides Directive relates to the authorisation and placing on the market of biocidal products such as pesticides, herbicides, or fungicides that may be harmful to the environment and in particular to groundwater quality.

Pesticides on the European market are already risk assessed by the European Chemicals Agency, so defining products to be targeted by the tax and grouping them into different rate bands would be relatively straightforward. For example, since 1999, Norway has had a pesticides tax with seven tax bands which depend on the environmental health- and ecosystem-related risks of the specific pesticide (OECD, 2017[2]; Böcker and Finger, 2016[39]) (see Box 6.7.).

Box 6.7. The Norway banded tax system for pesticides

In 1999, Norway passed a law to introduce a banded tax system for pesticides, in order to reduce the use of pesticides that represent the greatest risk to human health and the environment.

Pesticides are sorted into seven different categories, which depend on (i) risks for human health and (ii) environmental risks. All pesticides for professional use are tested according to several criteria and then categorised as a low, medium, or high risk. Each category is assigned a specific factor, which ranges from 0.5 to 150.

The tax applies per hectare, and is calculated by multiplying NOK 25 times the factor associated with the pesticide category.

Note: For further information, see Annex 6.A.
Source: Böcker and Finger (2016[40]) (2016[39]), Spikkerud (2006[41]) (2006[41]).

Regarding fertilisers, a tax on quantity used or purchased is not necessarily the best way to address their environmental externalities. In particular, tax rates could only constitute a proxy for their pollution potential, as the principal negative environmental impact of fertiliser use comes from an application in excess of plant needs or in unsuitable weather conditions (washout is particularly strong when applied just before rain episodes) (EPA, 2005[34]). Such situation-specific conditions are difficult to assess and reflect in a general tax rate.

The effect of taxes on fertilisers and pesticides will depend on the responsiveness of input use to increased price rates and again, their general equilibrium, affordability, yield and political feasibility[40] effects should be carefully assessed. Regarding pesticide use, a meta-analysis by Böcker and Finger (2016[39]) finds a median price elasticity of demand for pesticides of −0.28.[41] This means that a tax increase of 1% is estimated to reduce demand for pesticide inputs by 0.28%. These relative low price elasticities make it very improbable that dealing with water pollution from these inputs can be tackled through taxation only. It may also be a sign that farmers do not see (or have) alternatives to pesticide use given current constraints. This stresses the importance of complementary policies, which can help farmers reduce pesticide use without risking an important decrease in yield or at least income[42] or of a broader policy environment that is aligned with water protection objectives (such as policies that promote quality of agricultural production over quantity). Public awareness campaigns, stressing the risks for the environment as well as farmers themselves and their business may also help increase responsiveness levels.

It is also important to bear in mind that there is a risk associated to unilaterally introducing input taxes at a regional level with no concerted action with neighbouring regions. Indeed, the price-elasticity of input demand may also be low due to mobility of the base, which is not an issue encountered in the case for water usage. Indeed, it could be feasible for farmers to get their input provisions from other regions with no input tax. This issue would be worsened by a unilateral imposition of such a tax at a local level as opposed to the national level. It could however be dealt with through certain regulatory provisions.

From a political feasibility perspective, introducing taxes relating to inputs used by farmers may encounter resistance that needs to be managed well. Skou Andersen (2016[42]) for example, exposes the political difficulties surrounding the possibility of reintroducing a fertiliser tax in Sweden. These stem from an opposition to it by farmers, from a lack of a solid environmental impact assessment of this tax in the past and possibly from a shift of concern from water pollution to climate change.[43] Söderholm and Christiernsson (2008[43]) draw lessons from the European experience in fertilizer taxation and find that a form of earmarking of tax revenues can help increase the legitimacy of such taxes and that rates which achieve a close proportionality to damage done have a greater chance of being perceived as fair. Again, this highlights the importance of knowing about the effectiveness of these policies in order to make them politically acceptable.

Solutions to political frictions could be better communication on evidence-based results of pollution pricing mechanisms and earmarking of revenues. Use of revenues to encourage best-performing farmers in terms of pollution, through systems akin to a bonus-malus could increase effectiveness of the policy. These exist for vehicle taxation, for instance, and could be adapted to the context of water pollution. An example for the former is Italy, where vehicle taxation follows a bonus-malus system (also referred to as a system of *feebates* in other countries or regions) that penalises CO_2-intensive vehicles and subsidises cars emitting 60 grams of CO_2 or less per kilometre.

Advances in nutrient pollution modelling can provide an opportunity to tax diffuse pollution outputs directly, rather than taxing proxies such as fertiliser and pesticide inputs. Such models could allow setting pollution levies at levels that are directly proportional to the simulated amount of pollution generated by farm. For example, OverseerFM[44] is a New Zealand national model for farm-scale nutrient budgeting and loss estimation, which also identifies risks of environmental impacts through nutrient loss, including run-off and leaching (OECD, 2017[2]). New Zealand farmers are increasingly required by regional councils to use the model to develop nutrient management plans and budgets (OECD, 2021[24]), but it has not been used for direct pricing yet. The better alignment of taxes that would rely on such models with actual pollution effects could increase their efficiency, political acceptability and provide additional options for farmers to adapt, by helping them take agency on monitoring the polluting effects of the practices they choose to implement. Moreover, this could enable the introduction of an additional factor to tax rates, which would account for the state of potentially affected soils and aquifers.

As in the case of the French tax on water pollution (Box 6.5.), livestock could also be included in the base of water pollution taxation. The Wallonian tax on environmental impacts from farming (see Annex 6.A) seeks to water pollution through nitrogen, by targeting the number and type of livestock units as well as the land area and type of cultivation (organic or non-organic crops and organic and non-organic grassland). The design of the latter tax can encourage a switch to organic farming cultivation, to less polluting livestock or at least a smaller number of livestock units. Moreover, such taxes could target modes of production deemed to be more polluting. This would be the case for example of intensive as opposed to extensive livestock farming. However, again in this case, Policy Coherence is key – this issue comes up because for example of the push in the last decades towards more intensive modes of farming. Support for farmers in the transition is also key, as well as encouraging consumers to switch to more sustainable food consumption. These points are further discussed in section 6.3.

Another way of dealing with water pollution from agriculture would be similar to the collective responsibility mechanism described in Fuentes (2011[25]) for water quantity of aquifers. This could be used for pollution caused by input use to groundwater. Given that the chemical status of groundwater bodies is also measured for all river basins in Spain,[45] the idea would be to create financial incentives for Communities to keep pollution levels in aquifers at a reasonable level. The costs would then be passed on to farmers. Quality status of water bodies in Andalusia can also be informed by the Andalusian quality of water bodies atlas.[46]

Finally, water pollution in the agricultural sector does not only occur through the use of polluting inputs but also, as for the other two users, through the wastewater discharged in surface water. In that respect, the existing pollution control fee could also be extended to farmers – possibly through a separate instrument. Given that the irrigation fee depends on volumes used, the administrative capacity to implement such a tax exists.

6.2.5. Extending the coverage of the pollution control fee to contaminants of emerging concern

The increasingly pressing issue of CECs in wastewater was discussed at the beginning of this subsection, and it is not clear whether the pollution control fee accounts for these in one of its factors. The issue with CECs is that it is not necessarily possible nor advisable to call for a reduction in their use – the case of pharmaceuticals is particularly striking here. Countries have adopted a host of response packages to this rising phenomenon, which to date, however, focus on upgrading wastewater treatment plants (WWTP). In Switzerland, for example, the Waters Protection Act was revised in 2014 to further improve wastewater treatment for the removal of CECs, including pharmaceuticals. This included the introduction of a new technical wastewater treatment standard and public subsidies to fund technical upgrades of WWTPs. Some countries implement complementary measures, such as France, which has introduced financial incentives aimed at stimulating new innovative projects to manage CECs (OECD, 2020[5]). A recent OECD report (OECD, 2019[44]) outlines a policy mix of source-directed, use-orientated and end-of-pipe measures to manage pharmaceuticals harm to the environment.

6.3. Non pricing policies

Non pricing policies can be declined along two categories: (i) command-and-control measures that are used as substitutes or complements to pricing policies and (ii) accompanying measures to pricing policies. The first set of policies are generally deemed less efficient than pricing policies, which decentralises abatement choices, leaving the decision up to consumers as to the best way of reducing consumption and possibly encouraging technological innovation. However, they may help address targets more easily as responsiveness levels to prices are not always known or too low and they may also help deal with political frictions. The second set of policies contribute to the overarching Policy Coherence principle and are briefly discussed.

Command-and-control measures to ensure water conservation have long been used alongside or instead of pricing measures. Regulatory measures used for short-run water use and pollution management include restricting water usage during certain high-temperature periods or water shortage episodes or restricting the use of polluting inputs. Long-run water conservation policies are often technology standards. For example, in the United States, the National Energy Policy Act has, since 1992, required that all new construction install low-flow toilets, showerheads, and faucets (Olmstead and Stavins, 2009[17]).

While command-and-control measures may be easier to implement from a political point of view and may seem to be more straightforward than pricing policies in reaching certain targets,[47] they come with many drawbacks. The following discusses three of those. First, mandating technology standards that result in lower water flow can create rebound effects through income effects: lower water flow implies lower water bills, which in turn can encourage increased water use. Second, these measures are only effective if incentives such as monitoring and significant fines are implemented and implementable in case of non-compliance. Third, technology standards can actually dampen incentives to innovate and lock-in of current technologies once the standard is passed (Olmstead and Stavins, 2009[17]).

Accompanying measures to pricing policies are key to ensure their effectiveness and political feasibility. Three important policy instruments are discussed here. First, much evidence on low responsiveness of consumers to water prices (see Box 6.8) has been found to stem in particular from the low salience or high misperception of these prices. Gaudin (2006[45]) finds that demand responsiveness may increase when price information is posted on water bills. García-Valiñas et al. (2021[22]) find that policies aiming at promoting the careful reading of one's bill, providing additional detail about water consumption and tariffs or also promoting individual metering can significantly increase the impact of water pricing. Second, innovation support is key in enabling substitution possibilities and fostering adaptability. Innovation support has been discussed in the case of France for projects aiming at managing CECs. This has also been the case for new irrigation technologies which require much less water. Finally, measures to ensure affordability are key, from a political feasibility perspective certainly, but also because of the specific status of water as a necessary good for survival. Targeted measures to compensate poorer households for unaffordable water bills or to accompany small farmers through water price increases can become essential. Annex 6.A presents options to address affordability issues in the case of France.

Box 6.8. Water users' responsiveness to water pricing

Recent findings show that responsiveness depends on the type of user considered. In particular, industrial users' responsiveness to water pricing is generally found to be higher than that of agricultural users', higher in turn than residential users'.

The responsiveness of industrial users is heterogenous across firms and is generally found to be higher when the firm has the potential for in-plant recirculation of water as substitution for freshwater.

Recent evidence (Chakravorty, Dar and Emerick, 2023[46]) highlights that when farmers face a price per volume of water used, they have higher chances of adopting water-saving technologies, hence reduce their consumption than when facing a fixed fee for water access.

Responsiveness of agricultural users and households is generally higher when measuring long-run elasticities, when enough time is given to these users to adapt, change water-saving habits and adopt water-saving technologies. This has been shown in the case of New Zealand after volumetric water pricing was introduced and in Denmark, where water prices increased by 54% over two decades.

Moreover, low-value water use from households (e.g., gardening, swimming polls) is more elastic than high-value water use (e.g., drinking, cooking).

Finally, some studies find that the price elasticity of water use increases with higher prices – since higher water charges then account for a larger share of household expenditures. Given that water prices are generally low, this can help explain why responsiveness estimates are generally low as well.

Note: This Box is largely based on an OECD note (Leflaive, 2022[30]), pp. 17-18 and Box 3.2.
Source: Chakravorty et al. (2023[46]),Leflaive (2022[30]), Leflaive and Hjort (2020[47]), Reynaud (2015[48]).

Policy coherence is also called for by the OECD Council Recommendations on Water and is central to avoid conflicting signals and incentives. This holds particularly true in the case of farmers. For example, many policies that support agriculture production encourage greater land use change and intensive use of inputs, such as fertilisers, pesticides or irrigation (OECD, 2017[2]). In Andalusia, an example of potential misalignment of broader policy objectives with environmental policy relates to the recent rise in avocado production, a crop which is highly water-intensive. Policies or policy strategies that incentivise more avocado production – or do not disincentivise it – may be misaligned with water-conservation policies as under current circumstances (no appropriate water pricing and water pollution policies are in place for agriculture) it exacerbates environmental problems related to extreme water use.

Relatedly, changing agricultural and cultivation plans according to water sustainability can also help avoid lock-in effects. Given the predicted rise in temperature and water shortages in the South of Spain, the cultivation of many water-intensive crops is bound to become unsustainable in the years to come. For example, a recent study by the University of Cordoba and the Centre for Research in Geospace Science (Arenas-Castro et al., 2020[49]) found that the cultivation of many olives varieties in Andalusia was likely to become unstainable in the years to come. Certain varieties of olives, such as the Picual variety were found to be more resistant to climate change and to require less water. Switching to the cultivation of such varieties could help promote sustainability and avoid future lock-in effects.

Ensuring policy coherence but also setting clear policy goals and priorities is key in achieving water use and pollution sustainability and fairness without prejudice to other policy areas, including economic development. Indeed, all policies come with trade-offs. In the agricultural sector, national or regional agricultural planning policies may negatively affect water conservation (rice cultivation requires more water than certain vegetables) but ensure food security and competitiveness on the global market. Irrigated agriculture can also contribute to rural development (OECD, 2010[37]). For example, the recent rise in avocado cultivation in Andalusia enables farmers to make a decent living, since the global demand for this fruit is increasing and its price relatively stable (at EUR 2.4/kg on average in the recent years) (Campos, 2021[26]). However, given the urgency of the water scarcity issue in Andalusia and given the resulting long-term unsustainability of such agricultural practices, setting water conservation as a priority in policy decisions in Andalusia could be envisaged.

The White Book for Tax Reform in Spain has called "priority attention to the water-energy nexus and tax treatment that results in best practice for the whole" (Comité de personas expertas, 2022[10]). And indeed, climate change mitigation and adaptation policies may also come with trade-offs concerning water conservation and pollution issues. For example, desalination may help adapt to water scarcity but is very energy-intensive. On the other hand, setting higher energy prices for the pumping of water from wells could be an alternative way of restricting water abstraction from private wells (Ryan and Sudarshan, 2022[50]).

6.4. Key findings and strategic recommendations

Spain is characterised by a high temporal and spatial variability in water resources and the Southern part of the country regularly experiences water scarcity and long periods of droughts. The important heterogeneity in water resources has resulted in the historical use of a supply-side response, through the construction of dams, reservoirs, inter-basin water transfers and the intensive use of groundwater through the drilling of wells. However, the increasing risks linked to climate change and changing precipitation patterns calls for a focus on demand-side instruments (such as taxes and levies, or certain non-market-based instruments) – even if used in parallel with other supply-side strategies such as desalination and reuse.

In Andalusia many pricing instruments apply to the use of water. These are national-level, Andalusian-level and more local instruments. The national level instruments principally address service-cost recovery and environmental costs related to the installation of water extraction activities. The Andalusian and local charges address service-cost recovery as well as, to a certain extent, affordability and sustainable use criteria.

The main water uses in Andalusia are agriculture (about 80%, above the world average of 70%) and urban use. Industry not connected to the grid presents a much smaller share of water use, except in the Tinto-

Odiel-Piedras water basin. While many instruments are in place to directly price water use for urban users and only one is in place for pricing water use from agriculture. This instrument principally seeks to cover service-related costs and address equity between agricultural users. Urban users, however, face prices that do not only cover service-related costs but also seek to ensure affordability and sustainable use.

The equity principle between users could be improved in Andalusia. This holds between agriculture and urban uses as well as within urban uses. The differential rates and coverage observed, however, could be due to other factors, general equilibrium reasons (an increase in prices for farmers ultimately results in an increase in prices for households) and to different demand elasticities of users.

The improvement fee (though temporarily suspended by the Andalusian government from 1 January to 31 December 2023) could be better designed in terms of equity between households of different sizes and between households and firms. However, it holds the interesting feature of charging water suppliers for water losses in the network – an important issue in Spain. The low price levels for this fee, however, might imply that the resulting total water charge does not weigh much in households' budgets and hence does imply behavioural changes on their part. Accompanying measures, such as public awareness campaigns about water scarcity, information on water fees themselves aimed at increasing their salience or smart metering devices can contribute to increasing responsiveness.

In order to better balance cost-recovery and financial sustainability needs, equity, affordability and sustainable use, clear sustainable use objectives should be set and more information on costs and demand elasticities should be acquired. Environmental costs of water abstraction could then also be included in prices.

Given the special status of water as a good for which "the right to safe and clean drinking water and sanitation as a human right [...] is essential for the full enjoyment of life and all human rights" (United Nations General Assembly resolution, 28 July 2010), usual market forces that would increase prices and decrease demand when supply is tight are not in play. Hence, government intervention can be justified to ensure sustainable use.

For concessions extracting water for urban, industrial and agricultural uses, an Andalusian-level abstraction charge could be put in place, to align with the sustainable use goals of Andalusia. This is also a recommendation for the national level of the 2022 White Book for Tax Reform in Spain. For informal extraction of water (through wells, whether legal or not), which is a non-negligible share of agricultural water use, monitoring mechanisms could be put in place at the user association level. Monetary fines, for example, could be put in place if the groundwater body to which they are attached reaches poor quantitative status. This latter mechanism, however, would fall within the jurisdiction of Spain.

Water pollution comes from urban, industrial and agricultural users and may originate from point sources or diffuse sources. In the urban and industrial sectors, water pollution is mainly due to wastewater and direct industrial discharges. The usual water pollution sources from the agricultural sector include sedimentation and pesticides use as well as certain practices of nutrient use (applied in the form of chemical fertilisers, manure, and sludge), animal feeding, livestock grazing and irrigation. Contaminants of emerging concern (CECs) are an increasing issue for all users.

The main externalities from water pollution relate to health and ecosystems (environmental externalities) but are also economic, due to the need to sanitise polluted water for consumption. These externalities are addressed by a pollution control fee on discharges of water from urban and industrial use. However, no pollution tax or fee applies in the agricultural sector, even though it is the main sector responsible for aquifer pollution today.

While the lack of water pollution pricing for agriculture may be due to the diffuse nature of the pollution arising from this sector, the pricing of polluting inputs, such as pesticides and fertilisers – both responsible for an important share of water pollution – may be considered. Such taxes would target the quantities purchased of a specific product and rates would depend on their respective environmental impact.

Pesticides on the European market are already risk assessed by the European Chemicals Agency, so defining products to be targeted by the tax and grouping them into different rate bands would be relatively straightforward. Norway has such a tax. Taxes could also target number and type of livestock as well as area and type of cultivation, such as in Wallonia in Belgium.

Evidence points to low responsiveness of farmers to input taxes and high political barriers. This stresses the importance of complementary policies, which can help farmers reduce pesticide use without risking an important decrease in yield or income or of a broader policy environment that is aligned with water protection objectives (e.g., policies that promote quality of agricultural production over quantity). Coordination with other Autonomous Communities is also key, as farmers could get their input provisions from other regions with no input tax. Finally, political barriers may be addressed through better communication on evidence-based results of pollution pricing mechanisms and earmarking of revenues.

Advances in nutrient pollution modelling can provide an opportunity to tax diffuse pollution outputs directly, rather than taxing proxies such as fertiliser and pesticide inputs. This could increase the efficiency of water pollution taxes and might reduce political friction, by promoting a tax which would be closer to direct environmental damage and hence be perceived as fairer.

Finally, ensuring policy coherence but also setting clear policy goals and priorities is key in achieving water use, pollution sustainability and fairness without prejudice to other policy areas, including economic development. In this respect, long-term and short-term goals and sustainability should be carefully assessed. However, given the urgency of the water scarcity issue in Andalusia, setting water conservation as a priority in policy decisions in Andalusia could be envisaged.

References

2030 Water Ressources Group (2009), *Charting Our Water Future*. [6]

Agence de l'eau Loire-Bretagne (2017), *Une agence au service de l'eau depuis plus de 50 ans*. [59]

Andersen, M. et al. (2006), *The Use of Economic Instruments in Nordic and Baltic Environmental Policy 2001-2005*, TemaNord, Nordic Council of Ministers, Copenhagen K, https://doi.org/10.6027/TN2006-525. [28]

Arbués, F. and M. García-Valiñas (2020), *Water Tariffs in Spain*, Oxford University Press, https://doi.org/10.1093/acrefore/9780190632366.013.246. [20]

Arenas-Castro, S. et al. (2020), "Projected climate changes are expected to decrease the suitability and production of olive varieties in southern Spain", *Science of The Total Environment*, Vol. 709, p. 136161, https://doi.org/10.1016/j.scitotenv.2019.136161. [49]

Aznar-Sánchez, J. et al. (2019), "Aquifer Sustainability and the Use of Desalinated Seawater for Greenhouse Irrigation in the Campo de Níjar, Southeast Spain", *International Journal of Environmental Research and Public Health*, Vol. 16/5, p. 898, https://doi.org/10.3390/ijerph16050898. [1]

Böcker, T. and R. Finger (2016), "A Meta-Analysis on the Elasticity of Demand for Pesticides", *Journal of Agricultural Economics*, Vol. 68/2, pp. 518-533, https://doi.org/10.1111/1477-9552.12198. [39]

Böcker, T. and R. Finger (2016), "European Pesticide Tax Schemes in Comparison: An Analysis of Experiences and Developments", *Sustainability*, Vol. 8/4, p. 378, https://doi.org/10.3390/su8040378. [40]

Boxall, A. (2012), *New and Emerging Water Pollutants arising from Agriculture*. [35]

Campos, A. (2021), *Avocado cultivation depletes southern Spain's water resources*. [26]

Cardone, R. and C. Fonseca (2003), *Financing and Cost recovery*. [14]

Chakravorty, U., M. Dar and K. Emerick (2023), "Inefficient Water Pricing and Incentives for Conservation", *American Economic Journal: Applied Economics*, Vol. 15/1, pp. 319-350, https://doi.org/10.1257/app.20210011. [46]

Comité de personas expertas (2022), *Libro Blanco Sobre la Reforma Tributaria*. [10]

Cycoń, M., A. Mrozik and Z. Piotrowska-Seget (2019), "Antibiotics in the Soil Environment—Degradation and Their Impact on Microbial Activity and Diversity", *Frontiers in Microbiology*, Vol. 10, https://doi.org/10.3389/fmicb.2019.00338. [36]

Da Costa et al. (2015), *Public Water and Wastewater Services in France - Economic, Social and Environmental Data*. [62]

EPA (2005), "Protecting water quality from agricultural runoff", *United States Environmental Protection Agency*, https://www.epa.gov/sites/default/files/2015-09/documents/ag_runoff_fact_sheet.pdf. [34]

Estrela, T. and T. Sancho (2016), "Drought management policies in Spain and the European Union: from traditional emergency actions to Drought Management Plans", *Water Policy*, Vol. 18/S2, pp. 153-176, https://doi.org/10.2166/wp.2016.018. [8]

European Commission (2021), *Ensuring that Polluters Pay - Taxes, charges and fees*. [54]

European Commission (2019), "Fertilisers in the EU - Prices, trade and use", *EU Agricultural Markets Briefs*, Vol. 15, https://ec.europa.eu/info/sites/default/files/food-farming-fisheries/farming/documents/market-brief-fertilisers_june2019_en.pdf. [38]

European Court of Auditors (2021), *Sustainable water use in agriculture: CAP funds more likely to promote greater rather than more efficient water use*. [27]

European Environment Agency (2017), *Briefing: Water management in Europe: price and non-price approaches to water conservation*. [29]

European Environment Agency (2013), "Assessment of Cost Recovery Through Water Pricing". [52]

FAO & IWMI (2017), *Water pollution from agriculture: a global review*. [3]

Fuentes, A. (2011), "Policies Towards a Sustainable Use of Water in Spain", *OECD Economics Department Working Papers*, No. 840, OECD Publishing, Paris, https://doi.org/10.1787/5kgj3l0ggczt-en. [25]

García-Valiñas, M., R. Martínez-Espiñeira and M. Suárez-Varela Maciá (2021), "Price and Consumption Misperception Profiles: The Role of Information in the Residential Water Sector", *Environmental and Resource Economics*, Vol. 80/4, pp. 821-857, https://doi.org/10.1007/s10640-021-00611-8. [22]

Gaudin, S. (2006), "Effect of price information on residential water demand", *Applied Economics*, Vol. 38/4, pp. 383-393, https://doi.org/10.1080/00036840500397499. [45]

Grafton, R., L. Chu and P. Wyrwoll (2020), “The paradox of water pricing: dichotomies, dilemmas, and decisions”, *Oxford Review of Economic Policy*, Vol. 36/1, pp. 86-107, https://doi.org/10.1093/oxrep/grz030. [16]

Grafton, R. and S. Wheeler (2018), “Economics of Water Recovery in the Murray-Darling Basin, Australia”, *Annual Review of Resource Economics*, Vol. 10/1, pp. 487-510, https://doi.org/10.1146/annurev-resource-100517-023039. [32]

INE (2023), *Regional Accounts of Spain. Series 2016-2021..* [9]

Leflaive, X. (2022), *Background note: The economics of water scarcity*. [30]

Leflaive, X. and M. Hjort (2020), “Addressing the social consequences of tariffs for water supply and sanitation”, *OECD Environment Working Papers*, No. 166, OECD Publishing, Paris, https://doi.org/10.1787/afede7d6-en. [47]

Legifrance (2018), *Environmental Code*. [60]

Les entreprises de l’eau (2019), “Public water and wastewater services in France”. [63]

Luis Caparrós-Martínez, J. et al. (2020), “Public policies for sustainability and water security: The case of Almeria (Spain)”, *Global Ecology and Conservation*, Vol. 23, p. e01037, https://doi.org/10.1016/j.gecco.2020.e01037. [7]

Mayol, A. (2018), “Social and Nonlinear Tariffs on Drinking Water: cui bono? Empirical Evidence from a Natural Experiment in France”, *Revue d’économie politique*, Vol. Vol. 127/6, pp. 1161-1185, https://doi.org/10.3917/redp.276.1161. [61]

Möller-Gulland and Lago (2011), *Water Abstraction Charges and Compensation Payments in Baden-Württemberg (Germany)*, Ecologic Institute. [51]

Mottershead, D. et al. (2021), *Green taxation and other economic instruments. Internalising environmental costs to make the polluter pay*, European Commission, Directorate-General for Environment, Brussels, https://environment.ec.europa.eu/publications/green-taxation-and-other-economic-instruments-internalising-environmental-costs-make-polluter-pay_en. [15]

Norwegian Ministry of Agriculture and Food (2015), *Regulation on Pesticides*. [55]

OECD (2021), *Toolkit for Water Policies and Governance: Converging Towards the OECD Council Recommendation on Water*, OECD Publishing, Paris, https://doi.org/10.1787/ed1a7936-en. [24]

OECD (2020), “Belgium”, in *Taxation in Agriculture*, OECD Publishing, Paris, https://doi.org/10.1787/8195087d-en. [58]

OECD (2020), *Financing Water Supply, Sanitation and Flood Protection: Challenges in EU Member States and Policy Options*, OECD Studies on Water, OECD Publishing, Paris, https://doi.org/10.1787/6893cdac-en. [5]

OECD (2020), *OECD Economic Surveys: Israel 2020*, OECD Publishing, Paris, https://doi.org/10.1787/d6a7d907-en. [23]

OECD (2019), *Pharmaceutical Residues in Freshwater: Hazards and Policy Responses*, OECD Studies on Water, OECD Publishing, Paris, https://doi.org/10.1787/c936f42d-en. [44]

OECD (2017), *Diffuse Pollution, Degraded Waters: Emerging Policy Solutions*, OECD Studies on Water, OECD Publishing, Paris, https://doi.org/10.1787/9789264269064-en. [2]

OECD (2017), *Environmental fiscal reform - progress, prospects and pitfalls. OECD report for the G7 environment ministers..* [18]

OECD (2016), "OECD Council Recommendation on Water", https://www.oecd.org/environment/resources/Council-Recommendation-on-water.pdf. [4]

OECD (2015), *Water Resources Allocation: Sharing Risks and Opportunities*, OECD Studies on Water, OECD Publishing, Paris, https://doi.org/10.1787/9789264229631-en. [33]

OECD (2010), "Economic instruments for mobilising financial resources for supporting IWRM". [53]

OECD (2010), *Pricing Water Resources and Water and Sanitation Services*, OECD Studies on Water, OECD Publishing, Paris, https://doi.org/10.1787/9789264083608-en. [11]

OECD (2010), *Sustainable Management of Water Resources in Agriculture*, OECD Studies on Water, OECD Publishing, Paris, https://doi.org/10.1787/9789264083578-en. [37]

Olmstead, S. and R. Stavins (2009), "Comparing price and nonprice approaches to urban water conservation", *Water Resources Research*, Vol. 45/4, https://doi.org/10.1029/2008wr007227. [17]

Ostrom, E. (1965), *Public Entrepreneurship: A Case Study in Ground Water Basin Management*, University of California, Los Angeles. [31]

Reynaud, A. (2015), "Modelling Household Water Demand in Europe - Insights from a Cross-Country Econometric Analysis of EU-28 countries.", *Publications Office of the European Union*, https://publications.jrc.ec.europa.eu/repository/handle/JRC96268. [48]

Rogers, P. (2002), "Water is an economic good: How to use prices to promote equity, efficiency, and sustainability", *Water Policy*, Vol. 4/1, pp. 1-17, https://doi.org/10.1016/s1366-7017(02)00004-1. [13]

Rogers, P., R. Bhatia and A. Huber (1998), "Water as a social and economic good: How to put the principle into practice.", *Global Water Partnership Technical Advisory Committee (TAC)/Swedish International Development Cooperation Agency, Stockholm, Sweden*. [12]

Ryan, N. and A. Sudarshan (2022), "Rationing the Commons", *Journal of Political Economy*, Vol. 130/1, pp. 210-257, https://doi.org/10.1086/717045. [50]

Skou Andersen, M. (2016), *Fertilizer tax in Sweden*. [42]

Söderholm, P. and A. Christiernsson (2008), "Policy effectiveness and acceptance in the taxation of environmentally damaging chemical compounds", *Environmental Science & Policy*, Vol. 11/3, pp. 240-252, https://doi.org/10.1016/j.envsci.2007.10.003. [43]

Spikkerud, E. (2006), "Taxes as a Tool to Reduce Health and Environmental Risk from Pesticide Use in Norway", in *Evaluating Agri-environmental Policies: Design, Practice and Results*, OECD Publishing, Paris, https://doi.org/10.1787/9789264010116-21-en. [41]

Tooth, R. (2014), *LRMC Pricing for Water Services - Background Paper on LRMC Pricing*. [19]

UN Environment Programme (2020), *Study on the effects of taxes and subsidies on pesticides and fertilizers*. [56]

Van Dender, K. et al. (2022), *Why governments should target support amidst high energy prices*, OECD Publishing, Paris, https://www.oecd.org/ukraine-hub/policy-responses/why-governments-should-target-support-amidst-high-energy-prices-40f44f78/. [21]

Wallonie agriculture SPW (2017), *La taxe sur les charges environnementales générées par les exploitations agricoles*. [57]

Annex 6.A. Detailed case studies: water pollution and usage

This annex presents selected case studies in the domain of water pollution and usage across the OECD.

Germany: Baden-Württemberg Water Abstraction Charge

Annex Table 6.A.1. Baden-Württemberg water abstraction charge (Germany)

Legal bases	Baden-Württemberg's Water Act (1988 amended in 2010)
Objective	To raise awareness about water scarcity issues, to incentivise water-saving behaviour, to reduce the economic advantage of agents benefiting from the direct abstraction of water in comparison to those who do not benefit from it and to compensate investments of water bodies in charge of maintenance and cleaning.
Level of responsibility	Region (Baden-Württemberg)
Tax setter(s)	Region (Baden-Württemberg)
Revenue beneficiary(ies)	Region (Baden-Württemberg)
Tax payer(s)	Industries (including energy) and, indirectly, households receiving water from public water suppliers
Tax base (including main exemption(s), credits or deductions)	The tax base is the annual volume of water abstracted in cubic metres. Since the amendment of 2010, exemptions apply to the following cases: • Minor uses, such as abstractions below 4 000 m3/year, • The use of groundwater for heating and cooling buildings when the water is returned to the groundwater, • Use of groundwater to avert hazards in the context of groundwater remediation. • Reductions apply to the following cases: • Up to 90% for water-intensive industries that prove the abstraction charge undermines their competitive position, • Up to 75% of the abstraction charges can be offset by investments that reduce pollution related to heating, enhance waterbodies or enable the substitution of groundwater with surface water, • Up to 25% for specific industries when environmental management systems are in place.
Tax rate(s) (including their calculation)	The tax rate depends on the water source and type of extraction, ranging between EUR 0.01/m3 and EUR 0.051/m3, as shown in Annex Table 6.A.2.
Governance and implementation	The implementation of the charge involved a high level of openness and dialogue between different stakeholders (e.g. associations, committee meetings of the Länder parliaments, online stakeholder consultations) to design the legislation.
Environmental, social & health impacts	Water abstraction decreased by 34% between 1987 and 2007. Other factors may have also affected it, such as technology (Möller-Gulland and Lago, 2011[51]).

Annex Table 6.A.2. Tax rates (Baden-Württemberg Germany)

Sources	Cost categories	Tax rates (EUR/m3)
Surface water	Public water supply	0.051
	Others	0.010
Ground water	Public water supply	0.051
	Others	0.051

Source: (Möller-Gulland and Lago, 2011[51]), (European Environment Agency, 2013[52]; Möller-Gulland and Lago, 2011[51]; OECD, 2010[53]; European Commission, 2021[54]).

Norway: Banded Tax system for Pesticides

Annex Table 6.A.3. Banded tax system for pesticides (Norway)

Legal bases	Regulations on plant protection products (1999 amended in 2015)
Objective	To reduce the use of pesticides that represent the greatest risk to human health and the environment.
Level of responsibility	Central government (Norway)
Tax setter(s)	Central government (Norway)
Revenue beneficiary(ies)	Central government (Norway)
Tax payer(s)	Wholesalers
Tax base (including main exemption(s), credits or deductions)	The tax system is area-based and consists of seven tax bands based on patterns of use, human health, and environmental risk. Products allowed in organic farming are exempt.
Tax rate(s) (including their calculation)	The system is area-based as the tax rate is calculated according to the pesticides' specific standard doses per hectare and consists of seven tax bands based on the risks posed to human health and the environment. Each pesticide is evaluated against human health and environmental risk criteria. The former is based on intrinsic properties and exposure, which depends on the type of formulation and application method. The environmental criteria consider toxicity, bioaccumulation, persistence and leaching potential. All pesticides for professional use are tested according to several criteria and then categorised as a low, medium, or high risk. The categorisation of the factors can be seen in Annex Table 6.A.4, which gives rise to different tax bands. To extra tax bands arise for products for hobby use. The tax per hectare for each tax band is calculated by multiplying the base rate of NOK 25/ha (EUR 2.4/ha), which is set by the central government and the same for all products, by the tax band factor: Tax per hectare = base rate x factor for the given tax band This is converted to a tax per gram or milliliter considering as the Standard Area Dose (SAD), which is the maximum application rate for the main crop for which the pesticide is used, applying the formula Tax per kg or liter = (base rate x factor) x 1 000 / SAD
Governance and implementation	The Norwegian Crop Research Institute proposes the SADs and SAD guidelines. The Norwegian Food Safety Authority approves the SADs and determines the tax bands for each pesticide.
Environmental, social & health impacts	Between 1998 and 2011, a decline was reported in the sales of pesticides from categories 4 and 5, which are the most taxed, while an increase was observed in the sales of categories 1 and 2. In 2014, the Norwegian Food Authority reported no sale of pesticides in category 5. At the crop level, pesticides of higher categories have also been substituted for lower-category products. However, because of the hoarding effect on sales before the tax increase, the reduction of pesticide risks could only be observed years later (Böcker and Finger, 2016[40]).

Source: (Spikkerud, 2006[41]; Norwegian Ministry of Agriculture and Food, 2015[55]; Böcker and Finger, 2016[40]; UN Environment Programme, 2020[56])

Annex Table 6.A.4. Categorisation of pesticides factors (Norway)

Tax category		1	2	3	4	5	6	7
Pesticide characteristics	Human health risks Environmental risks	Both risks low	One low and one medium risk	One low and one high risk or both risks medium	One medium and one high risk	Both risks high	Concentrated products for hobby use	Ready-to-use product for hobby use
	Factor i (* NOK25/ha)	0.5	3	5	7	9	50	150
	Tax (NOK/ha)	12.5	75	125	175	225	1 250	3 750

Source: (Böcker and Finger, 2016[40]).

Belgium: Wallonia Tax on Environmental Impacts from Farming

Annex Table 6.A.5. Wallonia tax on environmental impacts from farming (Belgium)

Legal bases	MB 29.12.2014 of the Environmental Code of Wallonia
Objective	To internalise external environmental costs linked to agricultural activities' impacts on water resources and the use of fertilisers and phytosanitary products in crops
Level of responsibility	Region (Wallonia)
Tax setter(s)	Region (Wallonia)
Revenue beneficiary(ies)	Region (Wallonia)
Tax payer(s)	Farmers meeting specific criteria, of which the number of animals owned and farming area.
Tax base (including main exemption(s), credits or deductions)	The annual tax is based on two components: animals and land. Exemptions apply to each component as follows: • The livestock environmental component is nil when the farm holds a certificate of compliance for the livestock manure storage infrastructure or when the issue of this certificate is under examination. • The land component determines that the first thirty hectares of a farm are exempt from the tax.
Tax rate(s) (including their calculation)	The tax system is based on two components, which are summed to calculate the number of environmental load units. The two components are: • Livestock: measured by the number of livestock held or environmental charges generated by run-off from on-farm livestock effluent storage facilities that reach groundwater and surface water. • Land: measured in hectares for the area where agricultural activities require fertilisers. The environmental load unit is calculated by applying the formula: $N = 2 + N1 + N2$ Where N = the number of environmental load units, N1 is the livestock environmental load, and N2 is the land environmental load. The livestock component is calculated by summing the products of the multiplications of the number of animals by their nitrogen coefficient using Annex Table 6.A.6. The land component is calculated by adding the products resulting from the multiplication of crop and grassland areas by nitrogen coefficients, which reflect the average nitrogen residue in the soil, the average use of pesticides and the erosive potential of the crop. The coefficients are: • "Crop" coefficient = 0.3 • "Organic crop" coefficient = 0.15 • "Grassland" coefficient = 0.06 • "Organic grassland" coefficient = 0.03 These coefficients reflect the soil's average nitrogen residue, pesticide use, and the erosive potential of crops and grasslands. The base rate of the environmental load unit was EUR 10 in January 2015, and is indexed by inflation.
Governance and implementation	The tax supports the long-term plans outlined in Wallonia's First Strategy on Sustainable Development (2013). This strategy aims to internalise the external environmental costs of multiple activities, including the production of food products. Consultations with the Wallonia Council for Environment and Sustainable Development (CWEDD), the Wallonia Council for Economy, Society and the Environment (CESW) and the Wallonia High Council for Cities, Towns and Provinces were also held to develop the strategy.

Source: (Wallonie agriculture SPW, 2017[57]; OECD, 2020[58]; European Commission, 2021[54])

Annex Table 6.A.6. Animal categories and nitrogen coefficients (Wallonia, Belgium)

	Animal category	Nitrogen coefficient
Bovines	Milk cow	0.5538
	Suckler cow	0.4062
	Cull cow	0.4062
	Other cattle over two years old	0.4062
	Cattle with less than six months old	0.0615
	Heifer from six to 12 months	0.1723
	One to two years old heifer	0.2954
	Young bull from six to 12 months old	0.1538
	Young bull from one to two years old	0.2462
Sheep and goats	Sheep and goats under one year old	0.0203
	Sheep and goats over one year old	0.0406
Horses	Equine	0.3446
Pigs	Jerseys	0.0923
	Treason	0.0923
	Fattening pigs and gilt	0.0480
	Fattening pigs and gilt on bio-controlled litter	0.0277
	Piglets (four to 10 weeks old)	0.0117
Rabbits	Mother rabbits	0.0222
	Fattening rabbits	0.0020
Poultry	Broilers (40 days)	0.0017
	Laying or breeding hens (343 days)	0.0037
	Pullets (127 days)	0.0017
	Breeding roosters	0.0026
	Ducks (75 days)	0.0026
	Geese (150 days)	0.0026
	Turkeys (85 days)	0.0050
	Guinea fowl (79 days)	0.0017
	Quail	0.0002
	Ostriches and emus	0.0185

Source: (Wallonie agriculture SPW, 2017[57])

France: Water Management and Taxes

The French water taxation system aims to finance the protection of water resources and the aquatic environment. It is based on the polluter-pays and user-pays principles This system is implemented by the water agencies *(agences de l'eau)*, which aim to collect water levies from water users and to distribute water aids. They also determine tax rates within national statutory limits. There are six water agencies in France (Annex Figure 6.A.1). Each is composed of a board and a basin committee. The board defines action programmes and water levies that it submits to the basin committee and it determines how aids are allocated. The basin, which is made up of state representatives, subnational governments' elected officials and water users, is responsible for assessing the board's programmes and water levies, for planning actions related to water management in the basin and for evaluating regulations and projects with a direct effect on water and aquatic environment. This participatory model aims to facilitate the acceptance of taxes by liable entities. The model also includes the possibility for periodic adjustments to fairly represent water usage and retain its acceptability level. Tax revenue is earmarked for reinvestment in water quality and for dealing with scarcity at the basin level and is managed by the water agencies. The main beneficiaries of water aids are subnational governments to finance projects related to water protection. The other beneficiaries are businesses, farmers, associations, etc.

Annex Figure 6.A.1. Mapping of the basin committee in France

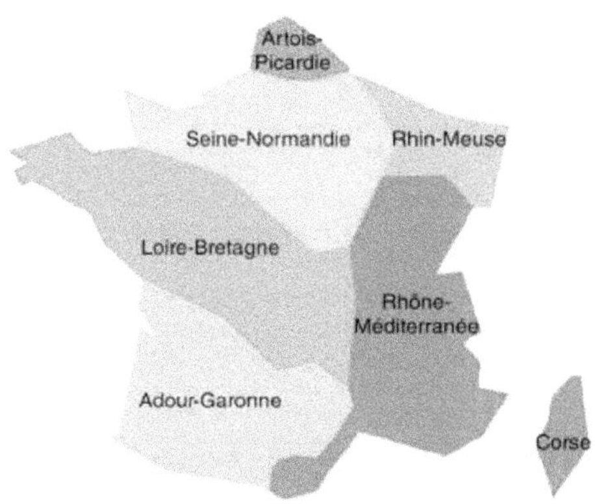

Source: Agence de l'eau Loire-Bretagne (Agence de l'eau Loire-Bretagne, 2017[59]) (2017[59])

Since the 1 January 2020, there are eight water-related levies (*redevances*) in France, defined under article L213-10 of the Environmental Code (Legifrance, 2018[60]), which aim at internalising the externalities of water uses by different user groups and to finance water protection activities. Four of the taxes are associated with water pollution (Legifrance, 2018[60]):

- **Tax on domestic water pollution** (*redevance pour pollution de l'eau d'origine domestique*) (art. L231-10-1 to 10-4) (Legifrance, 2018[60]): charges domestic users for their water consumption. The tax rates lie within the national limit of EUR 0.50/m^3.
- **Tax on non-domestic pollution** (*redevance pour pollution de l'eau d'origine non domestique*) (art. L231-10-1 to 10-4) (Legifrance, 2018[60]): charges any economic or industrial activity that discharges pollution. Pollution is assessed directly by monitoring systems and, when not possible, a theoretical level of pollution is calculated using benchmarks. The tax base is the annual pollution discharged above a threshold and the rates are different for each pollutant (Legifrance, 2018[60]).
- **Tax on sewer systems modernisation** (*redevance pour modernisation des réseaux de collecte*) (art. L231-10-5 to 10-7) (Legifrance, 2018[60]): charges all domestic or non-domestic users

connected to a public sewerage network for the volume discharged in the sewer network. The tax rates lie within the national limit of EUR 0.30/m^3.

- **Tax on diffuse pollution** (*redevance pour pollutions diffuses*) (art. L231-10-8) (Legifrance, 2018[60]): charges users of phytopharmaceutical products (i.e. pesticides) according to the substance class and quantity applied. The tax rates are different for each substance class.

The remaining four levies charge particular economic activities mostly related to water use (Legifrance, 2018[60]):

- **Water abstraction charge** (*redevances pour prélèvement sur la ressource en eau*) (art. L231-10-9) (Legifrance, 2018[60]): charges all users (e.g. households, industries and agriculture) for water withdrawal from the water resource. Exemptions apply to withdrawals of sea water, excavation of mines whose activity has ceased as well as withdrawals necessary for underground works and withdrawals during drainage to keep buildings dry, withdrawals related to aquaculture, geothermal energy, frost control for perennial crops, withdrawals outside the low water period or withdrawals intended exclusively for the supply of water to heritage fountains located in mountain areas and within the limit of a maximum of 5,000 m^3. The rates depend on the water use and the water basins. They are determined by the water agencies in EUR/m^3 within the national limits (Annex Table 6.A.7).
- **Hydroelectricity production charge** (a particular case of the water abstraction charge) (art. L231-10-9) (Legifrance, 2018[60]): it charges hydroelectric operators relatively to the volume of diverted water. The water agency sets the tax rate of the fee within the limit of a ceiling of EUR 1.8 per million m^3 and per meter of fall.
- **Tax on storage in low water level periods** (*redevance pour stockage d'eau en période d'étiage*) (art. L231-10-10) (Legifrance, 2018[60]): charges any person who has a storage facility of more than a million cubic meters of water and who stores the volume discharged into a watercourse during low water periods. The tax base is the water stored during the low water period, and the water agencies determine the rate within the limit of EUR 0.01/m^3.
- **Tax for the protection of freshwater environments** (*redevance pour protection du milieu aquatique*) (art. L231-10-12) (Legifrance, 2018[60]): it charges recreational fisheries per recreational fisher. The rates are set annually by the water agency, with the following limits: EUR1 per individual who fishes for one day, EUR4 per individual who fishes for seven consecutive days and EUR10 per individual who fishes for one year. An additional EUR20 per individual is required for fishing eel fry, salmon and sea trout.

Annex Table 6.A.7. National limits on water abstraction charge according to water uses (France)

Water uses	Category 1	Category 2
Irrigation (except gravity irrigation)	3.6	7.2
Gravity irrigation	0.5	1
Drinking water supply	7.2	14.4
Industrial cooling leading to a return of more than 99%	0.5	1
Feeding a channel	0.03	0.06
Other economic uses	5.4	10.8

Notes: Water resources of each basin are classified in category 1 when they are located outside the water distribution zones defined in the law and in category 2 otherwise.
Source: Author's own elaboration based on (Legifrance, 2018[60]).

Since 2010, the French Law has allowed the introduction of discriminating tariffs amongst consumers. These tariffs can be either based on social criteria, where special rates are offered to low-income families (social tariffs) or based on consumption (Mayol, 2018[61]). The form of the aid may vary from one municipality to another. An example is Dunkerque's "*éco-solidaire*" tariff, which differentiates the rate into three tiers reflecting the amount of water used: (i) vital consumption tier (consumption below 75 m^3/year), (ii) useful consumption (between 75 and 200 m^3/year) and (iii) comfort (above 200m^3/year). Under this system, the social beneficiaries of the universal health care coverage (*couverture maladie universelle*) had a 70% discount for the vital consumption tier (Mayol, 2018[61]).

Other mechanisms exist to support households with difficulties to pay housing expenses, such as the Solidarity Fund for Housing (*Fonds de Solidarité pour le Logement,* FSL). This fund is managed by departments (ie subnational governments in France). It may finance fully or partially water bills for financially distress households through a subsidy or a loan. Departments have also established measures to waive water arrears and provide pre-emptive support to families. In 2013, these measures were applied in 35,000 cases amounting to EUR 2.4 million (Da Costa et al., 2015[62]) (Da Costa et al., 2015[62]). Water voucher schemes are another form of support. They are issued by water operators and allocated to welfare recipients via local social welfare bodies (e.g., *Centres Communaux d'Action,* CCAS). In 2017, these vouchers were distributed to more than 19,000 customers, who received EUR 120 on average (Les entreprises de l'eau, 2019[63]) (Les entreprises de l'eau, 2019[63]).

Notes

[1] Hence, even though the improvement fee has been temporarily suspended by the Andalusian government (Decreto-Ley 7/2022, de 20 de Septiembre) from 1 January to 31 December 2023, it is still discussed in this section.

[2] https://www.epa.gov/sites/default/files/2015-09/documents/ag_runoff_fact_sheet.pdf.

[3] The Segura river basin is shared between Murcia (59%), Castilla La-Mancha (25%), Andalusia (9%) and Valencia (7%) (https://hispagua.cedex.es/en/instituciones/demarcaciones/segura, as accessed on 20/06/2022). The Guadalquivir river basin is shared between Andalusia (90.2%), Castilla La-Mancha (7.1%), Extremadura (2.5%) and Murcia (0.2%) (https://hispagua.cedex.es/en/instituciones/demarcaciones/guadalquivir, as accessed on 20/06/2022). Finally, the Guadiana river basin is also shared with Portugal. Within Spain, the division is as follows: 48% is within Castilla La-Mancha, 42% within Extremadura and 10% within Andalusia (https://hispagua.cedex.es/en/instituciones/demarcaciones/guadiana, as accessed on 20/06/2022).

[4] More precisely, this deficit is defined as the "aggregated global gap between existing accessible, reliable supply and 2030 water withdrawals, assuming no efficiency gains".

[5] https://ec.europa.eu/environment/water/water-framework/economics/pdf/Country%20fact%20sheet%20-%20SPAIN.pdf.

[6] https://inspain.news/extraordinary-drought-and-extreme-heat-in-andalucia-2021/.

[7] Runoff refers to all the water, such as rainfall or snowmelt that comes into a river water system.

[8] In 2016, Spain had around 1,200 dams, placing it as the ninth country in the world in terms of number of dams (Estrela and Sancho, 2016[8]).

[9] In 2010, similar shares as those presented in Figure 6.1 resulted in an Andalusian-level share for agricultural use of about 82%, for urban use of about 14%, for industrial use of about 3% and of less than

1% for other uses (see Table 106 in https://www.juntadeandalucia.es/medioambiente/portal/documents/20151/518740/PAMA2017_13febrero_portada.pdf/63069352-5250-aea3-3653-7000b57cdfeb?t=1368464000000).

[10] This is more than twice as high as the share of national agriculture GVA in Spain's GVA (2.9%).

[11] https://www.miteco.gob.es/es/agua/temas/planificacion-hidrologica/summaryrbmp2ndcycledraft_tcm30-379040.pdf.

[12] This distinction is further explained here: https://www.juntadeandalucia.es/medioambiente/portal/web/guest/areas-tematicas/agua/gestion-del-agua/recuperacion-de-costes/mapa-institucional-servicios-agua.

[13] In particular when allocation between users is well-defined.

[14] Economic externalities due to wastewater treatment are discussed in the subsection on water pollution.

[15] A/RES/64/292.

[16] https://ec.europa.eu/environment/water/water-framework/economics/pdf/Country%20fact%20sheet%20-%20SPAIN.pdf.

[17] https://ec.europa.eu/environment/water/water-framework/economics/pdf/Country%20fact%20sheet%20-%20SPAIN.pdf.

[18] Article 73 of Law 9/2010 of July 30.

[19] As a reminder, the Draft Hydrological Plan for the Mediterranean river basin defines urban use as uses through a connection to the urban grid – i.e., by households, regulated accommodation (e.g., hotels, rural tourism, campsites), non-regulated accommodation, industry connected to the urban grid, commercial and institutional uses, losses and uncontrolled uses.

[20] https://www.juntadeandalucia.es/medioambiente/portal/web/guest/areas-tematicas/agua/gestion-del-agua/recuperacion-de-costes/tarifas-y-canones-uso-agua-dominio-publico-hidraulico-dph-y-dominio-publico-maritimo-terrestre-dpmt, as accessed on 14 June 2022.

[21] https://www.juntadeandalucia.es/medioambiente/portal/web/guest/areas-tematicas/agua/gestion-del-agua/recuperacion-de-costes/tarifas-y-canones-uso-agua-dominio-publico-hidraulico-dph-y-dominio-publico-maritimo-terrestre-dpmt, as accessed on 14 June 2022.

[22] https://www.epdata.es/datos/graficos-situacion-agua-mundo-espana/333.

[23] Since 128L x 30 (days) = 3840 and $2m^3$=2000L.

[24] I.e., taking each user independently and not considering the impact of the increase of costs for one user on the final consumers, households.

[25] This does not refer to self-abstraction, which is not covered by any fee.

[26] This can prove to be more complicated when objectives vary with the time of year. For example, in times of drought, it can be urgent to reduce water consumption that in other times would not be problematic. Complementary policies such as temporary bans for certain uses are further discussed in the subsection on non-pricing policies.

[27] This is, however, not the focus of the present analysis.

[28] https://www.miteco.gob.es/es/agua/temas/planificacion-hidrologica/summaryrbmp2ndcycledraft_tcm30-379040.pdf.

[29] Based on Table 55 in the Guadalete-Barbate Draft Hydrological Plan for 2022-2027, https://www.juntadeandalucia.es/medioambiente/portal/documents/20151/497870/DI_MEMORIA_GB.pdf/caaf7b70-5ec4-61ee-795a-5377b35f8d73?t=1582033431000 where water used is defined as the sum of water consumed and waste water.

[30] Defined "how much abstracted water returns to the water body".

[31] Through soil that is washed off fields.

[32] Which can enter and contaminate water through direct application, runoff, and atmospheric deposition.

[33] https://www.who.int/news-room/fact-sheets/detail/antibiotic-resistance, as accessed on 25 January 2023.

[34] Eutrophication is the excessive richness of nutrients in a lake or other body of water, often caused by run-off from the land, causing the dense growth of plant life, in particular algal blooms.

[35] The lowest is at EUR 0.15/kg/year (Lithuania) and the highest at EUR 6/kg/year (Belgium).

[36] The lowest is at EUR 0.21/kg/year and the highest at EUR 4.93/kg/year (the Netherlands).

[37] The pollution control fee is also referred to as the discharge control charge in the Spanish Water Law.

[38] Meaning, here, industry not connected to urban grids.

[39] https://ec.europa.eu/commission/presscorner/detail/en/ip_21_6265, as accessed on 26 January 2023.

[40] https://ieep.eu/wp-content/uploads/2022/12/SE-Fertilizer-tax-final_REV.pdf, as accessed on 29 March 2023.

[41] Peer-reviewed studies, however, tend to find lower responsiveness levels than in grey literature.

[42] Such as income-support measures, that could promote decreased pesticides use, or the promotion of alternative practices such as ploughing.

[43] Indeed, when the Swedish tax on mineral fertilizers was abolished in 2009, the reduced diesel tax rate for farmers was increased in exchange.

[44] https://www.overseer.org.nz/. The New Zealand Ministry for Primary Industries, AgResearch, and the Fertiliser Association of New Zealand each hold one-third stake in the Overseer intellectual property.

[45] https://www.miteco.gob.es/es/agua/temas/planificacion-hidrologica/summaryrbmp2ndcycledraft_tcm30-379040.pdf.

[46] https://laboratoriorediam.cica.es/Visor_DMA/?urlFile=https://laboratoriorediam.cica.es/Visor_DMA/service_xml/capas_dma.xml.

[47] As their effect may seem more predictable than in the case of pricing policies where responsiveness estimates are key to predict impact.

Part IV Circular economy and waste management

7 Context

Spain ranks 26th out of 27 EU countries in terms of the weight of environmental taxes with respect to GDP. Green taxation accounts for 1.77% of Spain's gross domestic product (GDP), well below the European Union average of 2.37%, according to Eurostat data for 2019 (Eurostat, 2022[1]).

The lack of nationwide environmental taxes related to waste management in Spain has been one of the main causes of the poor performance of the country in this area (Castells-Rey, Pellicer-García and Puig-Ventosa, 2022[2]). For example, Spain in 2020 recycled only 40.5% of its Municipal Solid Waste (MSW) while the EU-27 average was 49.2% (Eurostat, 2022[3]), Figure A.A.1 shows the EU overview of MSW recycling rates. Thus, it has not accomplished the MSW recycling target of 2020 (50% of the MSW), and it will be a challenge to accomplish the recycling target of the upcoming years summarised in Table 7.A.1 in Annex 7.A. Traditionally, environmental taxes in Spain are mainly regional taxes that were promoted and implemented by some Spanish Autonomous Communities. Nevertheless, the situation has recently changed as two environmental taxes were included in the national Spanish law on waste and contaminated sites for a circular economy (Boletín Oficial del Estado, 2022[4]).

This chapter provides recommendations for policy reform at regional level to the Andalusian regional government (hereafter "Junta de Andalucía" or "Junta") on how to use its tax competencies to promote a Circular Economy (CE) and improve waste management. Such tax instruments could be used to address the material circularity within the region, as well as to increase waste prevention and improve its waste source separation and recycling rates. In addition to environmental taxes, other economic instruments (such as Extended Producer Responsibility) are vital to operationalise the Polluter Pays Principle (PPP) (European Commission, 2021[5]). The latter aims at changing the polluting behaviour and ensures that polluters compensate society for the costs caused by their activities.

The scope of the work includes reviewing existing tax instruments in Andalusia related to the circular economy and waste management, assessing their alignment with the goal to improve environmental quality while raising revenue and accounting for distributional concerns, and discussing possible additional tax interventions, with a view to identifying strategic reform options. The work includes a stocktake assessment of economic instruments used in this area at the EU, national, and regional levels. It identifies opportunities for complementary tax-based interventions at the regional level and analyses the environmental, economic (including revenue generation and distributional impacts), and behavioural implications of several possible taxes.

7.1. Main Economic Sectors

Andalusia is the most populous of the 17 Autonomous Communities (AC) of Spain, with 8.47 million inhabitants in 2021. Andalusia's Gross domestic product (GDP) was 160 billion euros in 2021, accounting for 13.3% of Spanish economic output. Although the Autonomous Community of Andalusia is the third largest contributor to the national GDP (after Madrid and Catalonia), the per capita GDP is the third lowest with EUR 19,026 in 2021 (INE, 2022[6]; INE, 2022[7]).

Andalusia has traditionally been an agricultural area, but nowadays the service sector predominates. The latter contributes with 22% of the Andalusian GDP, followed by public administration and defence (20%), real estate activities (12%) and extractive industries, energy and water supply, sanitation, and waste management and decontamination (11%). The agricultural sector nowadays accounts for approximately 6% of the GDP. Compared to the average GDP contribution in Spain, Andalusia's GDP relies more on the primary sector, and less on industry. Figure 7.B.1 in Annex 7.B shows the evolution of Andalusia's GDP and the sectors' contributions from 2000 until 2021. While some sectors have increased their economic output steadily over time until 2019, such as the service sector, public administration and defence, and real estate activities, most sectors have remained constant or have only varied slightly over time such as extractive industries, manufacturing industries and the primary sector. The construction sector is the only one that has experienced an important decrease in the last years.

7.2. Waste generation and management

7.2.1. Non-hazardous waste

According to the Junta de Andalucía (2019[8]), 18.34 million tonnes of non-hazardous waste were generated in Andalusia in 2018. As can be seen in Figure 7.1, category LER 19 had the highest generation in 2018 (31% of the non-hazardous waste generated), this category includes rejections from waste treatment facilities and sludge from wastewater treatment plants, which can be considered as a secondary contribution to waste generation.

LER 20 (Municipal waste) represented 27% of the total non-hazardous waste generated in 2018 and LER 17 (Construction and demolition waste) 22% of the total non-hazardous waste generated in 2018. The fourth and fifth positions on generation were LER 10 (Residues from thermal power plants and iron and steel industries) that accounted for 6% and LER 02 (Agricultural and livestock waste) that constituted 5.2% of the estimated total production. This group also includes waste generated by the agri-food industry. The rest of LER groups contributed less than 5% to the total generation of non-hazardous waste in Andalusia.

Figure 7.1 also illustrates that more than 38% of the waste generated in 2018 was subjected to the operation "R12 conditioning of waste prior to recovery", which includes the classification, disassembly, crushing, and conditioning, among other activities. This is mainly due to the contribution of the municipal waste treated at the recovery and composting plants, as well as in packaging selection plants. The second destination was landfill, more than 26% of the non-hazardous waste generated in 2018 was landfilled (D1 and D5). It should be noted that 3 Mt out of the 4.8 Mt landfilled corresponded to rejects from waste treatment facilities. Material recovery (R3, R4 and R5) accounted for 17.4% of waste treatment, which includes composting operations of the organic fraction of municipal waste, among other treatments. Finally, 9.6% of the waste is destined for intermediate storage in transfer facilities (D15 and R13).

Considering all recovery and disposal operation except intermediate storage, 68.3% of the non-hazardous waste generated in Andalusia in 2018 was valorised, while 31.7% was landfilled. According to (Sastre, Llopart and Puig Ventosa, 2018[9]), Andalusia had in 2014 a recycling rate of 31% similar to the average in Spain (30.8%).

The MSW generation (included in LER 20) accounted for 4.59 million tonnes in 2018 and only 10% was collected separately (Junta de Andalucía, 2019[8]). Separate collection included: 2% of light packaging, 3% of paper and cardboard, 2% of glass, and 3% of other selective collections. The amount of separately collected MSW has been kept an order of magnitude lower than the mixed collected waste since 2005 (Figure 7.2). Since 2013, most of the mixed waste collected in Andalusia went to recovery and composting plants (78%) including primary sorting within such facilities, while the rest was landfilled (Figure 7.3).

Figure 7.1. Generation of non-hazardous waste per LER group (1,000 t) and its management (%) in Andalusia in 2018

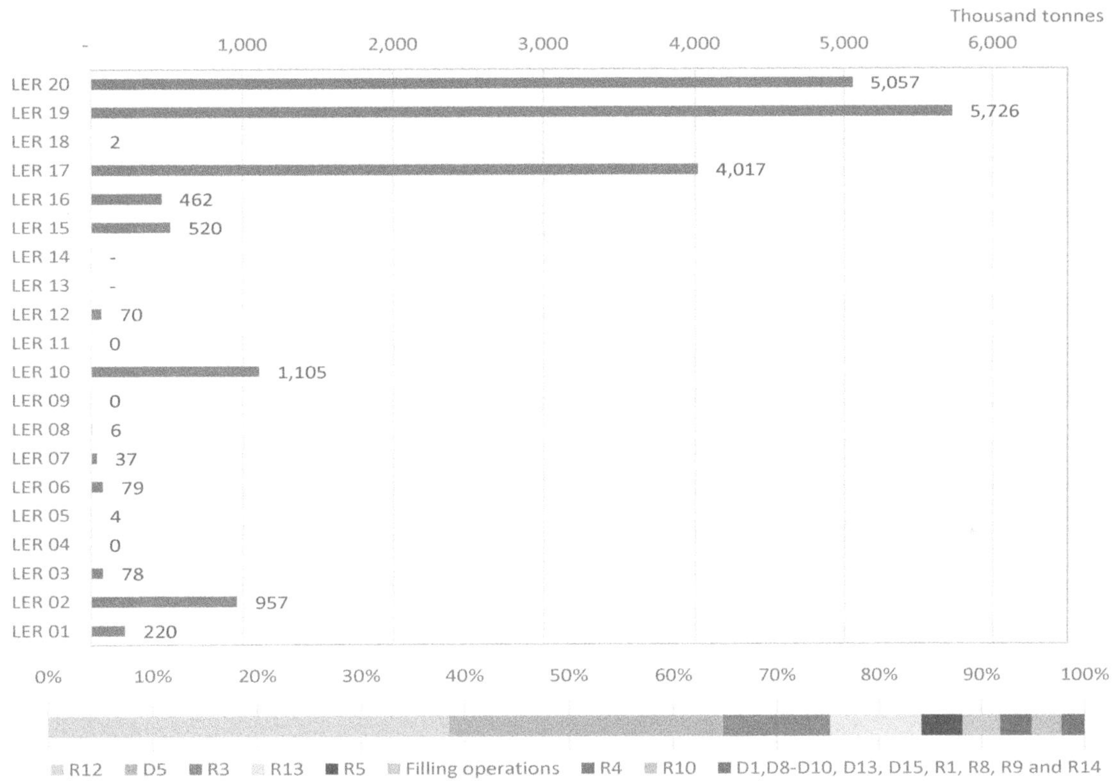

LER 01 Wastes resulting from exploration, mining, quarrying and physical and chemical treatment of minerals	**LER 11** Wastes from chemical surface treatment and coating of metals and other materials; non-ferrous hydrometallurgy
LER 02 Wastes from agriculture, horticulture, aquaculture, forestry, hunting and fishing, food preparation and processing	**LER 12** Wastes from shaping and physical and mechanical surface treatment of metals and plastics
LER 03 Wastes from wood processing and the production of panels and furniture, pulp, paper and cardboard	**LER 13** Oil wastes and wastes of liquid fuels
LER 04 Wastes from the leather, fur and textile industries	**LER 14** Waste organic solvents, refrigerants and propellants
LER 05 Wastes from petroleum refining, natural gas purification and pyrolytic treatment of coal	**LER 15** Waste packaging, absorbents, wiping cloths, filter materials and protective clothing not otherwise specified
LER 06 Wastes from inorganic chemical processes	**LER 16** Wastes not otherwise specified in the list
LER 07 Wastes from organic chemical processes	**LER 17** Construction and demolition wastes (including excavated soil from contaminated sites)
LER 08 Wastes from the manufacture, formulation, supply and use (MFSU) of coatings (paints, varnishes and vitreous enamels), adhesives, sealants and printing inks	**LER 18** Wastes from human or animal health care and/or related research
LER 09 Wastes from the photographic industry	**LER 19** Wastes from waste management facilities, off-site wastewater treatment plants and the preparation of water intended for human consumption and water for industrial use
LER 10 Wastes from thermal processes	**LER 20** Municipal wastes (household waste and similar commercial, industrial and institutional wastes) including separately collected fractions

Note: The complete names of the Operation codes can be found in the EU Waste Framework Directive (European Parliament, 2008[10]; European Commission, 2014[11])

Source: Own elaboration based on the data reported in (Junta de Andalucía, 2019[8]).

StatLink https://stat.link/r6p3fu

Figure 7.2. MSW generated and collected in Andalusia, 2005-2018

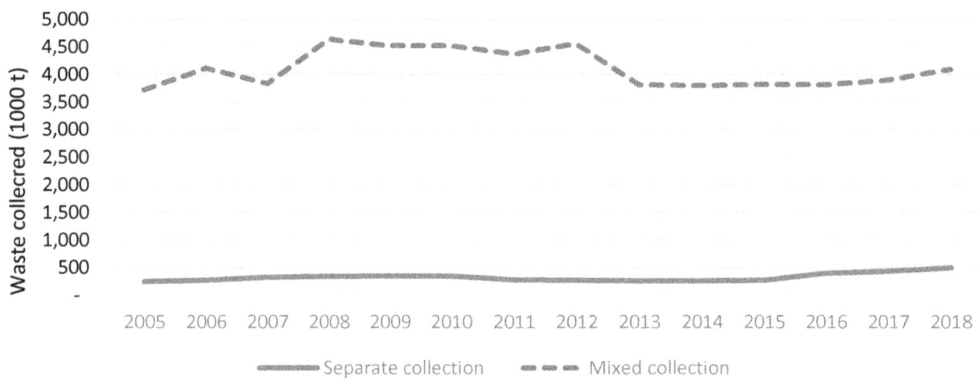

Note: Data from 2010 was not available and it has been assumed equal to 2009 data.
Source: Own elaboration based on the data obtained from (Junta de Andalucía, 2019[8]).

StatLink https://stat.link/05l8wk

Figure 7.3. Primary destination of the mixed collected MSW in Andalusia, 2005-2018.

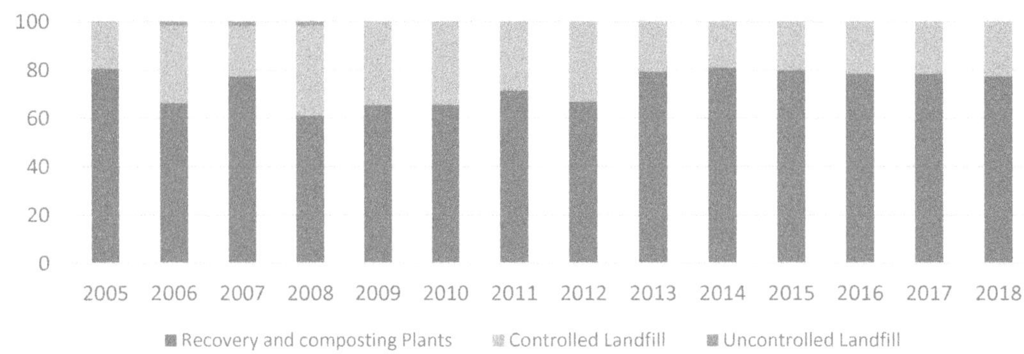

Note: Data from 2010 was not available and it has been assumed equal to 2009 data. Recovery and composting plant of mixed waste includes a primary sorting step before waste undertakes further treatments.
Source: Own elaboration based on the data obtained in (Junta de Andalucía, 2019[8]).

StatLink https://stat.link/4ife5v

7.2.2. Hazardous waste

In 2018, the generation of hazardous waste declared to the Junta de Andalucía was 327,646 tonnes (Junta de Andalucía, 2022[12]). The sectors contributing the most to such waste generation were waste recovery (26%), extractive and metallurgic industry (22%), energy sector (13%), chemical industry and associated products industries (12%), commercial services (11%) and waste elimination (7%). The rest of sectors contributed with less than 5% of the total generation (Figure 7.4).

It should be mentioned that while the generation of such waste summed up 327,646 tonnes in 2018, the amount of hazardous waste treated or/and disposed of in Andalusia equalled 842,499 tonnes in the same year (Junta de Andalucía, 2022[12]). Such difference can be explained by two facts: 1) some waste is treated more than once (there are primary and secondary destinations) and 2) some of the waste managed in Andalusia is imported from other ACs or countries for treatment and/or disposal.

Figure 7.4. Generation of hazardous waste per sector in Andalusia in 2018 (1,000 t)

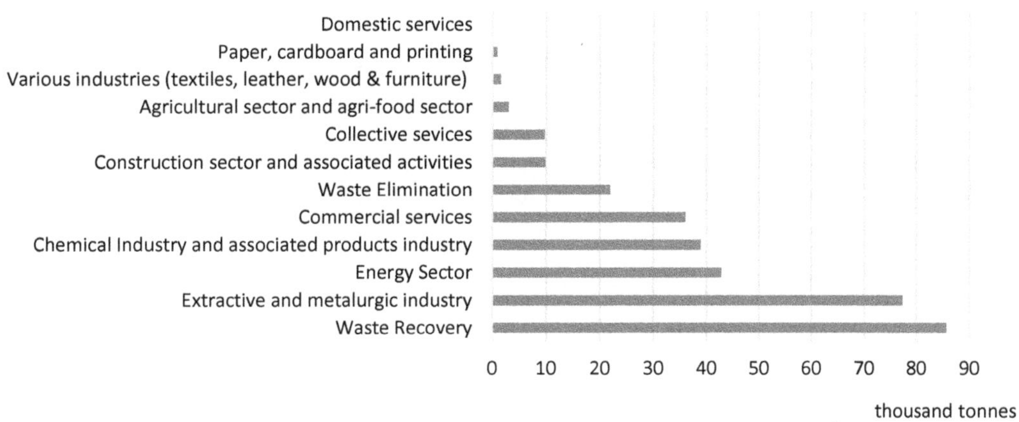

Source: Own elaboration based on the data reported in (Junta de Andalucía, 2022[12]).

StatLink https://stat.link/cmz3yo

References

Boletín Oficial del Estado (2022), *Ley 7/2022, de 8 de abril, de residuos y suelos contaminados para una economía circular*, https://www.boe.es/buscar/act.php?id=BOE-A-2022-5809 (accessed on 22 March 2023). [4]

Castells-Rey, I., P. Pellicer-García and I. Puig-Ventosa (2022), *Los instrumentos económicos y fiscales en la Ley de Residuos*, Retema, https://www.retema.es/revista-digital/marzo-abril-7 (accessed on 17 January 2023). [2]

European Commission (2021), "Green taxation and other economic instruments", http://Green taxahttps://ieep.eu/uploads/articles/attachments/134d9257-53c5-4a20-885b-9f6615452486/Green%20taxation%20and%20other%20economic%20instruments%20%E2%80%93%20Internalising%20environmental%20costs%20to%20make%20the%20polluter%20pay_Study_10.11.2021.pdf?v=63807385248tion and other economic instruments (accessed on 17 January 2023). [5]

European Commission (2014), *2014/955/EU: Commission Decision of 18 December 2014 amending Decision 2000/532/EC on the list of waste pursuant to Directive 2008/98/EC of the European Parliament and of the Council*, https://eur-lex.europa.eu/legal-content/EN/TXT/?uri=celex%3A32014D0955 (accessed on 18 January 2023). [11]

European Parliament (2008), *Directive 2008/98/EC of the European Parliament and of the Council of 19 November 2008 on waste and repealing certain Directives*, https://eur-lex.europa.eu/legal-content/EN/TXT/?uri=celex%3A32008L0098 (accessed on 18 January 2023). [10]

Eurostat (2022), *Environmental tax revenues*, https://ec.europa.eu/eurostat/databrowser/view/env_ac_tax/default/table?lang=en (accessed on 17 January 2023). [1]

Eurostat (2022), *Recycling rate of municipal waste*, https://ec.europa.eu/eurostat/databrowser/view/cei_wm011/default/table?lang=en (accessed on 17 January 2023). [3]

INE (2022), *Contabilidad regional de España*, https://www.ine.es/dyngs/INEbase/es/operacion.htm?c=Estadistica_C&cid=1254736167628&menu=ultiDatos&idp=1254735576581 (accessed on 17 January 2023). [6]

INE (2022), *Contabilidad Regional de España. Serie 2016-2021: P.I.B. a precios de mercado y valor añadido bruto a precios básicos por ramas de actividad: Precios corrientes por comunidades y ciudades autónomas, magnitud y periodo*, https://www.ine.es/jaxi/Tabla.htm?path=/t35/p010/rev19/l0/&file=01001.px&L=0 (accessed on 17 January 2023). [7]

Junta de Andalucía (2022), *Índices Medio Ambiente 2019. 5. Producción declarada de residuos peligrosos por grupo de actividad y constituyente, 2018*, https://www.juntadeandalucia.es/medioambiente/vem/?c=Menu/tema/580 (accessed on 18 January 2023). [12]

Junta de Andalucía (2019), "Informe sobre producción y gestión de residuos no peligrosos en Andalucía. Año 2018. Datos definitivos", https://surminas.org/webs/default/media/Alegaciones/Informe%20PyG%20RnoP%202018.pdf (accessed on 17 January 2023). [8]

Sastre, S., J. Llopart and I. Puig Ventosa (2018), "Mind the gap: A model for the EU recycling target applied to the Spanish regions", *Waste Management*, Vol. 79, pp. 415-427, https://doi.org/10.1016/j.wasman.2018.07.046. [9]

Annex 7.A. EU Waste Targets and current recycling rates

Annex Figure 7.A.1. Recycling rate (%) of municipal waste in 2020

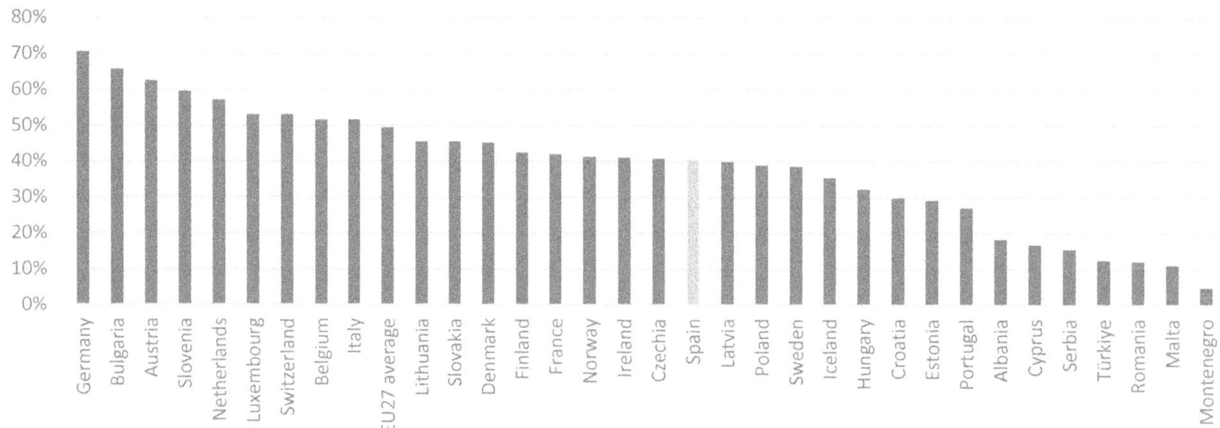

Source: (Eurostat, 2022[3]).

StatLink https://stat.link/fb1y52

Annex Table 7.A.1. EU targets on separate collection, preparation for re-use and recycling of MSW

	Material	2024	2025	2030	2035	Directive
Separate collection	Organic waste	Mandatory separate collection as of 31 December 2023				Directive (EU) 2018/851
	Textile		Mandatory separate collection as of 1 January 2025			
	Hazardous waste produced by households					
	Single-use plastic products*		77%	90%		Directive (EU) 2019/904
Preparation for re-use or recycling	Packaging		65%	70%		Directive (EU) 2018/852
	Plastic packaging		50%	55%		
	Wood packaging		25%	30%		
	Ferrous metal packaging		70%	80%		
	Aluminium packaging		50%	60%		
	Glass packaging		70%	75%		
	Paper and cardboard packaging		75%	85%		
	MSW		55%	60%	65%	Directive (EU) 2018/851

Note: (*) single-use plastic products listed in Part F of the Annex of Directive (EU) 2019/904.
Source: Own elaboration.

Annex 7.B. GDP by sector in Andalusia

Annex Figure 7.B.1. Sector contribution to Andalusia GDP 2000-2021

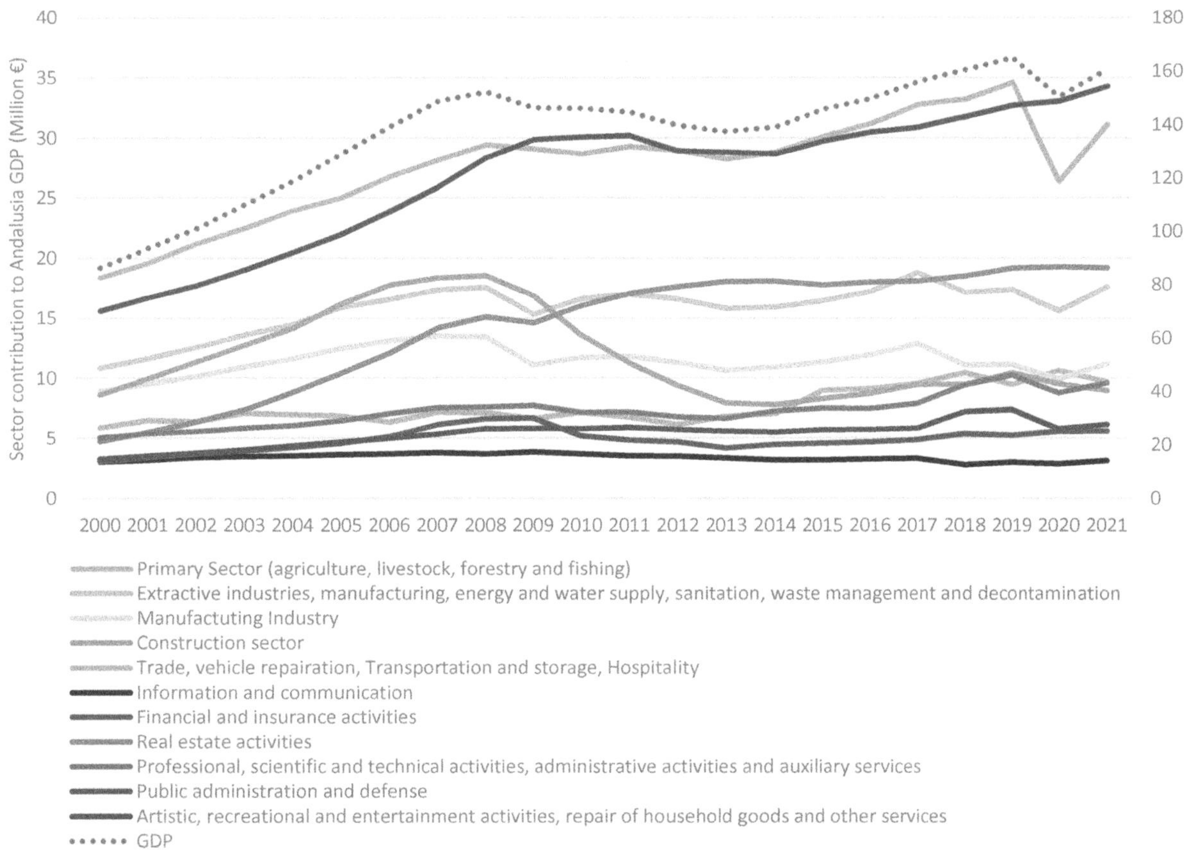

Source: Own elaboration based on the data available at (INE, 2022[6]; INE, 2022[7]).

StatLink https://stat.link/wstfdv

8 Legal stocktake: Circular Economy and Waste Management

This section provides a description of the most relevant legal instruments and existing environmental taxes at different governance levels that are relevant for waste management and the circular economy in Andalusia.

8.1. Legal framework, competencies and responsibilities on waste management

At EU-level, the first Circular Economy Action Plan (CEAP) in Europe was approved in 2015 (European Commission, 2015[1]) and, in 2020 the second CEAP was adopted (European Commission, 2020[2]). The CEAP includes 35 actions such as setting waste reduction targets for specific streams and other measures on waste prevention.

The Waste Framework Directive (WFD) is the main regulation on waste in Europe. The WFD sets targets for preparation for re-use and recycling of municipal waste of 55% by 2025, 60% by 2030 and 65% by 2035. In addition, the WFD establishes the basic requirements for Extended Producer Responsibility (EPR). The Packaging and Packaging Waste Directive strengthens the reuse of packaging, setting qualitative and quantitative objectives and the use of economic incentives.

The goal of the Single-Use Plastics (SUP) Directive is to prevent and reduce the impact on the environment of certain plastic products. It bans several single-use plastic products and for other single-use plastics it established design requirements (recycled content of plastic bottles) and set targets for separate collection and for recycled content for PET bottles.

At the Spanish level, Law 7/2022 on Waste and Contaminated Soils for a Circular Economy transposes the objectives of the directives.

Table 8.1. Main legislation and targets in the domain of waste and resources across different levels of government

	Legal framework / laws / instruments	Objectives and targets
EU-level	Waste Framework Directive [Directive 2008/98/EC]	• 50% of preparation for re-use and recycling of MSW by 2020, 55% by 2025, 60% by 2030 and 65% by 2035.
	Single Use Plastic Directive [Directive (EU) 2019/904]	• 70% of separated collection of bottles in 2023 and of 90% in 2029 • 30% recycled content for PET bottles by 2025
	Packaging and Packaging Waste Directive [Directive (EU) 94/62]	• 65% packaging waste recycling by 2025 and 70% by 2030
	Landfill Directive [Directive (EU)1999/31]	• By 2035, MSW deposited in landfills must be reduced to 10%
	Plastic Bags Directive [Directive (EU) 2015/720]	• annual consumption level does not exceed 40 light plastic bags per person by 2025

National level (Spain)	Spanish Law 7/2022 on Waste and Contaminated Soils for a Circular Economy	• Transposes and adopts the targets of the Waste Framework Directive and Single Use Plastic Directive. • 13% reduction in weight of waste generated by 2025 compared to 2010, 15% by 2030.
	Spanish Royal Decree 293/2018 on the reduction of the consumption of plastic bags	• By 2020, it bans thick plastic bags with less than 50% of recycled plastic • By 2021, it bans light plastic bags
	Spanish Circular Economy Strategy 2030	• By 2030, 30% reduction in national consumption of materials in relation to GDP • By 2030, 15% reduction of waste generation • By 2030, increase reuse up to 10% of municipal waste
Regional level (Andalusia)	Decree 73/2012, of the Andalusian Waste Regulation	• 50% of preparation for re-use and recycling of MSW by 2020.
	Draft bill of the Andalusian Law on Circular Economy (not yet approved)	• Implementing a separate collection of organic waste by 2023 • Implementing a separate collection of textile waste by 2025 • Implementing a separate collection of used cooking oil by 2025 • Implementing a separate collection of hazardous waste produced by households by 2025
	Andalusian Mining Strategy 2020[1]	• Environmental integration of mining activity and enhancement of the Andalusian mining heritage, as a cultural, social and economic resource that helps the sustainable development of the territory.

Table 8.2. Distribution of the main competences in the domain of waste and resources across different levels of government

Matter	National	Andalusia	Local (Municipal)
Waste	Basic legislation on waste management. The national legislation establishes minimum targets for reducing waste generation, as well as preparation for reuse, recycling and other forms of recovery. The national government approves the National Framework Plan for Waste Management, and it authorises shipments of waste to or from non-EU countries (art. 12.1,2,3 Law 7/2022)	Policy development of the basic national legislation and establishment of additional protection regulations. The regional government approves regional plans for waste prevention and management. It is also in charge of authorization, inspection and sanction of waste production and management activities. It also registers information on production and authorises waste management of the shipment of waste from or to EU countries (art. 12.4 Law 7/2022). The regional government also can increase the waste disposal tax rates (art. 93.2 Law 7/2022)	Municipalities are obliged to provide the collection, transportation, and treatment of household waste. Municipalities with more than 5,000 inhabitants are obliged to approve waste management programs. Municipalities can manage commercial waste. Municipalities must establish waste charge to finance the costs of the provided services (art. 12.5 Law 7/2022)
Resources	Basic mining legislation (article 149.1.25 of the Spanish Constitution and Law 2857/1978 which approves the general regulations for the mining regime)	Inspection and monitoring of mining activity. Management of mining resources, resource exploitation authorization and exploration permissions (Law 2857/1978 which approves the general regulations for the mining regime).	Urbanism competences (art. 25.2 Law 7/1985 of the bases of the local regime)

Andalusia is one of the first AC discussing in its Parliament a draft bill for a circular economy law. According to the legislative proposal approved by the Regional Government in February 2022 (BOPA, 2022), an Andalusian Office of Circular Economy would be created as an administrative unit for the development of advisory functions, dynamization, coordination and management of the actions provided for in the Law (chapter I). The legislative proposal includes the following references to economic instruments:

- Article 25.5 states that: "The taxes, charges or, where appropriate, other types of levies, established by Local Entities, in accordance with the provisions of the applicable national legislation, must reflect the real cost of the collection operations, transport and treatment of waste, including the monitoring of these operations and the maintenance and monitoring after the closure of landfills, and should allow progress in the establishment of pay-as-you-throw schemes, without prejudice to the financing obligations that correspond to the Collective Systems of Extended Producer Responsibility in accordance with the national regulations."

- Article 33.1 states that: "In accordance with the principles of respect for the environment and sustainability of the Andalusian port system, set forth in Law 21/2007, of December 18, on the Legal and Economic Regime of the Ports of Andalusia, tax incentives may be established in the rates regulated in such Law for those taxpayers who carry out marine litter collection activities."
- Article 52.3 states that: "Local Entities, within the scope of their competences, may adopt measures of deduction, reduction or discount in charges paid by to those companies, households, neighbourhood communities, or other users, who adopt biowaste composting systems."
- Article 64.1. states that: "The Administration of the Junta de Andalucía may take into account the obtaining of internationally recognised certificates in terms of environmental sustainability of buildings and urbanizations in order to propose rebates in municipal taxes or other tax incentives."

Due to the elections in Andalusia in June 2022, the Parliament was dissolved, and ongoing legislative processes were temporarily suspended. As such, this draft bill will likely not be approved in the near future.

8.2. Environmental taxes applied in other EU Member States relevant for the study

Based on the OECD database on Policy Instruments for the Environment (PINE) (OECD, 2022[3]), as well as on international studies such as two reports by the European Commission (2021[4]; 2021[5]), three types of environmental taxes were identified as relevant to promote circular economy: 1) waste disposal taxes, 2) taxes on raw materials extraction, and 3) taxes on specific products. This section reviews examples of these environmental taxes in place in other EU Member States.

8.2.1. Waste disposal taxes

Waste disposal taxes are justified by the environmental impacts of landfilling and incineration, compared to other options higher up in the waste management hierarchy established in the Waste Framework Directive (European Parliament, 2008[6]). Thus, such taxes are intended to favour waste prevention and increased recycling levels, and move towards the targets of the Landfill Directive. The incineration tax is often applied to prevent the diversion of waste from landfill to incineration.

According to the OECD database on Policy Instruments for the Environment (OECD, 2022[3]) and the latest version of the CEWEP database on landfill taxes and restrictions (CEWEP, 2022[7]), 26 countries out of 30 (27 EU member states plus Norway, Switzerland, and the United Kingdom) have landfill taxes and 5 have incineration taxes, see Table 8.A.1 for the details, as well as Annex D for case studies in Italy, Belgium and the United Kingdom. As can be seen in Table 8.3, disposal tax rates in European countries vary significantly between countries and types of waste.

In addition to the tax rates, tax policies also vary with modifications. In some countries:

- Disposal taxes are supplemented by additional limitations on the quantities that can be landfilled (more stringent than those indicated in Directive 31/1999 on landfill of waste), e.g., Belgium, Sweden, Denmark, Estonia, and Finland.
- Disposal taxes are earmarked, e.g., Lithuania, Hungary, Finland, and Austria.
- Specific waste types are exempt from disposal tax if no better treatment than landfill is available, such as asbestos in Flanders, Sweden, and the Netherlands, and for waste from waste recovery processes in Sweden, the Walloon region of Belgium and Portugal. In addition, in the United

Kingdom landfill operators can offset a maximum percentage of their tax liability by financing environmental projects through the Landfill Communities Fund.

- A differentiated tax rate applies. The tax rate can discriminate based on whether the input waste is pre-treated or not prior to landfill, as done in several Italian regions such as Piemonte, Calabria and Campania, or whether the municipality has implemented separate collection of the organic waste, as is the case in Balearic Islands. In other cases, the tax rate is determined based on the percentage of selective collection of the municipality, or on the quality of the waste landfilled, as in the region of Puglia (Italy) or in the Slovak Republic. These tax configurations would provide an extra incentive to improve selective collection and recycling.
- Disposal taxes apply to landfill and incineration and generally incineration tax rates are lower than landfill tax rates to incentivise energy recovery over disposal (in line with the waste hierarchy).

Table 8.3. Waste disposal tax rates in European countries, based on Table 8.A.1

Taxable Event	Mean (€/t)	Standard Deviation (€/t)
MSW Landfill	39.75	25.95
Industrial Waste Landfill	28.10	27.44
Inert and Construction Waste Landfill	12.33	17.33
MSW Incineration	12.19	9.04
Industrial Waste Incineration without energy recovery	30*	

Note: (*) Standard deviation could not be estimated because there was only one case.
Source: Own elaboration based on data published in (OECD, 2022[3]).

8.2.2. Taxes on raw material extraction

Taxes on the extraction of raw materials have been widespread in Europe since the early 1990s. This type of taxes can reduce demand of primary resources in favour of secondary raw materials while preserving the resource and the landscape.

One of the main raw materials extracted are aggregates. According to the OECD database on Policy Instruments for the Environment (OECD, 2022[3]), there are currently 88 different taxes applied on extractive activities of aggregates, gypsum and salt in OECD countries. More than half of these taxes (58%) are earmarked. 16% are *ad valorem*, and the remaining (84%) are *ad quantum*.

In relation to the tax base, 64% of taxes are levied on some specific type of aggregates (e.g., calcareous, marble or clay), 24% to all minerals in general (and therefore also on aggregates), 6% to aggregates in general (all equally), 5% to gypsum and 3% to salt. Table 8.C.1 summarises key aspects of the taxes on aggregates currently applied in 10 of the 30 countries analysed (EU 27 plus Norway, Switzerland, and the United Kingdom). The average tax rate in European countries is shown in Table 8.4.The high standard deviation indicates a great variability between countries.

Table 8.4. Tax rates on the extraction of aggregates in European countries

	Quantum tax				Ad Valorem tax	
	€/m3		€/t		Value (%)	Benefit (%)
	Mean	Standard Deviation	Mean	Standard Deviation		
Raw material extraction (including aggregates)	1.06	1.31	1.35	1.63	10.00*	2.60*
Specific aggregates extraction	1.34	0.84	1.58*			

Note: (*) Standard deviation could not be estimated because there was only one case.
Source: Own elaboration based on data published in (OECD, 2022[3]).

Denmark was one of the first countries to introduce a tax on the extraction of aggregates. In general, there has been a slight decrease in the extraction of these materials since the introduction of the tax in 1977, but it has not resulted in any reduction in the consumption of these materials (Söderholm, 2011[8]). This indicates a relative inelastic demand. Although an increase in the use of recycled materials was observed, this was mainly attributed to the introduction of the landfill tax on construction waste that was implemented in parallel (Söderholm, 2011[8]).

In Sweden, a tax on the extraction of natural gravel has been applied since 1996 to preserve groundwater. It started with a low tax rate, which was raised in 2003. Such increment implied a greater decrease in the consumption of this material. However, the decrease in the extraction of gravel was already significant before the introduction of the tax and could be associated with an increased demand for crushed rock due to its higher quality compared to natural gravel (although its extraction requires higher energy consumption). The decrease in gravel consumption led to an increase in alternative materials with a greater impact on emissions. Therefore, while groundwater quality has improved, emissions have increased. This example highlights the need for careful analysis and possibly additional instruments to avoid burden shifting. The Swedish case also shows that the gradual increase in the tax helps producers to organise themselves, contributes to increasing the elasticity of demand and allows for a better acceptance of the tax (Söderholm, 2011[8]).

In Italy a regional tax on the extraction of aggregates (sand, gravel, and rock) has been applied since the early 1990s. Each region or municipality applies a different tax rate that can vary between €0.41 and €0.57/m^3. Each regional authority defines its tax, which is complemented by national legislation. No substantial change in the demand for aggregates has been observed since the implementation of the tax, which indicates a relative inelastic demand that can be associated with the low tax rate (tax payments represent only 5% of the estimated profits of the industry) and the little preparation of the industry to produce and assimilate recycled materials of similar quality, combined with the absence of taxation on landfill of construction and demolition waste (European Environment Agency, 2008[9]).

In 2002, a tax on the extraction of aggregates was introduced in the United Kingdom and its current rate is 2 GBP (EUR 2.45) per tonne of sand, gravel, and rock (on average 20% of the value of the product). Although in this case there has been a decrease in the extraction of aggregates, this decrease began before the implementation of the tax and is related to factors such as the reduction in investment in infrastructure or the existence of a landfill tax on construction waste (European Environment Agency, 2008[9]). Part of the demand shifted towards non-taxed materials capable of substituting the materials subject to the tax, which have become competitive in the presence of the tax. There are some exemptions from the aggregates levy, such as aggregates which are returned to the ground in the same place and in the same form as they were extracted.[2]

8.2.3. Taxes on consumer products

As it can be seen on Table 8.D.1, there are several consumer products levied with environmental taxes in different OECD counties, e.g., tyres, pesticides, plastic products, disposal tableware. The Danish

Packaging tax merits specific analysis in this report since it could be an option to consider in the reform of environmental taxes in Andalusia to complement the Spanish Packaging EPR for specific packaging items, such as the beverage cartons. The Danish packaging tax has been levied since 1978 to reduce waste and increase packaging reuse and recycling rates. Denmark chose to internalise packaging waste management costs through this tax instead of setting up an industry-run producer-responsibility scheme (such as the Green Dot system) as done by many EU countries. The tax was initially divided into a weight-based part and a volume-based one. Exports were tax-exempt to avoid damaging the international competitiveness of Danish producers.

In 2001, the tax rates for the weight-based part of the tax were modified to consider the life cycle environmental impact of each type of packaging, per kilogram. The volume-based tax is a duty per unit of packaging for spirits, wine, beer, and carbonated soft drinks (Danish Ecological Council and Green Budget Europe, 2015[10]). Table 8.D.1 shows the tax rates divided by material and volume.

The management of the tax was difficult due to the large number of producers involved and the complexity of the tax definition (OECD, 2015). By January 2014, the Danish government abolished the weight-based part to reduce the production costs and administrative burdens of firms, but it is still valid on plastic bags, disposable tableware, and PVC foil (Danish Ecological Council and Green Budget Europe, 2015[10]).

Gsell *et al.*, (2022) also propose a Packaging beverage tax for Germany with differentiated tax rates based on the environmental impacts of the material used for the packaging. Latvia instead has a packaging tax, as part of the natural resource tax, which is used as an incentive to join producer responsibility organisations (PRO), as organisations that join a PRO are tax exempted (European Commission, 2021[5]). Norway has also an environmental tax applied to beverage packaging with differentiated rates per material, as well as a basic tax that applies to all single-use packaging.

8.3. Taxes and regulations at national level in Spain

This section describes two fiscal measures included in the Spanish Law 7/2022 on Waste and Contaminated Soils for a Circular Economy[3], namely a special tax on non-recycled plastic used in non-reusable packaging and a disposal tax. In addition, this section describes two existing measures currently being applied in Spain that can influence the proposal of fiscal reform for Andalusia. These are the national ban to provide single-use plastic bags and the current regulation and situation of Extended Producer Responsibility (EPR).

It is also important to mention that the Spanish Law 7/2022 on Waste and Contaminated Soils for a Circular Economy includes the implementation of a Deposit-Refund System for single-use beverage containers with volume up to 3 litres if Spain does not meet the target of 70% of separated collection of bottles in 2023 and of 85% in 2027 established in the Directive (EU) 2019/904.

In addition, the White Book for Tax Reform, published in March 2022, proposes the following measures in relation to circularity: intensification and extension of the taxes of the Waste and Contaminated Soil Law, reformulation of municipal charging of waste to link it to pay-as-you-throw systems, creation of a tax on the extraction of aggregates, creation of a tax on nitrogenous fertilizers and to extend and harmonise taxation on certain emissions from large industrial facilities.

8.3.1. Special tax on non-reusable plastic packaging

Article 67 of the Spanish Law 7/2022 on Waste and Contaminated Soils for a Circular Economy creates a special tax that levies production, importation, and acquisition of non-recycled plastic (i.e. virgin plastic) used in non-reusable plastic packaging. The objective of the tax is to incentivise the reduction of non-reusable plastic packaging as well as plastic recycling. The tax rate will be 0.45 euros per kg of non-recycled plastic used in non-reusable packaging (Article 78). The part of recycled plastic will have to be certified by an accredited body with the certification UNE-EN 15353:2008 (article 79). Although the tax is not earmarked, the rationale for its creation is to raise an amount of revenue similar to the cost for Spain of the EU contribution for non-recycled plastic (Castells-Rey, Pellicer-García and Puig-Ventosa, 2022[11]; Puig-Ventosa, 2021[12]). This contribution, known as the Plastics own resource, was introduced on January 2021 and consists as a national contribution based on the amount of non-recycled plastic packaging waste, which represents a new EU revenue source to the 2021-2027 EU budget (European Commission, 2021[13]; Council of the EU, 2020[14]).

Although the law entered into force on the 10th of April 2022, the measures included in Title VII, i.e., the special tax on non-reusable plastic packaging (described in this section) and the national waste disposal tax (described in next section), will enter into force on the 1st of January 2023 (13th final provision of the Spanish Law 7/2022).

8.3.2. National waste disposal tax on landfill and incineration

The national tax on the deposit of waste in landfills, as well as on the incineration and co-incineration of waste, included in the Spanish Law 7/2022 on Waste and Contaminated Soils for a Circular Economy (articles 84-97) aims to disincentivise these disposal operations in Spain.

The tax rate (Article 93.1) differs among waste type and disposal activity. Table 8.5 shows landfill tax rates, Table 8.6 shows incineration tax rates and a sole tax rate of 0 €/tonne applies to co-incineration, regardless of the type of waste co-incinerated (Article 93.1.f). Article 93.2 establishes the possibility for Autonomous Communities to increase the tax rates even though the tax collection will in principle be carried out by the State.

The National Tax Administration Agency or, the offices with analogous functions of the autonomous communities, has the competence for the tax management, liquidation, collection, and inspection (Article 95.1). According to article 97, the tax revenue will be distributed back to the Spanish regions according to the location where the taxable event happens. The Law 7/2022 on Waste and Contaminated Soils for a Circular Economy does not determine how regions must use the revenue generated.

Two additional provisions in this Law are important for Andalusia. The 7th additional provision establishes “1. To the extent that the taxes established by this law fall on taxable events levied by the autonomous communities and this produces a decrease in their income, the provisions of article 6.2 of Organic Law 8/1980” (the provisions of article 6.2 are compensation measures in favour of such AC). “2. The provisions of the previous section will only apply to those taxes of the autonomous communities that are in force prior to December 17, 2020”. “3. The compensation measures in favor of the autonomous communities established based on article 6.2 of Organic Law 8/1980, will be reduced by the amount of the collection received by the corresponding autonomous communities in accordance with the provisions of this law”.

The 21st additional provision establishes that ACs that at the entry into force of the Spanish Waste Law in 2022 had in place a regional tax on the deposit of waste in landfills, incineration, and co-incineration of waste, may maintain their management if the necessary agreements are established. There is strong uncertainty on the practical implications of the two mentioned additional provisions, which will need to be discussed in the future.

Table 8.5. Landfill tax rates included in the Spanish Law 7/2022

	Landfill (EUR per tonne)		
	Non-hazardous***	Hazardous****	Inert*****
1. Municipal Solid Waste (MSW)	40		
2. Rejects from MSW treatment	30		
3. Different than 1 and 2 (without pre-treatment required*): General character	15	8	3
4. Different than 1 and 2 (without pre-treatment required*): > 75% of inert waste	3 (15)**		
5. Different than 1, 2, 3 and 4: General character	10	5	1.5
6. Different than 1, 2, 3 and 4: With more than 75% of inert waste	1.5 (10)**		

Note: (*) in the terms established in article 7.2 of Royal Decree 646/2020; (**) The amount before the parenthesis is the tax rate for the inert part and the tax rate within the parenthesis applies for the rest of waste component; (***) Article 93.1.a); (****) Article 93.1.b); (*****) Article 93.1.c).
Source: Own elaboration based on the Spanish Law 7/2022 on Waste and Contaminated Soils for a Circular Economy.

Table 8.6. Incineration tax rates included in the Spanish Law 7/2022

	Incineration (EUR per tonne)		
	Disposal D10*	Recovery R01**	Different than D10 and R01***
1. Municipal Solid Waste (MSW)	20	15	
2. Rejects from MSW treatment	15	10	
3. Different than 1 and 2	7	4	
4. Different than 1, 2 which have not previously been subject to R02-09, R12, D8, D9, D13 or D14			5
5. Different than 1, 2, 3 and 4			3

Note: (*) Article 93.1.d); (**) Article 93.1.e); (***) Article 93.1.f).
Source: Own elaboration based on the Spanish Law 7/2022 on Waste and Contaminated Soils for a Circular Economy.

8.3.3. Ban of single-use light plastic bags

The EU Directive 2015/720 amending Directive 94/62/EC aims at reducing the consumption of lightweight plastic carrier bags from 90 light plastic bags per person at the end of December 2019 to 40 light plastic bags per person by the end of December 2025 (European Parliament, 2015[15]). It also establishes that by 31 December 2018, lightweight plastic carrier bags cannot be provided free of charge at the point of sale. Very lightweight plastic carrier bags may be exempted from those measures.

The Spanish Royal Decree 293/2018, of May 18, on the reduction of the consumption of plastic bags and by which the Registry of Producers was created transposes Directive (EU) 2015/720 into the Spanish legal system. The Decree (see Table 8.7), bans light plastic bags as of 1st of January 2021 and thick plastic bags with less than 50% of recycled plastic as of 1st of January 2020. Thus, after these dates, only providing very thin compostable bags (free of charge), thin compostable bags (prior payment), thick bags with more than 50% recycled plastic (prior payment), and thick bags with more than 70% of recycled plastic (for free) is still allowed. Annex I of the Royal Decree provides indicative prices to be used by establishments to be applied from the 1st of July 2018.

This annual consumption is of plastic bags in Spain is currently well above the maximum consumption levels and envisioned targets (Box 8.1.).

Table 8.7. Measures and deadlines established to reduce the consumption of plastic bags in the Spanish Royal Decree 239/2018

Deadline	Lightweight plastic bags*	Thick weight plastic bags**	Fragmentable plastic bags***
1 July 2018	Free delivery to consumers is prohibited		
	Exception: Very light plastic bags.	Exception: If they contain ≥ 70% recycled plastic, they can be delivered against payment.	
1 January 2020		Delivery to consumers is prohibited if it contains < 50% recycled plastic****. If they contain ≥ 50%, they can be delivered after payment.	Delivery to consumers is prohibited.
1 January 2021	Delivery to consumers is prohibited.		
	Exception: -Compostable lightweight plastic bags, which can be delivered upon payment. -Very lightweight compostable plastic bags		

Note: (*) with wall thickness below 50 microns; (**) with wall thickness equal or above 50 microns; (***) plastic bags made of plastic materials that include additives that catalyse the fragmentation of the plastic material into microfragments. The concept of fragmentable plastic includes oxofragmentable, photofragmentable, thermofragmentable and hydrofragmentable plastics (Article 3.e of the Spanish Royal Decree 293/2018); (****) The retailer must have documentation provided by the manufacturer that proves this percentage.
Source: (Junta de Andalucía, 2022[16]) and Real Decreto 293/2018, de 18 de mayo, sobre reducción del consumo de bolsas de plástico y por el que se crea el Registro de Productores.

Box 8.1. Evolution of plastic bag consumption in Andalusia

According to the Spanish Association of Plastics Manufacturers (ANAIP), the consumption of non-biodegradable single-use plastic bags per inhabitant in Spain was 300 in 2008, but this consumption dropped in the following years (MITECO, 2022[17]). In 2014, 6,730 million units of lightweight plastic carrier bags (with wall thickness below 50 microns, as defined in Directive (EU) 2015/720) were placed on the market, of which 23% were very lightweight plastic carrier bags (with a wall thickness below 15 microns, as defined in Directive (EU) 2015/720). This means that in Spain there was an average annual consumption of 145 light plastic bags per inhabitant that year (Junta de Andalucía, 2022[18]).

8.3.4. Extended Producer Responsibility Schemes

Extended Producer Responsibility (EPR) is defined as the environmental policy that intends to transfer responsibility of the post-consumer phase of the product to the producer (OECD, 2016[19]). The two main reasons for assigning responsibility to producers is: (1) to implement the polluter pays principle and ensure economically efficient recovery and treatment of End-of-Life (EoL) products, and (2) the capacity of producers to change products in the design phase to minimise their environmental impact throughout their entire life cycle.

Although there is evidence that EPR schemes can reduce public costs of municipal waste management while increasing prevention and recycling rates, currently there are only 4 waste streams for which EU directives establish the use of EPR policies (packaging, batteries, end-of-life vehicles (ELVs), electrical and electronic equipment (EEE)) (Eunomia, 2020[20]). Additionally, the EU Single-Use Plastics Directive will require Member Countries to implement EPR schemes for tobacco product filters (i.e. cigarettes) by 2023 and fishing gear by 2025 (European Parliament, 2019[21]) and harmonised EPR rules will be proposed for textiles.

In addition, in some EU countries there are national EPR schemes for products that are not yet addressed in EU-wide legislation (e.g., tyres, graphic paper, used oil and medical waste). In Spain, there are currently six waste flows where EPR is applied: packaging (including Medical Products Packaging and Expired

Medicines), batteries and accumulators, EoL vehicles, EoL tyres, used industrial oils and Waste Electric and Electronic Equipment (See Table 8.E.1 for details).

In addition, several additional waste streams are expected to have EPR schemes in Spain in the future:

- Article 60.1 of the Spanish Law 7/2022 on Waste and Contaminated Soils for a Circular Economy states that the Spanish government will develop, by regulation, EPR schemes for single-use plastic products listed in Annex IV, part F. These regulations must be established before 6 January 2023 for tobacco products and before 1 January 2025 for food containers, containers and wrappers containing food intended for immediate consumption in the container itself, containers for beverages up to three litres capacity including their caps and lids, drinking glasses including their lids and caps, light plastic bags, wet wipes and balloons.
- Article 60.5 of the Spanish Law 7/2022 on Waste and Contaminated Soils for a Circular Economy states that the Spanish government will develop EPR schemes for fishing gear, by regulation, before 1 January 2025. Such regulation will set: 1) a minimum national collection rate for waste fishing gear containing plastic for recycling and 2) the necessary measures to monitor the fishing gear containing plastic placed on the market as well as the waste collected.
- Article 60.3 of the Spanish Law 7/2022 on Waste and Contaminated Soils for a Circular Economy states that the producers of tobacco products will bear the costs of collecting the waste of said discarded products in public collection systems, including the infrastructure and its operation and the subsequent transport and treatment of the waste. The costs may include the establishment of specific infrastructure for the collection of the waste of said products, such as appropriate receptacles for waste in places where the dumping of scattered garbage of this waste is concentrated. Likewise, they may include costs associated with measures for the development of alternatives and prevention measures to reduce the generation of waste and increase material recovery.
- Finally, the seventh final provision of the Spanish Law 7/2022 states that the Spanish government will develop, by regulation, EPR schemes for textiles, furniture and household items, and non-packaging plastics for agricultural use before 9 April 2025. Besides, in the regulatory developments of the law 7/2022, the application of the EPR scheme to single-dose coffee capsules may be included.

8.4. Taxes used at regional level in Spain

This section describes existing waste disposal taxes applicable in different ACs (including Andalusia) and the regional tax on single-use plastic bags applied in Andalusia.

8.4.1. Waste Disposal Taxes

Eleven Spanish Autonomous Communities (AC) apply taxes on waste landfilling and four AC levy waste incineration. The nature of these taxes is quite heterogeneous regarding type of waste, waste activity, and tax rates (see Table 8.8).

Among ACs with waste disposal taxes, most of them levy industrial waste (all except Balearic Islands) and construction and demolition waste (all except Andalusia, Balearic Islands and Cantabria), fewer ACs levy municipal solid waste (Catalonia, Balearic Islands, Extremadura, Castile and León, and Navarra). Most ACs apply the same fees regardless of the recovery potential of the waste fractions. In some cases, tax rates are higher for recoverable waste in comparison to non-recoverable waste to incentivise waste recovery, where possible (Andalusia, Castile and León and Valencian Community). The tax rate on the Balearic Islands for MSW disposal is reduced by half if the municipalities have implemented separate collection of organic waste. A similar reduced tax rate scheme was also applied in Catalonia from 2009 to

2016. The Catalan Disposal Tax is described in detail in Annex H and the way it is designed and implemented is considered a best practice.

Along with the creation of their taxes, Catalonia and Navarra created specific bodies to manage them and specific waste management funds where the revenue goes.

Andalusian Waste Landfill Tax

Law 18/2003, of December 29, approved fiscal measures and administrative regulations in Andalusia. Chapter I of Title II is dedicated to environmental taxes. In this way, the taxes on carbon emissions, dumping into coastal waters, deposit of radioactive waste and deposit of hazardous waste were created. Article 14 establishes that the income proceeding from the abovementioned ecological taxes will be used to finance the actions of Junta de Andalucía in terms of environmental protection and conservation of natural resources, but the law did not set up a separate body to manage the tax and the funds generated, as Catalonia and Navarra did. The Andalusian Tax Agency is responsible for the tax management as well as for the determination and verification, where appropriate, of the environmental parameters that allow the quantification of said taxes (Article 16 of Law 18/2003).

Section V (art. 65 to 77) of the Law 18/2003 specifies the tax on hazardous waste, which came into force in January 2004. The taxable event (art. 67) is "the delivery of hazardous waste in public or private landfills" and "the temporary deposit of hazardous waste in the producer's facilities, prior to its elimination or recovery, when it exceeds the maximum period allowed by law and there is no special authorisation from the Ministry of Environment". Taxpayers are those delivering hazardous waste to a landfill for deposit, as well as those that exceed the temporal period allowed by law for temporary storage prior to elimination or recovery of waste. The tax base (art. 71) is the weight of the hazardous waste deposited and the tax rates depend on whether the waste can be recovered or not, in such a way that it is intended to stimulate preventive treatment (see Table 8.8).

Figure 8.1 shows the evolution of the tax revenue and the tax rate in the period 2008-2020.

Figure 8.1. Andalusian Waste Landfill tax. Revenue and rate for the period 2004-2020

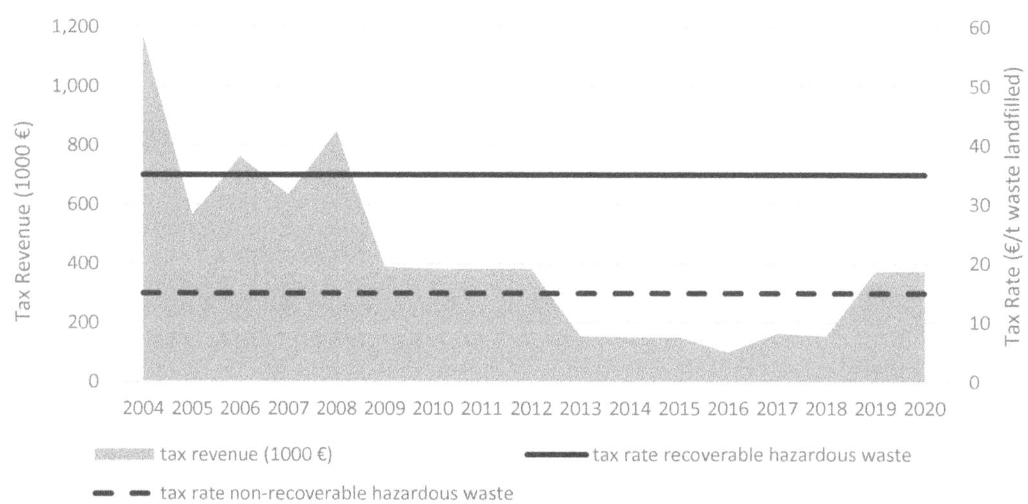

ource: Own elaboration based on the information available in the Portal of the Ministry of Finance and Public Function (2013-2020). Impuestos Propios (Secretaría General de Coordinación Autonómica y Local, 2022[22]).

StatLink https://stat.link/ug7rkd

Table 8.8. Tax rates (€/tonne) of existing waste disposal taxes in Spain, 2021

AC	Activity	Municipal Solid Waste		Industrial Hazardous Waste		Industrial Non-Hazardous Waste		Construction & Demolition Waste	Regulation
		Recoverable	Non-Recoverable	Recoverable	Non-Recoverable	Recoverable	Non-Recoverable		
Andalusia	Landfill			35	15				Law 18/2003
Balearic Islands	Landfill	40 (20)[1]							Law 3/2020
	Incineration	20 (10)[1]							
Cantabria	Landfill			2					Law 6/2009
Castile and León	Landfill	20	7	35	15	20	7	3	Legislative Decree 1/2013
Catalonia	Landfill	59.1[2]				15.8		3	Law 8/2008
	Incineration	29.6[2]							
Extremadura	Landfill	12		18		12		3.5[3]	Law 2/2012
La Rioja	Landfill			21	4[4]	12	4[4]		Law 10/2017
Madrid	Landfill			8		5		1[5]	Law 6/2003
Murcia Region	Landfill			15		7		3	Law 9/2005
Navarra	Landfill	20				20, 5, 1[6]		3[6]	Law 14/2018
	Incineration								
Valencia Community	Landfill			30		30	20	3	Law 21/2017
	(Co-)Incineration			30[7](20[8])	20[8]		20[7]		

Note: (1) The lower tax rate reported within brackets applies to municipalities that have initiated the separate collection of organic waste and Pay-As-You-Throw (PAYT) schemes. (2) Tax rates for 2022, see Table 8.F.1 for planned tax rates until 2024. (3) Rate applicable to Inert waste. For the non-inert part of the CDW, a non-hazardous waste tax rate is assumed to be applied. (4) Rate applicable to non-recoverable waste coming from waste treatment facilities. (5) Tax rate per cubic meter of CDW. (6) 20 €/t for non-hazardous waste in general, 5 €/t for industrial non-hazardous mineral residues with low lixiviation, 1 €/t for natural materials excavated (sand and rocks) and industrial inert waste. (7) Without energy recovery. (8) Hazardous waste in energy recovery operations.
Source: Own elaboration based on (Fundació ENT, 2022[23])

8.4.2. Single-Use Plastic Bag Tax in Andalusia

On 1 May 2011, the Single-Use Plastic Bag Tax came into force in Andalusia, regulated by Law 11/2010, on fiscal measures for the reduction of the public deficit through sustainability, which taxes the supply of plastic bags by commercial establishments located in Andalusia.

The taxpayers are the natural and legal entities owning the commercial establishments where single-use plastic bags are provided to customers. The law does not differentiate between type of plastics bags (e.g., thick, thin, and very thin), but compostable and reusable plastic bags are exempt from the tax. The main aim of the tax was reducing single-use plastic bags consumption, but additional tax revenues for the Junta de Andalucía also motivated the implementation of the tax.

The tax is fully passed on to consumers, and must be stated on the corresponding invoice, receipt, or voucher, as a separate item denoting the number of bags paid for. The tax revenue goes to the general funds of the AC. The tax base is the number of plastic bags provided by the retailer. The tax rate has been 5 cents per single-use bag since 2011 (See Figure 8.2 in Box 8.2.). An increase to 10 cents was planned, but never implemented

Bags supplied by commercial retail establishments in which the holders are registered exclusively under a heading of group 64 of the Tax on Economic Activities including for instance retailers of exclusively fruits and vegetables, meat, fish or bread. Not part of this exemption are retailers of the sub-groups 645,646 and 647, including for example retailers or wines and beverages, tobacco products or general grocery shops.

Box 8.2. Evolution of tax revenue from Andalusia's Single-use plastic bag tax

Based on the tax revenue of 2014 (EUR 388,380) and the tax rate (EUR 0.05 per bag), the taxable event in 2014 equalled to 7.77 million bags. Considering the population of Andalusia in 2014 (8.4 million inhabitants), the taxable event per inhabitant corresponded to 0.9 bags per inhabitant per year. This amount is very small compared with the Spanish average annual consumption of 145 light plastic bags per inhabitant reported by ANAIP for the same year (Junta de Andalucía, 2022[18]). This could mean either that the tax was highly effective and reduced almost completely the consumption of taxed single-use plastic bags, that exemptions applied to the establishments of group 64 of the Tax on Economic Activities commercial retail mean a large volume of bags that are not included in the tax revenue, or that some taxpayers are not fulfilling their obligations with regards to the bags delivered in their establishments.

Figure 8.2. Andalusian Plastic bags Tax revenue and rate for the period 2013-2020

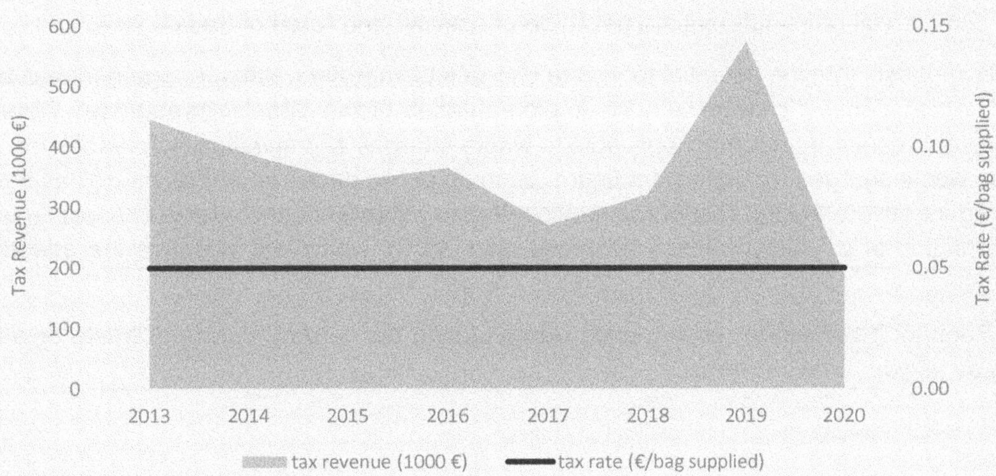

Note: According to Junta de Andalucía, in 2019 there was a peak in the supply of non-exempt bags, and it can be seen in the revenue of the same year. While the number of exempt bags decreased by 13%, the non-exempt ones increased by 40.7%. This could be related to the period of adaptation of businesses to the Spanish Royal Decree 239/2018 whose measures came into force in July 2018. In addition, the revenue of 2019 should be interpreted with caution, given that in December 2018 the Andalusian Budget Law for 2019 was not approved, for that the tax rate that applied from January 1 to July 24 of 2019 was 10-euro cents for each single-use plastic bag supplied instead of 5-euro cents. Subsequently, through the 4th Final Provision of Law 3/2019, of July 22, of the Budget of the Autonomous Community of Andalusia for the year 2019, with effect from January 1 of 2019 and indefinite validity (retroactive character), the tax rate was established at 5-euro cents for each single-use plastic bag supplied. Thus, part of the revenue was paid back to taxpayers and the actual revenue (after deductions) was less than the value shown in Figure 8.2.
Source: Own elaboration based on the data available from the Junta de Andalucía.

StatLink https://stat.link/xibf46

8.5. Charges at the municipal level in Spain

At municipal level, waste charges are used to finance waste collection and management services. Waste charges are regulated through the fiscal ordinances of each municipality and are often conceived as flat rates or depend on criteria different than waste generation. This lack of connection with the effective waste generation and source separation of each user represents a missed opportunity to incentivise waste prevention and separate collection at local level.

The Observatory on Waste Taxation carried out an assessment of the Spanish waste charges applied over five years (2015, 2018, 2019, 2020 and 2021) by evaluating qualitatively and quantitatively the waste fiscal ordinances of 125 Spanish municipalities (Fundació ENT, 2021[24]). The study concludes that:

- Great variability exists among the waste charges applied around the Spanish territory. This heterogeneity can be explained by the flexibility allowed by the Royal Decree 2/2004 on Local Treasuries when designing the charge and by the different configuration of waste collection services at municipal level, which translates into different costs.

- Waste charges have increased both for households and commercial activities between 2015 and 2021. However, some regression in the trends was observed in 2021, as some reductions were introduced to alleviate the economic effects of the COVID-19 pandemic.
- Most of the household waste charges are fixed rates (46.4% of the municipalities), while most of the commercial rates differentiate per "type of activity" and "area of trade".

The analysis suggests that the potential for waste charges to improve waste management has been hardly exploited. The situation will change with the implementation of the Spanish Waste Law 7/2022, since it contains relevant regulatory reforms applicable to waste charges and specifically foresees the mandatory nature of the waste charges (or equivalent figure, such as public prices or tariffs), as well as the obligation that such a figure covers the full cost of the service. It also establishes that waste charges "must allow for the implementation of pay-as-you-throw schemes" (art. 11.3), which will incentive the adoption of such schemes.

The White Book for Tax Reform recommends reformulating the current municipal waste taxation system to link it to pay as you throw systems.

References

Castells-Rey, I., P. Pellicer-García and I. Puig-Ventosa (2022), *Los instrumentos económicos y fiscales en la Ley de Residuos*, Retema, https://www.retema.es/revista-digital/marzo-abril-7 (accessed on 17 January 2023). [11]

CEWEP (2022), *Landfill taxes and bans overview - Last update: 28.10.2021*, https://www.cewep.eu/wp-content/uploads/2021/10/Landfill-taxes-and-restrictions-overview.pdf (accessed on 25 February 2019). [7]

Council of the EU (2020), *Council Decision (EU, Euratom) 2020/2053 of 14 December 2020 on the system of own resources of the European Union and repealing Decision 2014/335/EU*, https://eur-lex.europa.eu/legal-content/EN/TXT/?uri=CELEX%3A32020D2053 (accessed on 18 January 2023). [14]

Danish Ecological Council and Green Budget Europe (2015), "European expert platform on environmental taxation and green fiscal reform", https://green-budget.eu/wp-content/uploads/The-Danish-Packaging-tax_volume_FINAL.pdf (accessed on 18 January 2023). [10]

Ettlinger, S. (2017), "Aggregates Levy in the United Kingdom", https://ieep.eu/uploads/articles/attachments/5337d500-9960-473f-8a90-3c59c5c81917/UK%20Aggregates%20Levy%20final.pdf?v=63680923242 (accessed on 18 January 2023). [26]

Eunomia (2020), *Study to support preparation of the Commission's guidance for extended producer responsibility scheme*, https://op.europa.eu/en/publication-detail/-/publication/08a892b7-9330-11ea-aac4-01aa75ed71a1/language-en (accessed on 18 January 2023). [20]

European Commission (2021), *Ensuring that polluters pay: Taxes, charges and fees*, https://environment.ec.europa.eu/economy-and-finance/ensuring-polluters-pay/taxes-charges-and-fees_en (accessed on 18 January 2023). [5]

European Commission (2021), "Green taxation and other economic instruments", http://Green taxahttps://ieep.eu/uploads/articles/attachments/134d9257-53c5-4a20-885b-9f6615452486/Green%20taxation%20and%20other%20economic%20instruments%20%E2%80%93%20Internalising%20environmental%20costs%20to%20make%20the%20polluter%20pay_Study_10.11.2021.pdf?v=63807385248tion and other economic instruments (accessed on 17 January 2023). [4]

European Commission (2021), *Plastics own resource*, https://ec.europa.eu/info/strategy/eu-budget/long-term-eu-budget/2021-2027/revenue/own-resources/plastics-own-resource_en (accessed on 1 February 2022). [13]

European Commission (2020), *A new Circular Economy Action Plan For a cleaner and more competitive Europe*, https://eur-lex.europa.eu/legal-content/EN/TXT/?uri=COM%3A2020%3A98%3AFIN (accessed on 18 January 2023). [2]

European Commission (2015), *Closing the loop – An EU action plan for the circular economy*, http://eur-lex.europa.eu/resource.html?uri=cellar:8a8ef5e8-99a0-11e5-b3b7-01aa75ed71a1.0012.02/DOC_1&format=PDF (accessed on 18 January 2018). [1]

European Environment Agency (2008), *Effectiveness of environmental taxes and charges for managing sand, gravel and rock extraction in selected EU countries*, https://www.eea.europa.eu/publications/eea_report_2008_2/ (accessed on 18 January 2023). [9]

European Parliament (2019), *Directive (EU) 2019/904 of the European Parliament and of the Council of 5 June 2019 on the reduction of the impact of certain plastic products on the environment*, https://eur-lex.europa.eu/eli/dir/2019/904/oj. [21]

European Parliament (2015), *Directive (EU) 2015/720 of the European Parliament and of the Council of 29 April 2015 amending Directive 94/62/EC as regards reducing the consumption of lightweight plastic carrier bags*, https://eur-lex.europa.eu/legal-content/EN/TXT/?uri=celex%3A32015L0720 (accessed on 18 January 2023). [15]

European Parliament (2008), *Directive 2008/98/EC of the European Parliament and of the Council of 19 November 2008 on waste and repealing certain Directives*, https://eur-lex.europa.eu/legal-content/EN/TXT/?uri=celex%3A32008L0098 (accessed on 18 January 2023). [6]

Fundació ENT (2022), *Observatori de la fiscalitat dels residus. Impostos.*, https://www.fiscalitatresidus.org/impostos/ (accessed on 18 January 2023). [23]

Fundació ENT (2021), *'Evolución de las tasas de residuos en España 2015 - 2021*, https://www.fiscalitatresidus.org/wp-content/uploads/2022/05/EvolucionTasas_2015-2021.pdf (accessed on 18 January 2023). [24]

HMRC UK Government (2020), *Guidance: Exempt aggregate and reporting it to HMRC*, https://www.gov.uk/guidance/exempt-aggregate-and-reporting-it-to-hmrc (accessed on 18 January 2023). [25]

Junta de Andalucía (2022), *Bolsas de plástico*, https://www.juntadeandalucia.es/medioambiente/portal/areas-tematicas/residuos-suelos-contaminados-economia-circular/tipos-residuos/bolsas-plastico (accessed on 18 January 2023). [18]

Junta de Andalucía (2022), *Medidas y plazos establecidos para reducir el consumo de bolsas de plástico*, https://www.juntadeandalucia.es/medioambiente/portal/areas-tematicas/residuos-suelos-contaminados-economia-circular/tipos-residuos/bolsas-plastico/medidas-y-plazos-reducir-consumo-bolsas-plastico (accessed on 18 January 2023). [16]

MITECO (2022), *Las bolsas de plástico ya no podrán ser gratis*, https://www.miteco.gob.es/es/ceneam/carpeta-informativa-del-ceneam/novedades/bolsas-plastico-no-gratis.aspx (accessed on 18 January 2023). [17]

OECD (2022), *Policy Instruments for the Environment (PINE) Database*, https://pinedatabase.oecd.org/ (accessed on 28 June 2022). [3]

OECD (2016), *Extended Producer Responsibility: Updated Guidance for Efficient Waste Management*, https://doi.org/10.1111/jiec.12022. [19]

Puig-Ventosa (2021), *La fiscalidad en el Proyecto de Ley de Residuos y Suelos Contaminados*, https://www.retema.es/articulos-reportajes/la-fiscalidad-en-el-proyecto-de-ley-de-residuos-y-suelos-contaminados (accessed on 18 January 2023). [12]

Secretaría General de Coordinación Autonómica y Local (2022), *Capítulo III. Impuestos propios. Tributación Autonómica. Medidas 2009-2020'*, https://www.hacienda.gob.es/Documentacion/Publico/PortalVarios/FinanciacionTerritorial/Autonomica/Capitulo-III-Tributacion-Autonomica-2022.pdf (accessed on 18 January 2023). [22]

Söderholm, P. (2011), "Taxing virgin natural resources: Lessons from aggregates taxation in Europe", *Resources, Conservation and Recycling*, Vol. 55/11, pp. 911-922, https://doi.org/10.1016/j.resconrec.2011.05.011. [8]

Annex 8.A. Landfill and Incineration taxes in OECD countries

Annex Table 8.A.1. Landfill and incineration taxes for non-hazardous waste in OECD countries

Country	Region	Year	Taxable event	Tax Rate (€/t)	Earmarked
Austria		1989	MSW Landfill	9.2 - 87. Lower rates for landfills with modern technologies	Yes
		2006	Waste incineration	8	Yes
Belgium	Flanders	1990	Inert Waste Landfill	19.87	Yes
			Mineral Waste Landfill	9.03	
			MSW Landfill	59.33	
		1989	Waste Incineration and Co-incineration	8.18	Yes
	Wallonia	1991	MSW Landfill	66.89; non-valorisable waste: 36.12 €; Stabilised waste: 30.58 €	
Bulgaria			MSW Landfill	50.00	
Denmark		1987	MSW Landfill	79.00	
Estonia		1990	Mineral Waste Landfill	1.31	
			Inert Waste and MSW Landfill	29.84	
			Construction Waste Landfill	0.63	
Slovenia		2001	MSW Landfill	11.00	
Finland		1996	MSW & Non-hazardous industrial waste Landfill	70.00	Yes
France		1993	MSW Landfill	54.00	
Greece		2019	MSW Landfill	15.00	Yes
Hungary		2013	MSW Landfill	19.35	Yes
Ireland		2002	MSW & Non-hazardous industrial waste Landfill	75.00	
Israel			Construction Waste Landfill	0.94	
			Inert Waste Landfill	5.50	
			Mineral and Construction Waste Landfill	5.68	
		2001	Inert Waste incineration without energy recovery	0.2 - 2 depends on the region	
		2001	Non-inert Waste incineration without energy recovery	1.03 - 5.16 depends on the region	
	Abruzzo		MSW Landfill	25.00	
	Aosta Valley		MSW Landfill	18.00	
Italy	Apulla		MSW Landfill	MSW Separate collection: <65% = 20.69; >65% = 12.07; >90% =5.17	
Italy	Basilicata		MSW Landfill	20.00	
	Calabria		MSW Landfill	15.50; Pre-treated waste: 5.33; From outside the assigned area to the landfill: 25.82	
	Campania		MSW Landfill	10.30; Pre-treated waste: 5.2	
	Emilia-Romagna		MSW Landfill	19.00	
	Friuli-Venezia Giulia		MSW Landfill	25.82	
	Lazio		MSW Landfill	15.49; Well separated residue: 10.33	

Country	Region	Year	Taxable event	Tax Rate (€/t)	Earmarked
	Liguria		MSW Landfill	15.00	
	Lombardy		MSW Landfill	19.00	
	Marche		MSW Landfill	25.00	
	Molise		MSW Landfill	21.00	
	Piedmont		MSW Landfill	25.82; Pre-treated waste: 12.91	
	Sardinia		MSW Landfill	25.80; Stabilised waste: 18.	
	Trentino		MSW Landfill	12.86	
	South Tyrol		MSW Landfill	11.40	
	Tuscany		MSW Landfill	25.82; Stabilised waste: 21	
	Umbria		MSW Landfill	25.82	
	Veneto		MSW Landfill	25.82	
Latvia		1991	MSW Landfill & Non-hazardous industrial waste	65.00	
Lithuania			Inert Waste Landfill	30.41	Yes
			MSW Landfill	10.00	
Malta			Construction Waste Landfill	3.26	
Netherlands		1995	MSW Landfill	33.15	
Poland			MSW Landfill	46.00	
			Industrial waste Landfill	5.28	
Portugal		2007	MSW Landfill	22.00	
			Waste Incineration without energy recovery	7.70	
UK		1996	MSW Landfill	98.6	
Slovak Republic		2004	Industrial waste Landfill	7.00	
			Inert Waste Landfill	66.00	
			MSW Landfill	MSW separate collection: <10% = 33; <20% = 30; <30% =27; <40% = 22; >50% = 19; >60%=15; > 60%=11	
Czech Republic			MSW Landfill	20.00	
Romania		2019	MSW Landfill	17.00	
Sweden		2000	MSW Landfill	51.00	
Switzerland			Stabilised waste and Construction Waste Landfill	13.76	
			Inert Waste Landfill	4.30	

Note: The Spanish regions with waste disposal tax do not appear in this table. These are presented in Table 8.5.
Source: Own elaboration based on (OECD, 2022[3]) and (CEWEP, 2022[7]).

Annex 8.B. Case studies on landfill taxes in other OECD countries

Annex Table 8.B.1. Landfill tax in the United Kingdom

Legal bases	Finance Act of 1996
Objective	To encourage efforts to minimise the amount of material produced and the use of alternative waste management options, such as recycling or composting.
Level of responsibility	Central government (the United Kingdom)
Tax setter(s)	Central government (the United Kingdom)
Revenue beneficiary(ies)	Central government (the United Kingdom)
Tax payer(s)	Operators or controllers of landfill sites transfer the cost to waste producers, the waste industry and local authorities to dispose municipal waste.
Tax base (including main exemption(s), credits or deductions)	The landfill Tax is charged on material disposed of at a landfill or unauthorized waste site, and the tax base is tons of waste. Exemptions apply to the following waste fractions, disposed at authorised landfill sites: • Material removed from inland waterways and harbours by dredging,. • Material arising from mining and quarrying operations, • Burials of pets at certain authorised landfill sites (the site must be used solely for the burial of domestic pets), • Lower rated material used to fill existing or former quarries, • Waste from visiting military forces (e.g. North Atlantic Treaty Organisations).
Tax rate(s) (including their calculation)	The tax rate is based on the weight of waste, differentiated by two rates. As of 1 April 2022, the standard rate is GBP 98.6 (EUR 114.49) per tonne and the lower rate amounts to £3.15 (EUR 3.65) per tonne.
Governance and implementation	The landfill tax was the first UK tax to have an explicit environmental purpose. Nevertheless, it was considered "a popular tax" as it benefited from widespread support from industry, local authorities and NGOs due to the expected use of revenue to partially offset the burden to business (revised in 2003 to only 6%). The final instrument design was fine-tuned during a consultation period, which continued after its implementation. In 1998, it was suggested to increase the tax rates since the previous rates were shown to be insufficient to shift away from incineration towards more investment in recycling. This suggestion was implemented, and since then, the tax rates have been constantly updated (Eunomia, 2016[25]).
Environmental, social & health impacts	Combined with other policy measures, the tax has significantly contributed to reducing the quantity of waste sent to landfills: in 1996-1997, 50 million tonnes annually were sent to landfill, while it declined to around 12 million tonnes in 2015-2016 (Eunomia, 2016[25]).

Source: (Eunomia, 2016[25]), (The United Kingdom Government, 2021[26]), (The United Kingdom Government, 2022[27])

Annex Table 8.B.2. Belgium, Flanders: Tax on Landfilling and Incineration of Waste

Legal bases	Decree of December 23, 2011, on the sustainable management of material cycles and waste
Objective	To reduce or avoid altogether the landfilling of waste
Level of responsibility	Region (Flanders)
Tax setter(s)	Region (Flanders)
Revenue beneficiary(ies)	Region (Flanders)
Tax payer(s)	Landfill and incineration operators
Tax base (including main exemption(s), credits or deductions)	The tax base is the tons of waste.
Governance and implementation	Landfill taxes and bans on landfilling certain waste streams (e.g., separated waste and untreated municipal waste) have been used to shift from landfilling to incineration and recycling. These instruments are complemented with obligatory separated waste collection, pay-as-you-throw schemes, extended producer responsibility, as well as quotas on waste production per capita. Flanders also applies landfill taxes to waste exported for landfilling with the deduction of any taxes paid in the recipient country (a similar mechanism is used for waste exported for incineration) (OECD, 2021[28]).
Environmental, social & health impacts	The mix of waste policies in Flanders described above has contributed to reduce the average household waste from 555kg in 2007 to 490kg in 2017, resulting in only 1% of average household waste being directed to landfill sites. In 2012, Flanders only had 17 operational landfills in contrast with 118 in 1985 (Interreg Europe, 2018[29]).

Source: (OECD, 2021[28]), (Interreg Europe, 2018[29]), (European Environment Agency, 2013[30])

Annex Table 8.B.3. Tax rate(s) (including their calculation)

Depends on the waste fate and the type of waste:

	[EUR/t]
Landfilling of flammable waste	101.91
Landfilling of non-flammable waste	56.05
Incineration without permit	270.84
Landfilling of household waste that cannot be incinerated in an incinerator	36.12
Landfilling of flammable recycling residues (some categories have a lower tax rate = compensation factor)	101.91
Landfilling of non-combustible recycling residues (some categories have a lower tax rate = compensation factor)	56.05
Landfilling of dredging sludge on a specific site therefore permitted	0.19
Landfilling of residues from permitted treatment facilities of sewage sludge	5.42
Landfilling of residues from soil remediation	3.98
Landfilling of sludge residues from the cleaning of sieving sand	5.42
Landfilling of inert waste	19.87
Landfilling of ore residues	9.03
Landfilling of iron oxide of waste from zinc production	9.03
Landfilling of gypsum or calcium waste	1.81
Landfilling of immobilized non-flammable waste	30.58

Source: (Interreg Europe, 2018[29])

Annex Table 8.B.4. Regional landfill taxes in Italy

Legal bases	National law 549/1995 and all additional regional laws
Objective	To improve the waste management cycle by reducing the share of waste being landfilled, making landfills less convenient, supporting waste initiatives to reduce waste generation, and incentivising recycling and energy recovery alternatives.
Level of responsibility	Regions (Italy)
Tax setter(s)	Regions (Italy)
Revenue beneficiary(ies)	Regions and municipalities (Italy)
Tax payer(s)	Landfill operators
Tax rate(s) (including their calculation)	The tax rates vary regionally within the maximum threshold of EUR 25.8 per tonne, which is set by the central government. The rates are obtained by multiplying the unit amounts, differentiated by categories of waste, quality and conditions of delivery by the quantity, expressed in tons, of the waste delivered. The categories are the following: (i) urban waste and waste from urban treatment, (ii) inert waste, (iii) non-hazardous special waste, and (iv) special hazardous waste (European Commission, 2021[5]).
Tax base (including main exemption(s), credits or deductions)	The tax base is the tons of waste.
Governance and implementation	The landfill tax was introduced on 1 January 1996 to promote the separate waste collection and to support recycling and energy recovery plants. Although a landfilling reduction has been recorded since 1996, 22% of the total municipal waste was disposed of in landfills in 2018, which is far above the EU 10% target set for 2035. The main reason for this is the current relatively low rates of regional taxes. Since 2018, the Italian Regulatory Authority for Energy, Networks and Environment (ARERA) has been leading discussions and open consultation for tax enhancement and determining a "zero landfill" goal. Finally, to increase the effectiveness of the tax, the European Commission has recommended that Italy reforms the tax by increasing the rates and harmonising them across regions (European Commission, 2021[5]).
Environmental, social & health impacts	Since tax rates are determined at the regional levels, the effectiveness of taxes depend on the region considered. In Veneto, separate waste collection went from 34.4% in 2001 to 76.1% in 2020, whilst in Sicily, it varied from 3.3% to 43.3% in the same period (ISPRA, 2022[28]). Despite the reduction of landfilling, landfilling levels remain above the EU 10% target for 2035. Due to the tax's relatively low levels, it is also unclear whether landfilling decreased because of the tax or due to other mechanisms, such as pay-as-you-throw (PAYT) schemes, the improved sorting and recycling infrastructure, and other incentives (e.g., modulation of fees according to municipalities waste management performance) (European Commission, 2022[29]). For instance, Treviso in Veneto introduced a PAYT system in 2014 where 85 000 residents pay waste fees, which are 60% based on the number of people living in the same household, and 40% varies according to the amount of mixed waste. After the implementation of the tax, the separate collection in Treviso increased from 55% in September 2013 to 80% in December 2014, and the production of mixed waste decreased from 20kg/resident/month to 6kg/resident/month over the same period (European Commission, 2019[30]).

Source: (European Commission, 2021[5]), (European Commission, 2019[30]), (ISPRA, 2022[28]), (European Commission, 2022[30])

Annex 8.C. Taxes on Aggregates extraction in OECD countries

Annex Table 8.C.1. EU Environmental taxes on aggregates extraction in OECD countries

Country	Year	Material	Ad Quantum tax 2020	Ad Valorem 2020		Earmarked	Funds destination
			(€/m^3)	(% benefit)	(% market price)		
Estonia	1991	Clay-cement	0.79			Yes (partly)	Natural regeneration of resources, preserving the environment and repairing environmental damage. In 2014, 44% of the collection went to the general state funds.
		Clay-ceramic	0.75				
		Clay-Infusible	1.42				
		Dolomite-fill	0.94				
		Dolomite-high quality	2.36				
		Dolomite-low quality	1.40				
		Dolomite-technology	3.34				
		Gravel-construction	2.43				
		Gravel-fill	0.60				
		Limestone-fill	0.98				
		Limestone-finish	2.94				
		Limestone-high quality	2.36				
		Limestone-low quality	1.49				
		Limestone-technology	2.49				
		Sand-construction	1.55				
		Sand-fill	0.42				
		Sand-tech	1.64				
Lithuania	1991	Clay, Devonian period	0.86			Yes (partly)	20% of the revenue is transferred to the municipalities where the material is extracted and the funds are used to finance the Environment Protection Support Program of the municipality.
		Clay, others	0.51				
		Clay, Triassic	0.84				
		Dolomite	0.99				
		Limestone	0.84				
		Quartz sand	1.59				

		Sand	0.38				
		Construction sand	0.48				
		Sand used for silicone	0.44				
		Land used for construction	0.26				
Sweden	1996	Natural gravel	1.58 (2007)			No	State general fund.
Croatia	1959	Materials (without specifications)		2.6% (5% in protected areas) (2003)		Yes	Investments associated with economic development and environmental protection measures.
Cyprus	1990	Materials (without specifications)	0.26 (1999)			Yes	75% of the funds are used to regenerate the environmental damage in municipalities affected by extractive activity, the remaining 25% destined to projects for the restoration of abandoned quarries.
Czech Republic	1991	Materials (without specifications)	3.00(2011)		Up to 10%	Yes	25% allocated to projects for the restoration of abandoned quarries. Economic compensation for damages due to mining activity.
Denmark	2006	Materials (without specifications)	0.7 (2009)				
France	1999	Materials (without specifications)	0.20*				
United Kingdom	2002		2.50*				
Italy**	1998		It depends on the region		Up to 10.5% in Tuscany	Yes	50% goes to environmental recovery and remediation of disused quarries and degraded areas.

Note: In the OECD database these taxes appear as: mining charges, mineral extraction charges, natural gravel tax, quarrying charge, aggregates tax and general tax on pollution. The taxes of UK and Italy were not found in the OECD database. (*) Tax per tonne of material (**) Regional tax.
Source: Own elaboration based on the OECD database (OECD, 2022[3]).

Annex 8.D. Consumer products taxes in OECD countries

Annex Table 8.D.1. Consumer products levied with environmental taxes in OECD countries

Product	Country applying environmental tax
Household batteries	Croatia, Denmark, Hungary, Iceland, Italy, Poland, Portugal, Slovakia, Sweden
Disposable tableware	Belgium, Denmark, Latvia
Disposal cameras	Belgium
Aluminium foil	Belgium
Plastic carrier bags	Belgium, Denmark, Hungary, Ireland
Packaging items	Denmark, Latvia
Electric light bulbs	Denmark, Latvia, Slovakia
Motor vehicle batteries	Bulgaria, Iceland, Latvia, Lithuania, Poland, Portugal, Sweden
Car tyres	Bulgaria, Canada, Croatia, Denmark, Finland, Hungary, Latvia, Lithuania, Malta, Portugal, Slovakia
Paint, other solvent-containing products	Belgium, Canada
Pesticides	Canada, Denmark, Norway
Vehicles oils and lubricants	Canada, Croatia, Finland, Norway
Consumer electrical products	Canada, Hungary, Italy, Malta, Poland, Portugal, Slovakia

Source: Own elaboration using (OECD, 2015[25]) as main source of information.

Annex Table 8.D.2. Volume-based tax rate of the Danish packaging tax (in DKK/unit for 2022)

	Volume (cl)					
	<10	10-40	41-60	61-110	111-160	>160
Cardboard or laminate	0.08	0.16	0.26	0.53	0.79	1.05
Other (glass, plastic, metal, etc.)	0.14	0.26	0.42	0.84	1.27	1.69

Source: (Danish Ministry of Taxation, 2022[26])

Annex 8.E. EPR schemes in Spain

Annex Table 8.E.1. EPR schemes applied in Spain in 2022

Waste Flow		Producer Responsibility Organization	Spanish Regulation	EU Directive
Packaging	Light packaging (including plastic, metal, beverage carton and paper/cardboard)	ECOEMBES	Law 11/2997 Royal Decree 782/1998	Directive 2018/852
	Glass Packaging	ECOVIDRIO		
	Medical Products Packaging and Expired Medicines	SIGRE		
	Phytosanitary Products Packaging	AEVAE	Royal Decree 1416/2001	
	Agriculture Products Packaging	SIGFITO		
Batteries and accumulators		European Recycling Platform & Fundación Ecopilas	Royal Decree 106/2008	Directive 2006/66/EC
End-of-life vehicles		SIGRAUTO	Royal Decree 265/2021 that modifies Royal Decree 2822/1998	Directive 2000/53/EC
End-of-life tires		SIGNUS & TNU	Royal Decree 731/2020 that modifies Royal Decree 1619/2005	
Used Industrial Oils		SIGAUS & SIGPI	Royal Decree 679/2006 modified by Order ARM/795/2011	
Waste Electric and Electronic Equipment (WEEE)		>11 organizations (e.g., AMBILAMP)	Royal Decree 110/2015	Directive 2012/19/EU

Source: Own elaboration based on the information available at (MITERD, 2022[27]).

Annex 8.F. The Catalan Waste Disposal Tax

The **Catalan waste disposal tax** is one of the oldest of its kind that exist in Spain and the first one and still one of the few that levy not only landfilling but also incineration. The regulation on the management of the tax is defined in Law 8/2008 on the financing of waste management infrastructures and the fees on waste disposal.

The taxpayers are the local entities that own the MSW management service (or those that have it delegated) and the producers of the waste in the case of waste that is not the responsibility of the Local Authorities. The owners of the waste disposal facilities are substitute taxpayers.

Its design (object and tax rates) has evolved over time. In 2004 the landfill tax for municipal solid waste (MSW) was introduced, then it was extended to construction and demolition waste (CDW) in 2009 and to industrial waste (IW) in 2014. From 2009, the tax also included incineration of MSW. The tax rates have also been progressively increasing over time. In 2020, an increase in tax rates on municipal waste was approved until 2024 (see Table 8.F.1).

To encourage the implementation of separate collection of organic waste, from 2009 to 2016 differentiated rates were applied. The higher tax rate (reported within parentheses in 8.F.1) applied to municipalities without separate collection of organic fraction applied only to those that should have implemented separated collection of organic waste according to the separate collection deployment approved by the Catalan Waste Agency (ARC).

Annex Table 8.F.1. Tax rate evolution (€/t) of the Catalan waste disposal tax (2004-2024)

	2004-2008	2009-2010	2011	2012-2013	2014	2015	2016	2017	2018	2019	2020	2021	2022	2023	2024
MSW Landfill	10	10 (20)	12 (21)	12.4 (21.6)	15.8 (25.4)	19.1 (28.7)	19.1 (28.7)	30	35.6	41.3	47.1	53.1	59.1	65.3	71.6
MSW Incineration	-	5 (15)	5.5 (16)	5.7 (16.5)	7.4 (18.6)	9 (20.2)	9 (20.2)	14.5	17.8	20.6	23.6	26.5	29.6	32.7	35.8
IW Landfill	-	-	-	-	3.95	7.9	11.85	15.8	15.8	15.8	15.8	15.8	15.8	tbd	tbd
CDW Landfill	-	3	3	-	-	-	3	3	3	3	3	3	3	tbd	tbd

Note: tbd = to be defined; ( ) = Amounts within parentheses applies to municipalities without separate collection for organic waste that should have it according to the separate collection deployment project approved by the Catalan Waste Agency (ARC).
Source: Own elaboration.

The tax revenue goes to the Waste Management Fund regulated by Law 8/2008, of July 10, financing waste management infrastructure and the tax on the disposal of waste and attached to the Catalan Waste Agency (article 4). The Waste Management Fund is the body responsible for collecting and managing tax revenues, and it is the body that establishes the purpose of the funds collected and helps to prevent fraud. The fund is managed by two governing boards, the Governing Board for Municipal Waste and the Governing Board for Construction Waste. It is up to each of these boards to plan, decide and manage the destination of the funds. In the case of industrial waste, the funds are managed by a collegiate body made up of different organisations and entities.

Table 8.F.2 summarises the tax revenues from 2016 to 2019. The beneficiaries of the funds vary according to the type of waste. The beneficiaries of the revenues from MSW are municipalities and other local authorities in charge of collection and treatment MSW. The beneficiaries of the revenues from IW and CDW are the natural or legal persons who carry out actions for the prevention and material recovery of such waste.

Annex Table 8.F.2. Tax revenue (€) of the Catalan waste disposal tax, 2016-2019

Year	MSW Landfill	MSW Incineration	IW Landfill	CDW Landfill	Total Revenue
2016	28 109 717	6 197 830	4 414 252	3 701 139	42 422 938
2017	41 431 877	8 963 549	5 154 098	2 736 437	58 285 961
2018	56 558 520	11 030 233	6 375 591	4 871 586	78 835 930
2019	63 797 934	13 051 194	8 127 961	5 427 159	90 404 248

Source: Own elaboration based on data provided by the Catalan Waste Agency.

The distribution of the income collected with the MSW tax is of special interest. There is a procedure for revenue distribution, colloquially called "tax return", since it returns above 95% of tax revenues to taxpayers according to different concepts defined annually by the Municipal Waste Management Board. Concepts subject to tax return and the amounts returned to local entities in 2021 can be found in an guidance document published by the waste agency (Agència de Residus de Catalunya, 2021[28]). They depend on the performance of the Local Authorities in terms of waste management. The better the performance, the higher the tax return. One of the concepts is the amount of separately collected organic waste and the impurities content.

The funds collected from industrial waste must be allocated to prevention studies and new technologies for waste treatment (10%), abandoned waste management actions and other activities related to industrial waste management developed by ARC (40%), to green infrastructure actions and territorial environmental improvement (2%) and the remaining 48% to prevention actions. The funds collected with the tax on CDW must be used for actions to prevent and recover CDW, to optimise CDW management and to the promotion and research of recovered materials. As for IW, 2% of the funds collected must be used for actions of green infrastructure and territorial environmental improvement.

A statistical study carried out by ENT in 2021 demonstrated that the presence of a Catalan waste tax applied to MSW has significantly contributed to an increase in municipal separate collection. The design of the Catalan landfill and incineration taxes, with its earmarked character and with a detailed and dynamic income distribution system, has increased separate collection of organic waste and reduced landfill and incineration waste over time (see Annex Figure 8.F.1).

Annex Figure 8.F.1. Evolution of the separate collection of organic waste, landfilling and incineration in Catalonia, 2004-2020

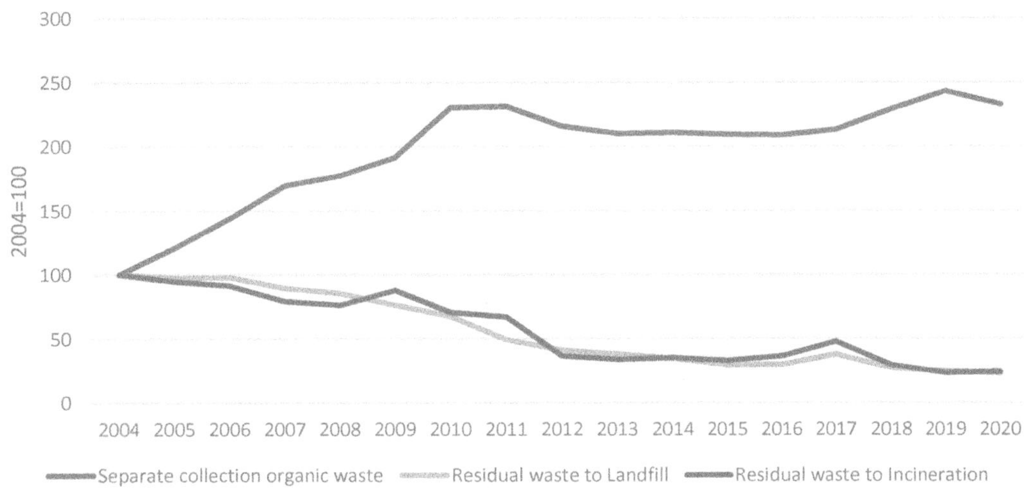

Source: Own elaboration based on the data available in http://estadistiques.arc.cat.

StatLink https://stat.link/erm418

Annex Table 8.F.3. Best practices: Summary of the Catalan Waste Disposal Tax

Catalonia (Spain)	**Title: Catalan Waste Disposal Tax** **Objective**: To discourage landfilling and incineration of waste and to create a twofold incentive for the separate collection and recovery of waste streams by making disposal more expensive and by channeling revenues to local authorities according to their results in separate collection and waste treatment. **Competence**: Catalan Waste Agency (ARC) **Legal basis**: Law 8/2008 on the financing of waste management infrastructures and the fees on waste disposal **Setter**: The Parliament of Catalonia **Beneficiary**: Municipalities and other local authorities in charge of collection and treatment of MSW through the Waste Management Fund created by Law 8/2008 **Payer**: Local entities that own the MSW management service (or those that have it delegated) and the producers of the waste in the case of waste that is not the responsibility of the Local Authorities. **Taxable event**: Landfilling and incineration of waste (Municipal solid waste, industrial waste and construction and demolition waste) **Calculation**: Progressive increase in tax rates **Strengths**: Distribution of the income collected with the MSW tax **Weaknesses**: Long time to reach significant tax rates

Annex 8.G. Tax on aggregates in Andalusia

Annex Table 8.G.1. Price and demand variation of the tax on aggregates proposed for Andalusia

	Price (including tax) Variation (%)		Demand Variation (%)	
	Ad Quantum	Ad Valorem	Ad Quantum	Ad Valorem
Andesite	59.2%	10.0%	-5.92%	-1.00%
Clay	91.0%	10.0%	-9.10%	-1.00%
Siliceous Sand	11.0%	10.0%	-1.10%	-1.00%
Sand and gravel	48.6%	10.0%	-4.86%	-1.00%
Sandstone	27.2%	10.0%	-2.72%	-1.00%
Basalt	23.1%	10.0%	-2.31%	-1.00%
Limestone	26.4%	10.0%	-2.64%	-1.00%
Diabase	26.4%	10.0%	-2.64%	-1.00%
Greenstone	25.7%	10.0%	-2.57%	-1.00%
Dolomite	27.1%	10.0%	-2.71%	-1.00%
Granite	27.3%	10.0%	-2.73%	-1.00%
Greywacke	52.0%	10.0%	-5.20%	-1.00%
Loam	91.8%	10.0%	-9.18%	-1.00%
Ophite	21.9%	10.0%	-2.19%	-1.00%
Trachyte	14.8%	10.0%	-1.48%	-1.00%
Gypsum	21.7%	10.0%	-2.17%	-1.00%

Note: Ad Quantum tax of 1.35 €/t, Ad Valorem tax of 10% of the market price, Elastic demand of 10%.
Source: Own elaboration.

Notes

[1] In June 2021, an Agreement was approved for the formulation of the Strategy for Sustainable Mining in Andalusia 2030 (Acuerdo de 1 de junio de 2021, del Consejo de Gobierno, por el que se aprueba la formulación de la Estrategia para una Minería Sostenible en Andalucía 2030 (EMSA 2030).

[2] See (HMRC UK Government, 2020[25]) for a complete description of the exemptions and (Ettlinger, 2017[26]) for a brief summary of the levy.

[3] Ley 7/2022, de 8 de abril, de residuos y suelos contaminados para una economía circular.

9 Assessment: Circular Economy and Waste Management

9.1. Identification of areas for strategic reform of tax instruments in Andalusia

A strategic environmental reform of tax instruments in Andalusia should be able to address circularity and to apply the polluter pays principle (PPP) to its main economic sectors, i.e., the service industry (including tourism), building industry, agriculture, and mining. In fact, the PPP is one of the governing principles of the draft bill of the Andalusian Law on Circular Economy (article 4d).

In addition, Andalusia could use such a tax reform to improve its source separation and recycling rate and meet the EU targets (Table 7.A.1).

Moreover, the Spanish Waste Law 7/2022 includes a national waste disposal tax, which will need to take into consideration the pre-existing regional taxes, as discussed later. In addition, the same Law requires local authorities to implement waste charges that cover collection and treatment costs and allows for the possibility of implementing pay-as-you-throw schemes.

Looking at the main economic sectors and the main waste management challenges of the region, existing environmental regional taxes will have to be modified (section 4.1) and some new instruments could be considered (section 4.2). Section 4.3 describes the selection of environmental taxes to be assessed in section 5. In addition to adjusting tax rates to higher polluting levels, one could also consider tax rebates and reimbursement measures for lower polluting activities, to incentivise a shift in the economy. These economic instruments lie however outside of the scope of this study.

9.1.1. Modification of existing environmental taxes in Andalusia

As previously described, there are two environmental taxes currently applied in Andalusia related to circular economy and waste management: the single-use plastic bag tax and the landfill tax on hazardous waste. Both are in its current form incompatible or in conflict with existing or forthcoming national regulation.

Single-use plastic bag tax

According to the measures and deadlines established by the Spanish Royal Decree 239/2018, of May 18, on the reduction of the consumption of plastic bags and by which the Registry of Producers was created, which transposes Directive (EU) 2015/720 into the Spanish legal system (see Table 8.7), it is only allowed in Spain to provide: 1) very thin compostable bags (free of charge), 2) thin compostable bags (prior payment), 3) thick reusable bags with more than 50% recycled plastic (prior payment), and 4) thick reusable bags with more than 70% of recycled plastic (for free).

Considering that thick bags are reusable, and that both reusable and compostable bags are exempted from the Andalusian Tax, the Andalusian Single-Use Plastic Bag Tax is incompatible with the Spanish Royal Decree insofar it levies bags that are banned by the Spanish regulation. Thus, it should either be repealed or reformulated to levy also compostable and/or reusable bags. Such reformulation could reduce the consumption of such bags and be compatible with the Spanish Royal Decree.

Regional tax on hazardous waste landfilling

After the entry into force of the national landfill and incineration tax, it is unclear what will happen to the existing regional taxes on waste disposal. In Andalusia, the existing tax affects the landfilling of hazardous waste. For cases like this, the 21st additional provision of Law 7/2022 states that the regional taxes can be maintained while the necessary agreements are being made between the State and the Autonomous Communities. But the new Spanish disposal tax will affect several other waste types and will levy not only landfill, but also incineration and co-incineration. For all these cases, the national tax will apply directly.

According to the Spanish Waste Law (art. 93.2), the AC will have the capacity to increase the tax rates above the minimum level established at national level. The tax revenue will be assigned to the Autonomous Communities based on the place where the taxable event occurs (article 96).

The current waste disposal rate in Andalusia for hazardous waste is 35€/t for recoverable waste and 15€/t for non-recoverable waste while the rate for waste disposal of hazardous waste in the national law is 8€/t for waste without pre-treatment required and 5€/t for the rest of the waste.

9.1.2. Potential new environmental taxes for Andalusia

Different products and activities could be levied in Andalusia to avoid or reduce their environmental impacts. Some of the externalities that are not yet part of the scope of national tax laws and where environmental taxes on sub-national level may thus lead to improvements include:

- **A regional tax on aggregates** extraction could incentivise reuse and material recovery of construction and demolition waste and could generate revenue to mitigate some of the impacts of these activities in Andalusia. The White Book for Tax Reform suggests the introduction of a tax on the extraction of aggregates (Proposal 14).
- **A regional tax on tourism** with differentiated rates depending on the environmental impact associated with different types of tourism could promote a change of preference in terms of tourism activities and/or compensate for the costs associated with tourism. However, as tourism affects different environmental fields (water, energy, mobility, waste, etc), such a tax should not merely focus in promoting circular economy and waste management and may be better suited to address a combination of different externalities, as is done in Part IV.
- **A regional tax on agricultural plastics** could be implemented to avoid their use or to reduce mismanagement of EoL agricultural plastic in Andalusia. Spain is the EU Member States with greatest consumption of agricultural plastics and Andalusia is the region with the largest consumption within Spain (Eunomia;, Deloitte; and ENT, 2021[1]). Due to the strong agricultural sector in Andalusia, end-of-life agricultural plastics is a significant waste stream and mismanaged plastics pollute the environment and degrade over time into microplastics. There is a need to address this issue through economic instruments. However, according to some sources,[1] there seems to be plans to create a mandatory national EPR scheme for agricultural plastic, which is considered the more cost-effective option as opposed to a tax.
- **A regional tax on textiles** could be introduced to disincentivise consumption of fast fashion products while revenues could be used to increase source separation for high quality recycling of end-of-life textiles. However, a national EPR scheme may be better suited to address textiles waste, due to the complexity of the textile value chain and the diversity of products within the sector. The EU textile strategy foresees that the European Commission provides guidance on the implementation of an EPR schemes for textiles.

- **A regional tax on beverage cartons.** Beverage cartons are often not part of the scope of a DRS and were initially considered to be excluded from the planned Spanish Beverage Deposit Refund System (Laubinger et al., 2022[2]). In addition, beverage cartons are hardly affected by the special tax on non-reusable packaging or by the target on reusable packaging, both measures that are included in the Spanish Law 7/2022 on Waste and Contaminated Soils for a Circular Economy. EPR fees for DRS are likely to be higher than EPR fees for kerbside collection. As such, there may be a risk of producers shifting from beverage containers that are required to participate in the more costly DRS (e.g. plastic bottles) to beverage cartons if these are excluded. A regional tax on beverage cartons would level the playing field between producer contributions to the DRS for bottles and overall producer fees for beverage cartons, to prevent a possible migration away from bottles. *This tax scenario was assessed during the analysis, however, due to new developments in the Spanish DRS law, which eventually lead to the inclusion of beverage cartons in its scope, the analysis was excluded from the final version of this report.*

9.2. Selection of taxes to be assessed

To concretise and delimit the proposed tax reform in Andalusia three of the seven taxes were selected due to its potential to incentivise a more circular economy. Some of the identified areas are in line with the suggestions of the White Book for Tax Reform: intensification and extension of the taxes of the Waste and Contaminated Soil Law and creation of a tax on the extraction of aggregates. Table 9.1 summarises the selection of the environmental taxes that will be assessed in more detail in the following chapter of this project and the associated justification for their selection:

Table 9.1. Selection of environmental taxes to be assessed in this project.

Regional tax on:	Selected	Reason for selection or exclusion
Modifications of existing taxes		
Hazardous waste landfill	Yes	As the tax rates on the landfilling of hazardous waste included in the Spanish Law 7/2022 are lower than the current rates applied in Andalusia, a regional tax rate increase for such waste fraction is highly recommended to prevent additional hazardous waste imports.
Single-use plastic bags	No	Modifications of this tax (based on the options currently allowed by the Spanish Royal Decree 239/2018) can have little environmental impact associated, as it would be only possible to levy compostable and/or reusable bags.
New environmental taxes		
Aggregates	Yes	A tax on aggregates can incentive the circular economy in the construction sector, one of the economic sectors with larger environmental impacts. In addition, competitiveness of the local industry will not be affected by the tax, if tax rate is defined below transport cost, due to the low price of aggregates.
Beverage Cartons	No	Beverage cartons is the packaging option for beverage with lower recycling rates due to its composite nature and at the same time is the beverage packaging option least affected by the measures included in the Spanish Waste Law for packaging, such as the plastic tax and reusable targets. A tax on beverage cartons to avoid unwanted shifts from bottles to beverage carton packaging due to specifics of the DRS design as assessed, but eventually excluded from the analysis, as the final version of the Spanish DRS aims to include beverage cartons in its scope. The low recyclability of beverage cartons is best reflected through modulation of EPR fees for different beverage container options.
Tourism	No	Tourism is an economic activity that puts pressure on different environmental areas simultaneously, i.e., not only waste management. Thus, a tax on tourism with a holistic environmental focus seems more suitable to reduce tourism environmental impact than a tax with a narrower focus on circular economy and waste management (see Part IV for a discussion on the taxation of tourist stays).
Agricultural plastics	No	These two groups of items are believed to be better addressed through EPR schemes able to capture the complexity of their value chain and the diversity of items included in them. The definition of taxes on these types of items may become obsolete when the EPR schemes are established. As mentioned in section 4.2, they are expected to be created within the next few years.
Textiles	No	

Source: Own elaboration.

References

Eunomia;, Deloitte; and ENT (2021), *Conventional and Biodegradable Plastics in Agriculture*, https://www.eunomia.co.uk/reports-tools/conventional-and-biodegradable-plastics-in-agriculture/ (accessed on 18 January 2023). [1]

Laubinger, F. et al. (2022), "Deposit-refund systems and the interplay with additional mandatory extended producer responsibility policies", *OECD Environment Working Papers*, No. 208, OECD Publishing, Paris, https://doi.org/10.1787/a80f4b26-en. [2]

Note

[1] Personal communication with Junta de Andalucía.

10 Evaluation of tax instruments

10.1. Waste Disposal Tax

According to Article 93.2 of the Spanish Waste Law, ACs will be able to increase the national tax rates on landfilling, incineration, and co-incineration of waste. This is likely to happen in ACs that border ACs with similar taxes and where the already established tax rates for specific waste types are higher than the rates required by the new Spanish Waste Law. These cases are represented in the blue cells of Table 10.1. It should be noted that the definitions used in the Spanish Law 7/2022 for each tax rate are, in most cases, not directly comparable to the terms used in the different laws defining waste disposal taxes on regional level the table is therefore a simplification of the real picture (see Table 8.5 and Table 8.6).

The Spanish waste disposal tax is not on specific waste types, but instead on waste that is deposited in one of three types of legally established landfills (i.e., landfill for non-hazardous waste, landfill for hazardous waste, and landfill for inert waste) and depending on whether it has been subject to prior treatment or not (see section 3.2.2). Contrary, most of the regional disposal taxes apply to specific waste types. For example, the Andalusian waste disposal tax is on hazardous waste regardless of the type of landfill where it is disposed of. This means, for example, that for asbestos, which is considered hazardous waste, the same tax rate applies in the Andalusian tax regardless of the type of landfill where this waste is disposed of (they can be disposed of in landfills for hazardous waste or for non-hazardous waste after some pre-treatment). On the contrary, different rates will apply to asbestos in the Spanish tax depending on the landfill type where it is disposed of. The current Andalusian waste tax on the other hand distinguishes between recoverable and non-recoverable wastes, which is not differentiated for in the Spanish waste tax.

Table 10.1. Landfill tax rates applied in different ACs and in Spain for different waste types

		Municipal Solid Waste (MSW)		Hazardous Waste (HW)		Industrial non-Hazardous Waste (INHW)		Construction and Demolition Waste (CDW)	
		Recoverable (R)	Non-recoverable (NR)	R	NR	R	NR	R	NR
Landfill	Andalusia			35	15				
	Balearic Islands	40 (20)							
	Cantabria					2			
	Castile & León	20	7	35	15	20	7	3	
	Catalonia	59.1				15.8		3	
	Extremadura	12		18		12		3.5	
	La Rioja			21		12			
	Madrid			8		5		1	
	Murcia Region			15		7		3	
	Navarra	10				5 (1)			
	Valencia Community			42	35	30	25	3	
	Spain	40 (30)		8 (5)		15 (3) or 10 (1.5)		3 (1.5)	
Incineration	Balearic Islands	20 (10)							
	Catalonia	29.6							
	Navarra	20							
	Valencia Community			42	35	30	25	3	
	Spain	20, 15 or 7 in D10 15, 10, 4 in R01		30*		8**			

Note: Blue cells represent the cases where existing regional waste disposal tax rates are above the national disposal tax as required by the Spanish Waste Law and a regional surtax on top of the national tax to match the incumbent rate is thus more probable.
(*) This rate applies to "rejects from MSW treatment" and we assumed such rejects are "non-recoverable MSW"
(**) This rate applies to residues different than MSW, rejects from MSW treatment, and without previous pre-treatment required. We assume this is equivalent to "recoverable hazardous waste".
Source: Own elaboration.

Andalusia has 79 legally established landfills: 2 landfills for hazardous waste, 30 landfills for non-hazardous waste (20 are landfill facilities associated with MSW treatment plants), and 47 landfills for inert waste (Junta de Andalucía, 2021[1]).

Two regional increases of the national tax rates could be considered in Andalusia: 1) on hazardous waste disposal and 2) on construction and demolition waste.

10.1.1. Hazardous waste:

In 2018, 10,771 tonnes of hazardous waste were disposed of in Andalusia. 43% of them (4,597 t) correspond to hazardous waste containing asbestos that can be deposited in non-hazardous waste landfills after treatment (Junta de Andalucía, 2021[1]).[1]

Although there are two landfills for hazardous waste in Andalusia (in Nerva and in Jerez de la Frontera), the latter has not received hazardous waste since 2005. It should be noted that even with the Andalusian waste disposal tax, there are already significant imports of hazardous waste to be landfilled in Andalusia, i.e. 60% of the hazardous waste landfilled in Nerva in 2018 came from outside Andalusia (Junta de Andalucía, 2021[1]).

If Andalusia does not increase the national tax rate on the disposal of hazardous waste, the imports of hazardous waste are likely to increase, since the landfill fees of the Nerva landfill would decrease relative to previous landfill tax rates. In addition, the landfill of Jerez de la Frontera could consider again to accept hazardous waste. The Andalusian government could decide not to increase the national tax rate and get the compensation of the revenue loss with the national tax introduction. This would imply that La Junta could get the same funds with or without increasing the national tax rates, but while increasing them will have a political cost associated, not doing it will likely derive in an increase of waste imports that will result in Andalusia receiving additional revenues but also having to deal with the environmental impact associated with increased waste imports. Increasing the tax rate entails an environmental benefit associated with a decrease in waste imports, subsequently reducing the environmental burden. Thus, to prevent an increase of hazardous waste imports to the region, increasing the tax increase to at least current rates is highly recommended.

It should also be considered that the amount of waste imports is limited by the Waste Regulation of Andalusia, as approved in March 2012, which establishes limits to the direct entries of hazardous waste from outside Andalusia to hazardous waste landfills in Andalusia. The Plan for the Prevention and Management of Hazardous Waste in Andalusia defined the following limits: 13,337 tons per year for the Nerva landfill and 681 tons per year for the Jerez de la Frontera landfill. The total limit for direct entries of hazardous waste from outside is thus 14,018 tons, compared with currently 10,771 tons being imported and disposed of in Andalusia. Increasing the tax rate to initial levels could help prevent imports from rising, which may likely be the consequence if landfill tax rates would be lowered from initial levels to the proposed levels by the Spanish Waste Law.

10.1.2. Construction and demolition waste:

Andalusia could also consider increasing the national tax rates on activities and waste fractions for which the tax rate is low compared to the regional taxes applied in other ACs, such as disposal of CDW.

The composition of CDW is mainly inert, although hazardous and non-hazardous non-inert waste can also be found within CDW. In 2018, 4,042 thousand tonnes of CDW were generated in Andalusia and only 0.6% (i.e. almost 25 thousand tonnes) corresponded to hazardous waste (Junta de Andalucía, 2021[1]), with the majority of waste being sand and stones containing hazardous substances (LER 170503*) and construction materials containing asbestos (LER 170605*), with a total of 19,378 tons and 5,296 tons, respectively.

Of the total construction and demolition waste managed in Andalusia, more than 90% is generated in the territory itself. During 2018, 92% of construction and demolition waste was subject to recovery operations, which include recycling (75%) and other recovery operations such as the restoration of degraded and filled spaces (17%), compared to 8% whose final destination was landfill (326 kt).

It should be considered that even if the authorised facilities have high recycling rates, a significant percentage of CDW generation remains unknown and deposited in unauthorised places (30% of the production can be considered uncontrolled according to the latest CDW Production and Management report in Spain). Considering the known CDW generation data, Andalusia would be reaching the objective of 70% of non-hazardous CDW destined for preparation for reuse, recycling and other recovery operations established in the National Waste Framework Plan (PEMAR) 2016-2022 for the year 2020 (and also stated in Law 7/2022 on Waste and Contaminated Soils for a Circular Economy). However, if uncontrolled CDW data were considered the objective would not be met.

The recovery of CDW in authorised facilities had an upward trend until 2015, although in recent years there has been a decline in recycling operations. This could be due to the low demand for recycled materials that could be motivated by several issues such as: 1) low prices of virgin material, 2) low prices of the deposit of CDW in authorised landfills, 3) illegal dumping of CDW and 4) insufficient promotion of the market of the products resulting from the treatment.

Thus, the national tax rate increase on the disposal of CDW will likely help to increase demand for recycled aggregates and revert the trend observed in recycling since 2015. Andalusia could consider increasing this landfill tax even further, in order to encourage material recovery. Accompanying enforcement and control measures would however be necessary to avoid an increase in illegal dumping, due to higher disposal taxes.

10.1.3. Tax definition

The **taxable event** is defined at Law 7/2022, i.e. "the delivery of waste for its disposal in authorised landfills, publicly or privately owned, located in the Spanish territory", but the increased tax would only apply to hazardous waste and CDW disposed of in landfills located in Andalusia, particularly to:

1) Hazardous waste disposed in landfills for hazardous waste
2) Hazardous waste disposed in landfills for non-hazardous waste (only applicable to asbestos) and includes CDW with asbestos content.
3) CDW disposed in landfills for inert waste

The definition of the **tax rate increase** in the case of hazardous waste would aim at compensating for the difference between the Spanish tax rate and the current Andalusian tax rate. The breakdown of the differences between the current Andalusian tax rate and the rate of the Spanish waste tax for the different waste types, as well as the resulting surtax that is proposed to re-establish current levels is listed in Table 10.2.

For CDW, since Andalusia currently has no disposal tax specifically for CDW, the difference between the Andalusian tax (null) and the Spanish tax rates are negatives. Nevertheless, also in the case of CDW, as the Spanish tax rate is low, and such residue is unlikely transported to other regions with lower disposal taxes, the tax rate is proposed to be increased to 5 €/t for CDW disposed of in landfills for inert waste without prior treatment and 3 €/t for CDW with previous treatment.

The revenue from both the national tax rate and from the suggested regional increase can be earmarked for transparency, to increase acceptability, to correct distributional impacts or to fund for instance waste management services in local authorities. It is however beyond the scope of the study to assess possibilities and purposes of earmarked tax revenues that arise from the recommended taxes. Overall, it should be considered that management of earmarked taxes is more complex and entails higher administrative costs than general taxes. They are also considered less economically efficient since budgetary flexibility is reduced (Dechezleprêtre et al., 2022[2]; Kallbekken, Kroll and Cherry, 2011[3]).

Table 10.2. Definition of the national waste disposal tax rate increase

	Landfill type	Recoverable	Pre-treatment	Current Andalusian tax rate	Spanish tax rate	Tax rate difference*	Proposed surtax on Spanish tax rate	**Proposed new tax rates for Andalusia**
Hazardous waste	Landfill for Hazardous waste	Yes	Without	35	8	27	27	**35**
			With	35	5	30	10	**15**
		No	Without	15	8	7	27	**35**
			With	15	5	10	10	**15**
	Landfill for non-Hazardous waste (asbestos)	Yes	Without	35	15	20	20	**35**
			With	35	10	25	5	**15**
		No	Without	15	15	0	20	**35**
			With	15	10	5	5	**15**
CDW	Landfill for inert waste		Without	0	3	-3	2	**5**
			With	0	1.5	-1.50	1.5	**3**

Note: * tax difference between current Andalusian tax rate and Spanish tax rate.
Source: Own elaboration.

10.1.4. Environmental implications

The environmental implications of increasing the tax rates at regional level would mainly affect hazardous waste and CDW management in Andalusia. The objective of such tax rate increase on hazardous waste management would be to prevent higher imports of such type of waste in Andalusia. As mentioned previously, with the current Andalusian tax rate, 60% of the waste disposed in Nerva Landfill is from outside of Andalusia, and a large part of the imports come from outside Spain, mainly from Italy and Montenegro (European Parliament, 2022[4]). Without such national tax rate increase, the disposal of hazardous waste in Nerva would become cheaper, and imports could increase.

The environmental risk of the hazardous waste disposal in Nerva Landfill has been raised by multiple actors in the region. The landfill is located 700 metres away from the town of Nerva (Huelva) and discharges its waters into the River Tinto, a Special Area of Conservation. In addition, waste imports coming from Italy and Montenegro arrive through the Port of Seville and transported in lorries across the River Guadiamar Special Area of Conservation (European Parliament, 2022[4]).

The environmental implications of the regional increase of the national tax rate on CDW would relate to increasing the circularity of such type of waste and thus preventing its disposal. In this case, the damage caused by this type of waste disposal is less important than the hazardous waste disposal, since most of it is inert waste, but it should be noted that building materials (such as concrete) have energy and waste intensive productions (EEBA, 2021[5]). Thus, if such a surtax on the national tax rate would incentivise the use of secondary building materials, it may also lead to reductions in the use of primary building materials and associated impacts on the environment and resource depletion.

10.1.5. Economic implications

Table 10.3 summarises the potential revenue of the waste disposal national tax rate increase for Andalusia and of the national waste disposal tax for the two types of waste discussed in this section. This calculation is based on two main assumptions:

- For hazardous waste, it has been assumed that most of the hazardous waste landfilled in Andalusia is non-recoverable, since dividing the tax revenue by the amount of waste disposed of in 2018, the resulting average tax rate was 14.31 €/t, which is closer to the non-recoverable tax rate (15€/t) than the recoverable tax rate (35 €/t).

- For construction and demolition waste, it has been assumed that all of the waste is pre-treated before being disposed of in landfills for inert waste in Andalusia, since according to the good practices for the management of CDW in Andalusia,[2] disposal of CDW without pre-treatment is banned in the region.

The revenue associated to the national tax rate increase would come mainly from the disposal of CDW (6 million €) while only 1% would come from the disposal of hazardous waste (66 thousand €). This uneven contribution between the two waste fractions can be explained by the difference in waste generation, the amount of CDW landfilled is a thousand times larger than the amount of hazardous waste. The same occurs for the national disposal tax, 98% of the tax revenue (associated only to these two waste fractions disposed of in Andalusia) would come from the disposal of CDW in inert landfills (6 million €) and 95 thousand € from hazardous waste disposal.

Table 10.3. Estimated revenue from the national waste disposal tax and from the suggested regional increases for hazardous waste and CDW landfilled in Andalusia

	Landfill type	Recoverable	Pre-treatment	Amount disposed of (t)	Tax increase revenue (€)	Spanish tax revenue (€)	Total revenue (€)
Hazardous waste	Landfill for Hazardous waste	Yes	Without		-	-	-
			With		-	-	-
		No	Without				
			With	6 174	43 218	49 392	92 610
	Landfill for non-Hazardous waste (asbestos)	Yes	Without		-	-	-
			With		-	-	-
		No	Without		-	-	-
			With	4 597	22 985	45 970	68 955
CDW	Landfill for inert waste		Without		-	-	-
			With	4 017 000	6 025 500	6 025 500	12 051 000
Total				4 027 771	6 091 703	6 120 862	12 212 565

Source: Own elaboration

10.1.6. Behavioural implications and distributional impacts

The proposed regional increase of the national tax rate on hazardous waste disposal has been calculated to maintain the same level of hazardous waste disposal in Andalusia. It is assumed that the hazardous waste generation sources would generate and dispose of the same amount, as they would bear the same costs as currently. Thus, no major changes are expected on taxpayer behaviour nor on the distribution impacts of the tax, both aspects would remain the same as currently.

Contrary, the national disposal tax on CDW disposal, together with the suggested regional increase of the corresponding tax rates, is expected to induce a change in the behaviour of CDW generators. The expected behavioural implications would be an increase on prevention and sorting efforts to reduce the amount of CDW disposed of in inert landfills. Even if the taxpayer would be the construction sector, it is expected to pass most of the tax costs to consumers (as in the case of the tax on aggregates discussed in section 5.2.5).

10.2. Regional Tax on Aggregates Extraction

Forty percent of the Spanish mining production value comes from Andalusia (MITERD, 2020[6]), including fuels, metallic minerals, industrial minerals, ornamental rocks and quarry products. Regarding quarry products, even if they are extracted in all ACs, Andalusia, Catalonia, Castile-Leon, Valencian Community and Aragon produce together around 60% of the Spanish production value. Worth noting that the province of Almeria (located in Andalusia) concentrates around 60% of the gypsum extracted in Spain (both in weight and in value) (MITERD, 2020[6]). Figure 10.1 summarises the amounts of aggregates extracted in Andalusia per type of material in 2019, as well as the market price in Andalusia (where available), neighbouring ACs, and in Spain in the same year. The most extracted aggregates in Andalusia are limestone, gypsum, sand and gravel, and dolomite.

The main motivation behind an environmental tax on aggregates would be to reduce the consumption on virgin aggregates in favour of recycled aggregates to incentivise their use in the construction sector. The White Book for Tax Reform also proposes a national tax on the extraction of aggregates with a tax rate equal to that of the UK Aggregates Levy of 2.35 €/t (2 pounds per tonne, 2021), with the aim to encourage the reduction of the consumption of aggregates and increase the use of recycled aggregates.

Even if metal mining represents around 60-80% of the mining production value in Andalusia (MITERD, 2020[6]), an environmental tax on metals extraction in Andalusia has not been analysed because of competitiveness concerns. As metals are traded in global markets, a regional tax on metals extraction could be detrimental for the local industry, if no carbon border adjustment mechanisms are implemented at the same time. More analysis would be needed in this respect.

Figure 10.1. Aggregates extraction in Andalusia and average price in Spain in 2019

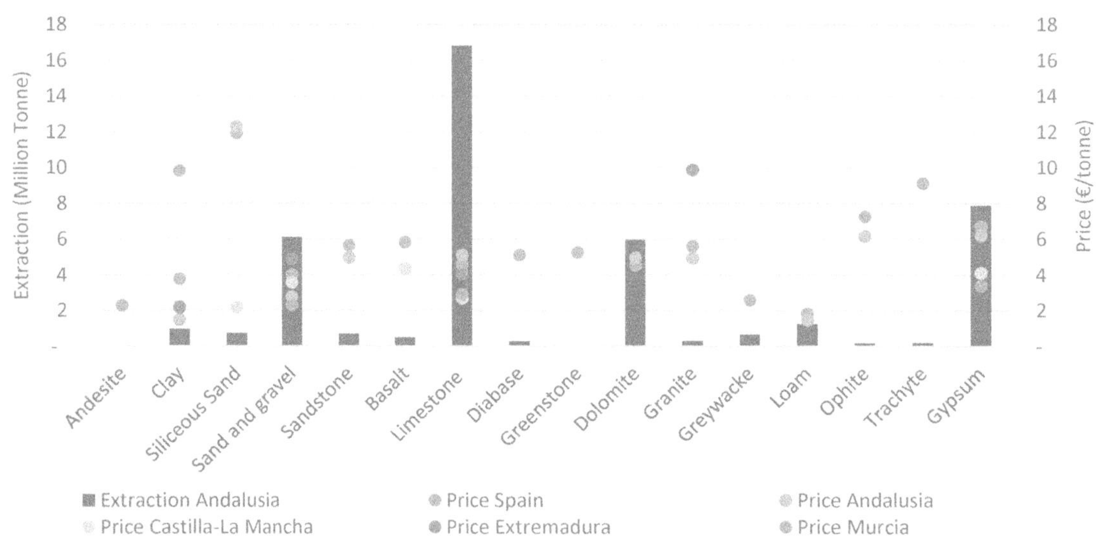

Source: Own elaboration based on MITERD (2020[6]).

StatLink https://stat.link/m8rylh

10.2.1. Tax definition

For the tax to have an environmental effect, the recommended **taxable event** is the affectation of ecosystem services and the environmental impact of the extractive activity of aggregates. Regarding the scope, it is recommended to tax all aggregates to avoid substituting one for another and thus favour the use of recycled construction and demolition materials.

The **tax base** depends on when the taxable event occurs (during consumption or extraction) and on the physical magnitude to be taxed (quantity extracted in weight or volume, affected area, affected ecosystem service, etc.).

A levy on the consumption of aggregates, even if they come from another territory, would prevent issues around the competitiveness of the local industry, but materials extracted but not consumed in Andalusia would not be subject to the tax. On the other hand, levying the extraction of aggregates in Andalusia would be a better option for the specific tax event, but it could incentivise imports, shifting part of the problem to other ACs.

Given that the sale of aggregates over long distances practically does not occur and therefore the risk of imports is minimal, it is recommended to tax extraction because its link with environmental impacts is clearer. In addition, a tax on production is easier to implement and to enforce than a tax on consumption.

The decision of the physical magnitude on which to apply the tax will depend mainly on the availability of data. If data is available, the recommended setting of the tax base would be a combination of the affected area and the extracted amount.

The tax base can be physical (e.g., quantity of product extracted) namely *ad quantum* tax, or it can be monetary (e.g., the sale price), namely *ad valorem* tax. Both options have their advantages and disadvantages (see Table 10.4). There is also the possibility of conceiving a tax that has mixed characteristics.

Table 10.4. Advantages and disadvantages of Ad quantum and Ad valorem taxes applied on aggregates

	Ad Quantum	Ad Valorem	Comment
Market impact	-	+	Ad valorem taxes have less economic impact on the market than ad quantum taxes since they cause less market distortion.
Revenue stability and predictability	+	-	Ad quantum taxes have more revenue predictability than ad valorem taxes.
Environmental impact representation	+	-	Ad quantum taxes represent better the environmental impact of the taxable event than ad valorem taxes.
Economic impact	-	+	Ad quantum taxes place a proportionally higher tax on cheaper products than ad valorem taxes.
Administrative characteristics	+	-	Ad quantum taxes are often simpler to administrate than ad valorem taxes.

Source: Own elaboration.

Regarding the tax rates, the economic theory does not rule unequivocally in relation to how to define them. On the one hand, environmental economics indicates that the tax rate should reflect the magnitude of the environmental externalities generated (measured in monetary terms). The estimation of ecosystem services can be a methodology to determine environmental costs and the latter can be considered to define the tax rate (*Pigouvia*n approach). The functioning of ecosystems provides, directly or indirectly, services to humans (Haines-Young and Potschin, 2012). These services can be for the provision of materials and resources, the regulation and support of the basic structure of the ecosystem (e.g., water regulation, climate regulation, erosion control or pollination), and cultural services (i.e., availability of natural spaces to develop activities). Environmental services generally do not have a market price and, therefore, do not have an associated monetary value, which is why a wide range of techniques have been developed to attribute a monetary value to them that allows the value to be compared with other goods or services that do have a market price. Different studies have estimated the environmental costs of aggregate extraction in different locations, see Table 10.5, but according to the authors' knowledge such studies have not been done for Andalusia and thus do not represent the Andalusian situation currently. An in-depth study to estimate environmental costs of extraction in Andalusia, which is outside of the scope of this study, would be required to consider this aspect in the tax rate.

Table 10.5. Summary of studies performing economic assessment of the environmental externalities of aggregate extraction

Year	Place	Quarry type	Externality value	Reference
2000	Aycliffe (UK)	Hard rock quarries	0,46 – 1,18 €/t	(Garrod and Willis, 2000[7]; Willis and Garrod, 1999[8])
1999	Yorkshire Deals and Peak District (UK)	Quarries of rock, gravel and sand	0,38 – 11,82 €/t rock, gravel and/or sand	(London Economics, 1999[9])
2003	Athens (Greece)	Abandoned marble quarry	0,88 €/m^2 – 92,44 €/m^2	(Damigos and Kaliampakos, 2003[10])

Source: Own elaboration.

On the other hand, ecological economics suggests that the desired volume of activity should be defined from outside the market and environmental taxes established to reduce activity to the desired levels. In the analysed case, with inelastic demand and without specific objectives for reducing activity, it can be difficult to put into practice this approach. In both cases, explicitly or implicitly, the tax rate should reflect the environmental impact of the activity and therefore can vary from one material to another and from one location to another. However, there are no studies available for Andalusia that indicate a differentiated environmental impact by type of material extracted. Therefore, a straightforward alternative would be to apply the same tax rate to all aggregates.

However, applying an *ad quantum* flat rate to all aggregates would represent a greater relative impact of the tax on cheaper materials. Differentiated rates based on groups of material prices could be preferable in terms of sector acceptability, i.e., lower tax rates for cheaper materials. Therefore, an ad quantum tax with differentiated rates based on 2-3 groups for aggregates of different values could be preferable, but this would complicate the management of the tax and would be difficult to apply. Another alternative would be to differentiate the tax rates not according to the type of aggregate, but according to the location of the extractive activity, singularly, that the rate be higher for activities located in protected natural spaces.

Earmarked taxes are recommended with tax revenues used to incentivise recycled aggregates and reduce extraction impacts. It is recommended that the fund has a certain compensatory effect on the loss of environmental services, for example, providing income for the environmental improvement of the municipalities closest to the extraction areas, dedicated to improving/creating recreation areas for the inhabitants of the area. Progressive and predictable tax rates are also recommended, i.e., the tax rate increases gradually, and the increments are known in advance by the affected agents so that they can adapt.

The effects on imports must be considered to define the tax rate. The possibility of importing aggregates directly influences the elasticity of local demand and, therefore, the effectiveness of the tax. The import of aggregates will depend mainly on two factors: availability in nearby areas, since aggregates are materials that are generally consumed less than 50 km away, and the cost of transportation from these nearby areas, which is relatively high in relation to the low price of the material.

However, it is important to highlight that some neighbouring ACs also extract important amounts of some of the same aggregates as Andalusia, e.g., Castilla-La Mancha extracts more sand and gravel than Andalusia (around 8 Mt in total), and the production of limestone in Castilla-La Mancha and Murcia together is around 2/3 of Andalusia's production (each AC produces around 5 Mt).

In general, the prices of aggregates in Andalusia (and other ACs) are low, up to 10 €/t, except for silicious sand with an average price of around 12 €/t. This means that the tax rate cannot be too high. It is observed that limestone, the most extracted aggregated in the Andalusia, is around 41% and 55% cheaper in the neighbouring ACs than in Andalusia. This must be considered when defining the tax, in order to avoid a tax rate that could favour imports.

The price difference between territories will determine the distance from which the transport of material will be profitable. Assuming that: 1) the transport of minerals is done with a truck with a maximum load of 24 useful tonnes, 2) with a cost of 1.42 €/km[3] (including fuel, vehicle depreciation, maintenance costs, personnel and industrial profit) (MITMA, 2021[11]), 3) the transport radius of 80% of the aggregates is a maximum of 50 km, 4) the truck runs at full capacity, the unit cost of transport would be 2.96 €/t. Tax rates higher than such amount would incentivise aggregates imports from border areas. The imposition of average tax rates lower than the calculated transport cost is recommended, to stimulate the substitution of raw materials by domestic recycled materials, rather than by imported raw materials from other regions. However, a large difference between the transport cost and the tax rate is not recommended since very low tax rates are not expected to incentivise reductions in aggregate extraction.

For the estimation of the tax implications two possible tax rates have been considered, both representing mean values of the taxes found in the EU (see Table 8.4 and Table 8.C.1): 1) An Ad Quantum Tax of 1.35 €/t, and 2) An Ad Valorem Tax of 10.00% of the aggregate price.

10.2.2. Price Elasticity of Aggregates

From a theoretical point of view, both the supply and the demand of aggregates in the short/mid-term are quite inelastic (European Environment Agency, 2008[12]), mainly because:

- The physical-chemical properties of all aggregates cannot be found in other materials (e.g., wood, synthetic materials) and therefore the only viable short-term substitute would be recycled aggregates for certain uses. But this option is only possible if recycled aggregates are found in the form and quantity required by the market. Further research is necessary to assess to available supply of secondary materials as substitute good.
- The low price of aggregates makes imports unprofitable due to transport costs.

It is likely, however, that each type of aggregate has a different elasticity, as their potential for substitution and their availability in territories close to Andalusia differ. Also there the tax design could benefit from a detailed study to look into price elasticity of different aggregates in Andalusia.

It is also important to consider that a disposal tax on CDW applied simultaneously to the extraction tax with a relatively high tax rate can incentivise the substitution of taxed raw materials with recycled material instead of importing materials from other regions and thus increase the elasticity. With the entry in force of the fiscal measures included in the Spanish Law 7/2022 on Waste and Contaminated Soils for a Circular Economy, there will be a national tax on disposal of CDW. However, the national tax rate included is low (3 €/t), therefore only a limited substitution of raw materials for recycled materials is expected.

As detailed data on empirical aggregates elasticities in Andalusia are not available, the implications of the tax have been estimated for two scenarios with different elasticities for all aggregates: 1) Price Elasticity of 0%, and 2) Price Elasticity of 10%.

10.2.3. Environmental implications

Aggregate exploitations and extraction has environmental impacts, such as soil degradation, damage to ecosystem functions and air pollution from fine particles, as well as greenhouse gas emissions from energy use. During the exploration and extraction phases, the use of explosives and/or heavy machinery is common to break the bedrock, with the consequent generation of dust, gases, and noise. Depending on the location of the activity and given that the subsoil is drilled, there may be cases of groundwater contamination and severe disruptions of ecosystem functions. In cases where the activity is carried out in mountainous massifs or open pit mines, it can also cause significant alterations to the landscape. In addition, the generation of waste that the activity entails must be considered, since during the extraction process sludge, dust and other non-useful materials may be generated. In addition, aggregates are a non-renewable resource and therefore, extraction will tend to be increasingly costly, economically and environmentally.

The main environmental implication of the tax on aggregate extraction would be a decrease in aggregates demand in favour of recycled aggregates or alternative products. Table 10.6 summarises the potential demand variation obtained in the four tax scenarios assessed. No demand variation is expected when price inelasticity is assumed. In Sweden, the tax on aggregates found an increase in elasticity due to the substitution effect of other types of materials, in the UK the elasticity increased due to the combined effect of the CDW disposal tax and the increase in other types of recycled materials and aggregates (Söderholm, 2011[13]). Conversely, when a 10% elasticity is assumed, for the tax rates stated above (1.35€/t and 10% of the price), the reduced demand associated with the *ad quantum* tax was more than three times larger than the one associated with the *ad valorem* tax.

According to European Environment Agency (2008[12]), earmarking of the revenue can help reinforce the impact of the tax if specific market failures are addressed and the revenue is used to improve environmental outcomes. For example, the United Kingdom used a proportion of the tax revenue to develop quality standards for recycled aggregates, which gave companies confidence in purchasing and using these materials.

An indirect environmental impact of the aggregate tax would be the decrease of C&DW landfilling when the decrease in demand of aggregates is compensated with an increase in the demand of recycled aggregates. These impacts would be likelier and larger if the aggregate tax would be accompanied with an increased tax rate on C&DW disposal (compatible with the Spanish Law 7/20220 on Waste and Contaminated Soils for a Circular Economy), as has been seen in Denmark (section 3.1.2). As earmarked tax, other environmental impacts related to the tax would come from the use of the revenue. Most of the earmarked mining taxes applied in Europe are used to restore old mines and regenerate ecosystems.

Table 10.6. Aggregates' Demand Variation in tonnes in the scenarios with 10% Demand Elasticity.

	Scenario Ad Quantum and 10% Demand Elasticity	Scenario Ad Valorem and 10% Demand Elasticity
Andesite	-88.78	-15
Clay	-85 878.75	-9 440
Siliceous Sand	-7 922.01	-7 224
Sand and gravel	-296 622.17	-61 096
Sandstone	-18 573.49	-6 837
Basalt	-11 564.43	-5 000
Limestone	-443 294.82	-168 231
Diabase	-7 351.68	-2 787
Greenstone	-744.51	-290
Dolomite	-161 986.56	-59 688
Granite	-7 908.74	-2 893
Greywacke	-34 110.17	-6 556
Loam	-113 596.85	-12 374
Ophite	-3 681.29	-1 683
Trachyte	-2 866.12	-1 935
Gypsum	-170 887.69	-78 698
Total	-1 367 078.00	-424 747

Note: Ad Quantum tax of 1.35 €/t, Ad Valorem tax of 10% of the market price, Demand elasticity of 10%.
Source: Own elaboration.

10.2.4. Economic implications

As presented in Table 10.7, the potential revenue of the *ad quantum* tax is more than double the potential revenue of the *ad valorem* tax. For the 4 scenarios, limestone is the aggregate contributing the most to the total revenue of the tax (40-41% of the total revenue), followed by gypsum (19-24% of the total revenue depending on the scenario), dolomite (14%), and sand and gravel (8-14% depending on the scenario).

In the *ad valorem* tax, the levy would represent 10% of the market price for all the aggregates, but in the case of the *ad quantum tax*, the tax represents up to 91% of the price of the cheapest aggregates (e.g. clay and loam) and only 11% for the most expensive aggregated, the siliceous sand, see Table 8.G.1. The same occurs with the variation of the demand, the most affected aggregates by the *ad quantum tax* are the most economical ones. Contrarily, the demand of all aggregates decreased by 0.13% with the *ad valorem tax.*

The total revenue expected with the *ad quantum* tax would represent around 28% of the aggregate sector turnover[4] in Andalusia while the total *ad valorem* tax revenue would represent 10% of the turnover.

Table 10.7. Aggregates' Tax Revenue in € per Scenario.

	Scenario Ad Quantum and 0% Demand Elasticity	Scenario Ad Quantum and 10% Demand Elasticity	Scenario Ad Valorem and 0% Demand Elasticity	Scenario Ad Valorem and 10% Demand Elasticity
Andesite	2 025	1 905	342	339
Clay	1 274 372	1 158 435	140 079	138 678
Siliceous Sand	975 258	964 563	889 344	880 450
Sand and gravel	8 247 984	7 847 544	1 698 863	1 681 874
Sandstone	922 929	897 855	339 711	336 314
Basalt	675 000	659 388	291 843	288 925
Limestone	22 711 127	22 112 679	8 618 882	8 532 693
Diabase	376 299	366 374	142 674	141 248
Greenstone	39 150	38 145	15 250	15 097
Dolomite	8 057 892	7 839 210	2 969 136	2 939 445
Granite	390 577	379 900	142 880	141 451
Greywacke	885 098	839 049	170 124	168 422
Loam	1 670 428	1 517 072	181 951	180 132
Ophite	227 216	222 246	103 883	102 844
Trachyte	261 267	257 398	176 417	174 653
Gypsum	10 624 285	10 393 587	4 892 771	4 843 843
Total	57 340 906	55 495 350	20 774 150	20 566 408

Note: *Ad quantum* tax of 1.35 €/t, *Ad valorem* tax of 10% of the market price
Source: Own elaboration.

10.2.5. Behavioural implications and distributional impacts

Several factors should be considered to achieve an appropriate level of **tax compliance** by the aggregates sector, and thus to prevent tax evasion. According to Harford (1978[14]), Macho-Stadler and Pérez-Castrillo (2004[15]) and Bontems and Bourgeon (2005[16]) polluting firms tend to evade environmental taxes, unless a high probability of audit is established. Harford (1978[14]) also concluded that tax evasion is more likely for larger tax rates. Alm (2011[17]) studied evasion of taxes, without focusing on environmental taxes, and found that fines and possible audits reduce tax evasion, but effects of tax rate on compliance was unclear. Alm acknowledged that rewards could be more effective than punishment to increase compliance. It is also known that earmarked and progressive taxes are better accepted in general and thus tax compliance would be likelier under these conditions.

Regarding the **response of the sector** with respect to the tax implementation, the aggregates demand could:

1. keep constant, but the sector (suppliers and consumers) would cover part of the environmental externalities of the extraction,
2. be reduced in favour of an increased demand of alternative raw materials (e.g., woods), but this substitution is only feasible in certain cases,
3. be reduced in favour of an increased demand of recycled products, but this substitution is only possible if the recycled aggregate market can supply quantities and qualities requested. According to the Spanish National Association of Entrepreneurs Manufacturers of Aggregates (ANEFA), 184.7 million tons of aggregates were produced in Spain in 2019, of which only 1.4% were recycled aggregates (ANEFA, 2022[18]).

To know who will ultimately bear the tax burden and to what extent, i.e., **the distributional impacts of the tax**, the relative price-elasticity of supply and demand of aggregates should be considered. When demand is more elastic than supply, producers bear most of the tax cost, and the opposite occurs when supply is more elastic than demand (buyers bear most of the tax burden)

Although the relative price-elasticity of supply and demand of aggregates in Andalusia is not available, supply is supposed to be more elastic than demand because: 1) the extraction rate can be, to some extent, adjusted, 2) high transportation costs, 3) few aggregates have substitution options.

Assuming such relative price-elasticities (i.e., demand more inelastic than supply), then most of the tax would be passed to the consumer and thus the aggregates industry in Andalusia would not be much affected by such a tax.

References

Alm, J. (2011), “Measuring, explaining, and controlling tax evasion: lessons from theory, experiments, and field studies”, *International Tax and Public Finance*, Vol. 19/1, pp. 54-77, https://doi.org/10.1007/s10797-011-9171-2. [17]

ANEFA (2022), *El sector de los áridos en 2019 y perspectivas 2020*, https://www.aridos.org/el-sector-de-los-aridos-en-2019-y-perspectivas-2020/ (accessed on 18 January 2023). [18]

Bontems, P. and J. Bourgeon (2005), “Optimal Environmental Taxation and Enforcement Policy”. [16]

Damigos, D. and D. Kaliampakos (2003), “Environmental Economics and the Mining Industry: Monetary benefits of an abandoned quarry rehabilitation in Greece”, *Environmental Geology*, Vol. 44/3, pp. 356-362, https://doi.org/10.1007/s00254-003-0774-5. [10]

Dechezleprêtre, A. et al. (2022), “Fighting climate change: International attitudes toward climate policies”, *OECD Economics Department Working Papers*, No. 1714, OECD Publishing, Paris, https://doi.org/10.1787/3406f29a-en. [2]

EEBA (2021), *The Environmental Impact of Building Materials*, https://www.eeba.org/the-environmental-impact-of-building-materials (accessed on 18 January 2023). [5]

European Environment Agency (2008), *Effectiveness of environmental taxes and charges for managing sand, gravel and rock extraction in selected EU countries*, https://www.eea.europa.eu/publications/eea_report_2008_2/ (accessed on 18 January 2023). [12]

European Parliament (2022), *Parliamentary question | Nerva landfill (Huelva) | E-000945/2022*, https://www.europarl.europa.eu/doceo/document/E-9-2022-000945_EN.html (accessed on 18 January 2023). [4]

Garrod, G. and K. Willis (2000), “Economic approaches to valuing the environmental costs and benefits of mineral and aggregate extraction”, *Minerals & Energy - Raw Materials Report*, Vol. 15/4, pp. 12-20, https://doi.org/10.1080/14041040009362569. [7]

Harford, J., Harford and J. D. (1978), “Firm behavior under imperfectly enforceable pollution standards and taxes”, *Journal of Environmental Economics and Management*, Vol. 5/1, pp. 26-43, https://doi.org/10.1007/s00028-003-0095-x. [14]

Junta de Andalucía (2021), *Plan Integral de Residuos de Andalucía. Hacia una Economía Circular en el Horizonte 2030 (PIRec 2030)*, https://www.juntadeandalucia.es/medioambiente/portal/landing-page-planificacion/-/asset_publisher/Jw7AHImcvbx0/content/plan-integral-de-residuos-de-andaluc-c3-ada/20151 (accessed on 18 January 2023). [1]

Kallbekken, S., S. Kroll and T. Cherry (2011), “Do you not like Pigou, or do you not understand him? Tax aversion and revenue recycling in the lab”, *Journal of Environmental Economics and Management*, Vol. 62/1, pp. 53-64, https://doi.org/10.1016/j.jeem.2010.10.006. [3]

London Economics (1999), "The environmental costs and benefits of the supply of aggregates : phase 2", p. 208. [9]

Macho-Stadler, I. and D. Pérez-Castrillo (2004), "Optimal Enforcement Policy and Firms' Emissions and Compliance with Environmental Taxes", http://www.CESifo.de (accessed on 18 January 2023). [15]

MITERD (2020), "Estadística Minera de España 2019", http://publicacionesoficiales.boe.es/ (accessed on 18 January 2023). [6]

MITMA (2021), *Observatorio de costes del transporte de mercancías*, https://www.mitma.gob.es/transporte-terrestre/servicios-al-transportista/observatorios-del-transporte/observatorios-del-transporte-de-mercancias-por-carretera/observatorios-costes-transporte-mercancias (accessed on 18 January 2023). [11]

Söderholm, P. (2011), "Taxing virgin natural resources: Lessons from aggregates taxation in Europe", *Resources, Conservation and Recycling*, Vol. 55/11, pp. 911-922, https://doi.org/10.1016/j.resconrec.2011.05.011. [13]

Willis, K. and G. Garrod (1999), "Externalities from extraction of aggregates", *Resources Policy*, Vol. 25/2, pp. 77-86, https://doi.org/10.1016/s0301-4207(99)00012-4. [8]

Notes

[1] The regulations on waste allow asbestos residues to be disposed of in non-hazardous waste landfills, without prior testing, provided that the requirements of Annex II of Royal Decree 646/2020, of July 7, which regulates the disposal of waste by landfill, in accordance with article 7 of the same Royal Decree, are met.

[2] https://www.juntadeandalucia.es/medioambiente/portal/documents/20151/371576/gestion_tratamiento_residuos_RCD_buenas_practicas.pdf/305bc319-6265-0369-4f32-05bb0961fda6?t=1606380697444

[3] This value applied before the oil price surge in the context of the Ukraine war.

[4] The turnover has been calculated with the extracted amounts and the material price of 2019 from (MITERD, 2020[6]).

11 Policy recommendations

Considering the Andalusian context (main economic sectors, current environmental taxes, and current waste sectors) and a stocktake assessment of economic instruments used to incentivise circular economy (at EU, national and regional level), this chapter identified two environmental taxes with the potential to incentivise circularity in Andalusia while raising revenue: 1) an increase in the tax rates for two subcategories of the national waste disposal tax and 2) an aggregates extraction tax. These taxes would address the circularity of different economic sectors while increasing waste prevention and improving waste source separation and recycling rates. In addition, such taxes would help to operationalise the Polluter Pays Principle in the region.

The recently published Spanish Law 7/2022 on Waste and Contaminated Soils for a Circular Economy has reconfirmed the interest of these two proposals. Firstly, the regional increase of the national tax rate on hazardous waste would be necessary in the region to prevent an increase of hazardous waste disposal, as the amount of hazardous waste treated and/or disposed of by Andalusia is already around two and a half times the hazardous waste than generated in the region. Additionally, the regional increase of the national tax rate on CDW disposal, together with an aggregates extraction tax, could increase the circularity and prevention in the construction sector, which is one of the sectors contributing the most to global warming.

Looking at economic implications of each instrument, the highest revenue would be obtained with the aggregates extraction tax whose revenue could range from 20 to 57 million EUR. The tax rate increase on CDW disposal is estimated to generate a potential revenue of 6 million EUR. Finally, the instrument with the lowest potential revenue would be the regional increase of the national tax rate on hazardous waste disposal. In addition to the environmental advantages and revenues derived from the two regional taxes proposed, there might also be a strategic argument as occupying these tax bases on the level of the AC may make the Junta eligible for compensation by the national government if national taxes were to be introduced for these tax bases.

While different economic sectors would be affected by these instruments – namely, the building sector, the industry generating hazardous waste and households generating general waste –, consumers are expected to bear most of the tax costs in the three instruments.

This chapter has estimated the implications of each of the instruments using a ceteris paribus approach, thus results should be taken as approximations. More detailed studies would be needed to get more accurate results. A more detailed study could also consider alternative tax designs, such as a progressive tax that would ease the administrative and/or financial burden on small and medium enterprises. Nevertheless, the adoption of the suggested instruments in the proposed form or slightly modified would certainly provide a better incentive structure for moving towards more resource efficiency and material circularity.

Table 11.1. Key aspects for tax instruments

Tax instrument	Description	Tax rate	Expected revenue	Comments (environmental, economic and behavioural implications)
Waste Disposal Tax	Increase of the national tax rates on landfilling of hazardous waste and CDW located in Andalusia.	• For hazardous waste, increase the basic national tax rate to make it equal to the current Andalusian rate (from EUR 8 to 35 per tonne for waste without pre-treatment required and EUR 5 to 15 per tonne for the rest of the waste). • For CDW, the tax rate could be increased to 5 €/t disposed of in landfills for inert waste without pre-treatment and 3 €/t with pre-treatment.	• Hazardous waste without pre-treatment: EUR 92 610 • Hazardous waste with pre-treatment: EUR 68 955 • CDW: EUR 12 051 000 • Total revenue: EUR 12 212 565	• Increases the circularity of waste: incentivises the use of recycled CDW that would prevent the use of primary building materials. • The revenue associated to the national tax rate increase would come mainly from the disposal of CDW. Only 1% would come from the disposal of hazardous waste. • The proposed regional increase of the national tax rate on hazardous waste disposal has been calculated to maintain the same level of the existing tax on hazardous waste disposal in Andalusia, so no major changes are expected. Contrarily, the national disposal tax on CDW disposal, and its suggested regional increase, is expected to induce a change in the behaviour of CDW generators.
Regional Tax on Aggregates Extraction	• Setting a regional tax on aggregates extraction that could incentivise reuse and material recovery of construction and demolition waste and could generate revenue to mitigate some of the impacts of extractive activities. • It is recommended to tax all aggregates to avoid substituting one for another and thus favour the use of recycled construction and demolition materials.	• An Ad Quantum Tax: 1.35 €/t • An Ad Valorem Tax: 10.00% of the aggregate price.	• Scenario Ad Quantum and 0% demand elasticity: EUR 57 340 906 • Scenario Ad Quantum and 10% demand elasticity: EUR 55 495 350 • Scenario Ad Valorem and 0% demand elasticity: EUR 20 774 150 • Scenario Ad Valorem and 10% demand elasticity: EUR 20 566 408	• An environmental impact of the aggregate tax would be the decrease of CDW landfilling when the decrease in demand of aggregates is compensated with an increase in the demand of recycled aggregates. • An additional impact would be reduced extraction of aggregates. • For the 4 scenarios, limestone is the aggregate contributing the most to the total revenue of the tax (40-41%) • Assuming relative price-elasticities (i.e., demand more inelastic than supply), most of the tax would be passed to the consumer and thus the competitiveness of the aggregates industry in Andalusia would not be much affected by such a tax.

Part V The taxation of tourist stays

12 The taxation of tourism activities in Andalusia

While tourist taxes exist in several European countries, there is currently no specific tax on tourists in Andalusia The tourism industry is one of the major economic sectors in the Andalusian economy, contributing to 13% of the region's GDP and to 14% of employment (Villegas, Del Carmen Delgado and Cardenete, 2022[1]). There currently is no specific tax on tourist stays in the region. Taxes on tourist stays exist in several European countries at the regional, local or city-level, such as in Switzerland, or in the cities of Amsterdam or Lisbon (Responsible Travel, 2022[2]). In Spain, the Autonomous Communities of Catalonia and the Balearic Islands have established tourist taxes (see Annex 12.A). In most of the cases, such taxes cannot be considered environmental taxes as their main goal is not to account for the environmental impact of tourism, but rather to raise funds for municipal expenses. Typically, these taxes do not incorporate environmental considerations explicitly but are fixed charges per night ranging from EUR 0.50 to EUR 5 or charged as a percentage of the price of the accommodation (up to 7% in Amsterdam). Whilst several tourist taxes use revenues to relieve some of the (environmental) pressures caused by tourism, there are few cases in Europe where the tax rate itself varies with explicit environmental criteria. This is the case of the Balearic Islands Tourist Tax (see Box 12.1 and Annex 12.A).

This chapter lines out few dimensions that may need to be considered in reflections on tourist taxes, without providing concrete recommendations on their implementation or design. Additional research on the actual extent of the problem and potential policy approaches to resolve them, reflections and discussions would need to be conducted (e.g. on the impact of specific design and management features), which is not in the focus of this report.

12.1. Setting the Scene

Spain is one of the most popular destinations for international tourists in the world. In 2019, Spain was ranked among the top foreign destinations for EU residents in terms of number of trips (Figure 12.1) as well as number of nights spent or expenditure.

Figure 12.1. Destinations for outbound trips made by EU residents within the EU (Nights spent in 2019)

Source: Eurostat (2022)

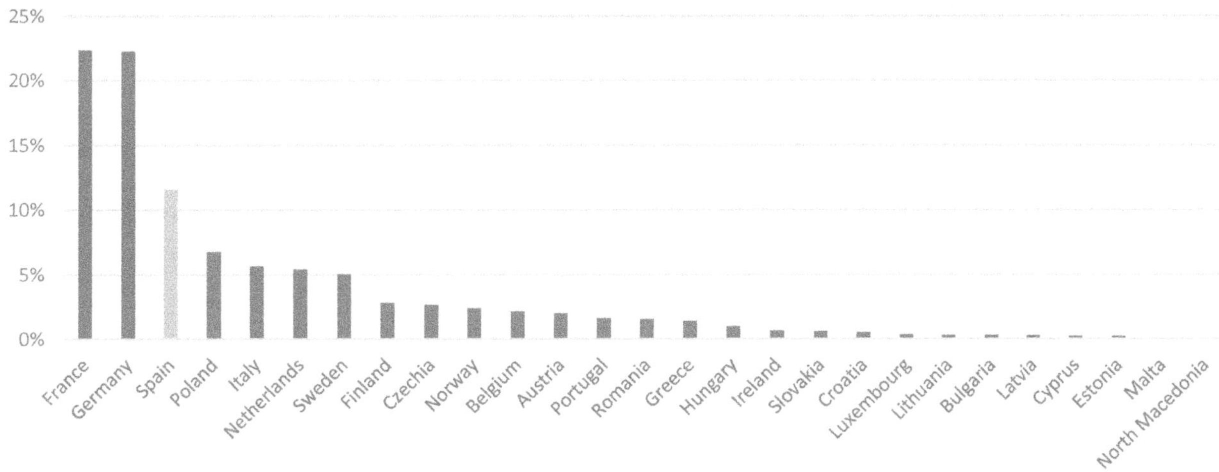

Source: Eurostat (2022[3]).

StatLink https://stat.link/7a8ejs

Within Spain, Andalusia accounted for approximately 14% of all visits in 2021 (Instituto Nacional de Estadistica, 2022[4]). Prior to the COVID-19 pandemic, Andalusia received between 11 and 12 million visitors annually. In 2021, Andalusia received approximately 4.2 million visitors, though as pandemic related restrictions are being lifted, visitor numbers are expected to increase again to near pre-pandemic levels. Recent low levels of tourism and the stress it exerted on the sector may affect support for the introduction of a new tourist tax.

Several economic arguments can justify the taxation of tourism, such as external cost management, revenue raising or rent extraction. Tourism activities involve different types of external costs, e.g. pollution, noise and congestion that may not be reflected in prices, they may also raise the costs of living. For example, extensive tourism activity in a region puts additional stress on ecosystems such as national parcs and beaches due to visitor increases and littering. Tourism also affects infrastructure needs: the seasonal inflow of tourists requires investments in road, water and waste infrastructure etc. to be designed to carry a larger capacity than required for local residents only. This infrastructure is however typically financed from funds to which temporary visitors do not necessarily or only partially contribute. Finally, in cases where the tourism sector receives preferential tax treatment (e.g. exemptions from waste taxes or water charges, preferential treatment through the VAT or CIT), equity concerns can arise between the treatment of local residents and visitors. This section provides an initial reflection on ways how the tax system in Andalusia could account for such impacts.

12.2. Dimensions to consider when reflecting on a tax on tourist stays

Andalusia currently does not levy a tax on tourist stays, as is the case in other autonomous communities such as Catalonia and the Balearic Islands. Importantly, if the objective of a tourist tax is to internalise external costs, "double taxation" should be avoided. For example, in instances where environmental impacts or other external costs by tourists are already sufficiently priced through consumption-based taxes (such as an air pollution tax, a carbon tax, or levies on road use, waste and water), an additional tourist tax on these external costs would involve double taxation. It is advisable to include tourism in such general consumption-based taxes and use and design a tourist tax only to account for additional external costs that cannot be covered through reforms of existing taxes or levies.

12.2.1. Removing preferential treatment in existing taxes or levies granted to the tourism sector

Equity and economic efficiency considerations call for a removal of preferential treatment of tourism related to external costs. Tourists use infrastructure, consume resources and energy, generate waste and emit pollutants to water and air just like local inhabitants. In a Pigouvian tax scenario, impacts caused by the consumption through tourism activity would thus ideally be charged at equal tax rates than impacts caused by locals. For some consumption-based taxes and levies, however, the tourist sector is currently taxed at lower rates.

For water consumption, hotels currently face a lower water abstraction charge compared to local households. While both tourists, as part of non-residential urban use, and local residents pay the improvement fee (currently temporarily suspended from 1 January to 31 December 2023), non-residential users do not face a fixed charge as locals do and the variable charge is not progressive (see Part III). Municipal fees also apply, but these are city-dependent and can vary significantly (Arbués and García-Valiñas, 2020[5]). This unequal treatment across users is even more important when considering that tourists tend to have a higher per-capita water consumption due to regular cleaning and towel washing in hotels and the existence of private pools in accommodation. Moreover, contrary to the increasing rate of the variable charge faced by local residents, the fixed rate of the variable charge for hotels does not make it more expensive to use water above certain thresholds. This in turn, does not include a sustainable use criterion in water pricing for the tourism industry.

In the context of vehicle taxes, rental cars are currently exempt from the national registration tax which results in locals paying a different price for using a car compared to tourists and tourist car drivers not contributing to costs associated with the construction and maintenance and public streets and highways.

A differentiation needs to be made between consumption of formal services of tourists, such as formal hotel and car hires, and consumption of services provided by private households, such as peer-to-peer lodging services or car rentals or other services provided through a sharing economy. Services of the latter are not subject to preferential treatment, as the owner, usually a private resident, is being charged the residential use rate, whereas formal services receive abovementioned exemptions.

Where the removal of preferential treatment is difficult, a tourist tax per night that aligns with a rough approximation of these consumption-based costs (e.g. for water consumption or waste generation) may be a second-best choice. Such a tax could be levied by the accommodation and could vary at the accommodation level, which would require reviewing the water and waste profile of the accommodations and apply higher rates for those with more harmful effects on the environment or to apply approximations for this. A regulation on the resource-efficiency of tourist accommodation may be another option to tackle these points.

To what extent a tourist tax would be passed on to tourists and what behavioural reactions to expect would need to be the subject of a more detailed study as it is not the focus of this report. Using tax and pricing policies jointly with non-pricing instruments (such as educational campaigns, nudging, encouraging the adoption of resource efficient technologies) could further improve the environmental performance of the tourism sector. In this respect, informing tourists about their water and energy consumption during their stay (and related costs) could be an interesting strategy to explore as a complement and/or alternative policy.

Box 12.1. The Balearic Islands Tourist Tax

The Balearic Islands Tourist Tax applies since 2016 with a tax rate varying from EUR 1 to EUR 4 per day, depending on the type of accommodation. More luxurious lodgings are being charged a higher rate as a proxy for a higher environmental footprint. Tax revenues are earmarked and dedicated to promoting sustainable tourism and mitigating environmental impacts associated with tourism.

Insights for Andalusia

A differentiated tax allows to account for differences in environmental footprints of different tourist accommodations. However, a more direct association with the specific environmental footprint could strengthen the incentive of lodgings to reduce their impact. For instance, in the Balearic case, a 4-star hotel does not have an incentive to implement water efficiency or waste reduction measures, because even if it did, its fixed charge would remain at EUR 4.

Note: See Annex 12.A for further details on the tourist tax in the Balearic Islands.

12.2.2. Accounting for infrastructure investment from tourism activity that exceeds needs of residents

Seasonality and fluctuation in tourist and visitor numbers to a destination is a common phenomenon linked to tourism activity. Seasonal fluctuation is well visible in the number of visitors that Andalusia is receiving, with monthly visitors being nearly three times as high in the summer months compared to the winter months (Figure 12.2). The net fluctuation in the number of visitors amounts to nearly 1 million tourists (compared to 8.4 million inhabitants in Andalusia). In certain smaller cities and villages, in particular along coastal areas, the seasonal fluctuation may be much greater than the Andalusian average shown in these figures (Cisneros-Martínez and Fernández-Morales, 2013[6]).

Figure 12.2. Number of tourist visits in Andalusia (in millions)

Panel A: Monthly

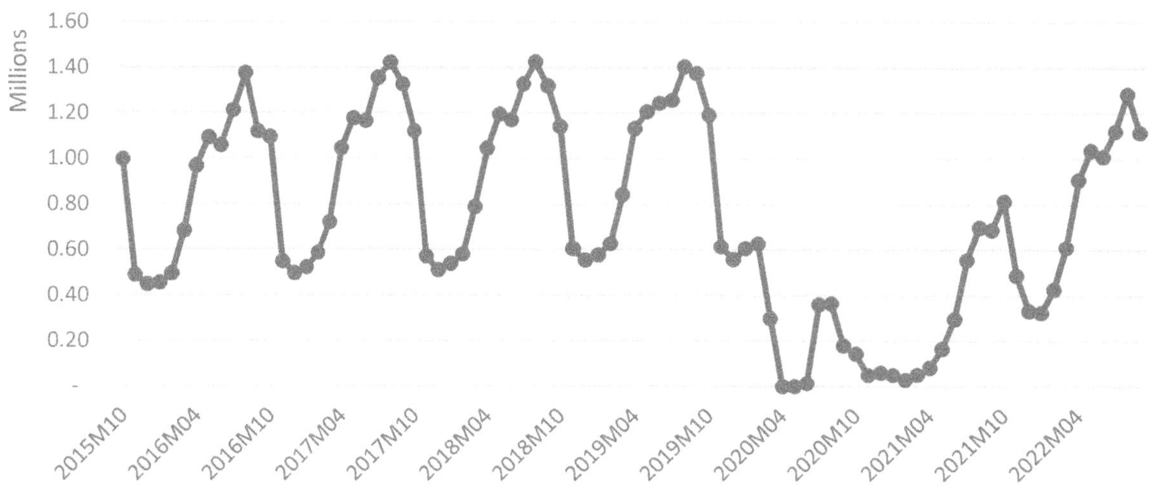

Panel B: By quarter

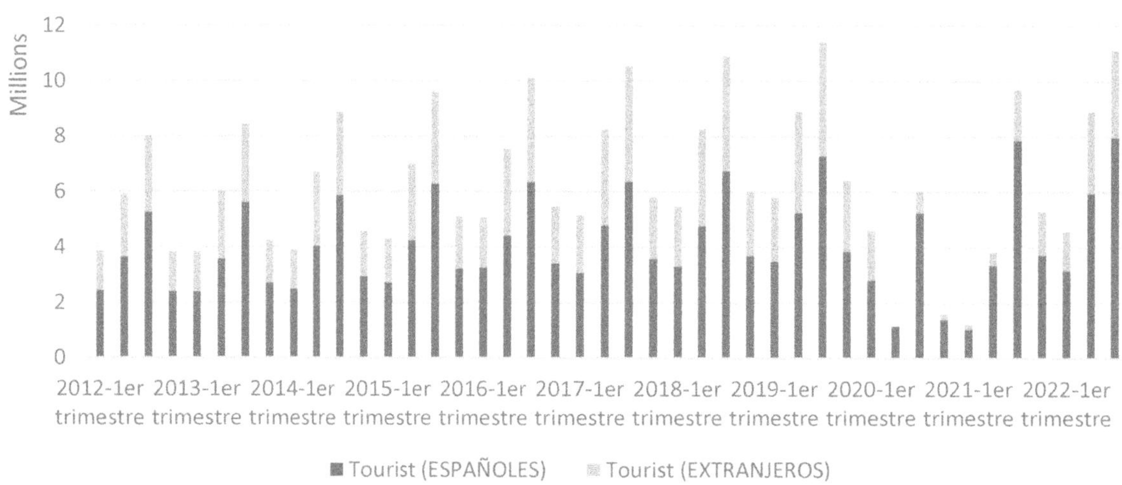

Source: Panel A: Instituto Nacional de Estadística (2022[4]); Panel B: Junta de Andalucía.

StatLink https://stat.link/jkygov

Destinations with high fluctuations in seasonality face various challenges. One main issue caused by tourism seasonality is the need to provide infrastructure capacity, such as roads, housing, water systems and sewage pipes and waste services, for the maximum capacity of users in peak-season, including regular inhabitants and tourists. When local residents pay for the additional infrastructure needs due to seasonal fluctuation of tourists (e.g. because the general budget finances the construction and maintenance of this additional infrastructure capacity) a fee on tourists may help alleviate equity concerns. A tourist charge could account for the costs of constructing and maintaining additional infrastructure capacity, which is only used during tourism peaks.

The under-use of tourism facilities during low-season and their maximum occupancy at full capacity during high-season, creates various additional challenges for the local economy, such as, among others, overcrowding, high prices, environmental degradation, a lack of services and job opportunities in shoulder and low seasons. Some tourism goods and services are more elastic than others to adjust for these fluctuations. For the inelastic goods and services, seasonality can be a significant external factor, which negatively affects the private sector, but also public services and environmental amenities.

For example, in the case of wastewater treatment, infrastructure must be able to adapt to the high variation of wastewater flow. To be able to accommodate larger effluent flows in peak-season, larger capacities for sewer and treatment facilities must be built. Whilst these additional treatment capacities remain idle for prolonged periods during low season, their construction and maintenance in low-season times, creates costs for wastewater treatment services, which can be directly linked to tourism seasonality. These additional costs, however, are usually integrated in and distributed across general wastewater treatment charges and thus borne by tourists, as well as year-round local inhabitants. Similar to wastewater treatment infrastructure, capacity for some waste services may also be required to accommodate demand changes during seasonal peaks. This would include additional kerbside pick-up trucks, as well as cleaning machines for sidewalks and beaches, which would be required to handle increased waste generation and littering during peak tourist seasons.

12.2.3. Policy recommendations

The tourism industry is one of the major sectors in the Andalusian economy, but generates external environmental costs and places pressure on local infrastructure. While tourism makes important contributions to the regional economy, it can weigh on resources and infrastructure and can thereby cause environmental and economic impacts. At the same time preferential tax treatments exist for tourists, such as reduced water abstraction charges or exclusions from national-level vehicle taxes for rental cars. Where the impacts of tourism are not priced appropriately through fiscal measures, there is a risk that the external and infrastructure costs are ultimately born by local inhabitants. The Junta may want to consider tools to align pricing with the external costs of the tourism sector, where possible.

Andalusia could adjust existing taxes to align with the external costs generated by the tourism sector and could consider additional tools to internalise costs and fund infrastructure. To internalise external environmental costs, Andalusia could reform existing taxes on use or consumption, such as the water improvement levy, which currently provides for a preferential tax rate for hotels. Mobilising existing taxes helps avoid double taxation, where costs associated with tourism would potentially be priced through both consumption-based taxes and specific tourist taxes. However, as Andalusia does not have direct control over certain taxes (e.g. the national vehicle registration tax), the scope for action at the regional level will be limited in some policy areas. In addition, existing taxes may be insufficient to fully cover costs, such as the additional infrastructure capacity needed during tourism peaks. In these cases, Andalusia could explore the introduction of a tourist tax. Two considerations are therefore proposed in the following order:

1. **Consider reducing preferential treatment in existing levies granted to the tourism sector.** Due to preferential tax treatment, some environmental costs (e.g. water consumption, road use) are only partially priced for tourists. The Junta may consider fully or partially repealing exemptions that apply to the tourism sector. Other reforms, such as aligning water use charges across residential and non-residential users, would indirectly reduce the preferential tax treatment that applies to the tourism sector.
2. **Explore alternative options to fund additional infrastructure investment and maintenance and to internalise the external costs associated with tourism.** While there is a strong seasonality and fluctuation of tourism activity over the year, infrastructure such as wastewater pipes needs to be built to carry the capacity of peak months and requires maintenance during off-peak months. Andalusia could consider whether a tourist tax would be appropriate to ensure tourists contribute to the costs of constructing and maintaining this additional infrastructure capacity. Additionally, where reducing the preferential tax treatment of the tourism sector is not feasible, a tourist tax would be a second-best option to internalise the externalities associated with tourism.

References

Arbués, F. and M. García-Valiñas (2020), *Water Tariffs in Spain*, Oxford University Press, https://doi.org/10.1093/acrefore/9780190632366.013.246. [5]

Cisneros-Martínez, J. and A. Fernández-Morales (2013), "Cultural tourism as tourist segment for reducing seasonality in a coastal area: the case study of Andalusia", *Current Issues in Tourism*, Vol. 18/8, pp. 765-784, https://doi.org/10.1080/13683500.2013.861810. [6]

Eurostat (2022), *Tourism statistics - top destinations*, https://ec.europa.eu/eurostat/statistics-explained/index.php?title=Tourism_statistics_-_top_destinations#Spain_was_the_top_foreign_destination_for_EU_residents_in_2019 (accessed on 30 November 2022). [3]

Instituto Nacional de Estadistica (2022), *Número de turistas según comunidad autónoma de destino principal*, https://www.ine.es/jaxiT3/Datos.htm?t=10823#!tabs-tabla (accessed on 30 November 2022). [4]

Responsible Travel (2022), *Tourist taxes map*, https://www.responsibletravel.com/copy/tourist-taxes-map (accessed on 29 November 2022). [2]

Villegas, P., M. Del Carmen Delgado and M. Cardenete (2022), "The economic impact of a tourist tax in Andalusia examined through a price effect model", *Applied Economics Letters*, pp. 1-6, https://doi.org/10.1080/13504851.2022.2128167. [1]

Annex 12.A. Detailed case studies: Balearic Islands Tourist Tax

The tourist tax in the Balearic Islands applies since 2016 as defined in Law 2/2016 Annex Table 12.A.1. The tax rate varies from EUR 1 to EUR 4 per day and depends on the type of accommodation (Annex Table 12.A.2). The tax is earmarked and funds are used to incentivise sustainable tourism. Since 2016, the revenue has been used to carry out 163 projects, equivalent to EUR 258 million. Most of the funds have been dedicated to environmental projects (42%), followed by social renting projects (20%), sustainable tourism (17%), research and development (11%), education and labour (7%) and historical heritage (3%). Most of the projects within the environmental area focussed on biodiversity, but some were also on water treatment plants and waste management (Govern Illes Balears, 2022).

Annex Table 12.A.1. Best practices: Summary of the Balearic Islands Tourist Tax

Balearic Islands	**Title: Balearic Islands Tourist Tax** **Objective**: The purpose of the tax is to reduce the environmental impact of tourism on the Balearic Islands. The tax is earmarked, and funds are used to incentivise sustainable tourism. **Competence**: Government of Balearic Islands **Legal basis**: Law 2/2016 on the taxation of tourist accommodation on the Balearic Islands and measures to boost sustainable tourism **Setter**: Parliament of the Balearic Islands **Beneficiary**: Fund to promote sustainable tourism regulated by Law 2/2016 **Payer**: Persons who stay in hotels and other types of tourist accommodation **Taxable event**: The stays made in the Balearic Islands in tourist establishments **Calculation**: The tax rate varies from 1 to 4 EUR per day and depends on the type of accommodation

Source: Own elaboration.

Annex Table 12.A.2. Tax rates of the Balearic Island Tourist Tax

Type of accommodation	EUR per day
Hotels, city hotels and apartment hotels of 5 stars, 5 stars grand luxury and 4 stars superior	4
Hotels, city hotels and apartment hotels of 4 stars, 4 stars grand luxury and 3 stars superior	3
Hotels, city hotels and apartment hotels of 3, 2 and 1 stars	2
Tourist apartments with four keys and four superior keys	4
Tourist apartments with three superior keys	3
Tourist apartments with one, two and three keys	2
Non-residential lodging establishments of tourist-residential companies	4
Holiday tourist homes, homes subject to commercialization of tourist stays and dwellings object of tourism marketing	2
Rural hotels, agrotourism, inns and tourist accommodation inland	2
Hostels, hostels-residence, pensions, inns and houses of guests, tourist camps or campsites	1
Shelters	1
Other tourist establishments or dwellings	2
Tourist cruise boats	2

Source: Ley 2/2016, de 30 de marzo, del impuesto sobre estancias turísticas en las Illes Balears y de medidas de impulso del turismo sostenible (BOIB núm. 46, de 2 de abril de 2016).

www.ingramcontent.com/pod-product-compliance
Ingram Content Group UK Ltd.
Pitfield, Milton Keynes, MK11 3LW, UK
UKHW051457290926
15471UKWH00008B/1266

9 789264 462830